D1702242

Edited by
Jürgen Schmidt

Process and Plant Safety

Related Titles

Stoessel, F.

Thermal Safety of Chemical Processes

Risk Assessment and Process Design

2008

ISBN: 978-3-527-31712-7

Reniers, G. L. L.

Multi-Plant Safety and Security Management in the Chemical and Process Industries

2010

ISBN: 978-3-527-32551-1

Nemerow, N. L., Agardy, F. J., Salvato, J. A.

Environmental Engineering

Environmental Health and Safety for Municipal Infrastructure, Land Use and Planning, and Industry

2009

ISBN: 978-0-470-08305-5

Center for Chemical Process Safety (CCPS)

Inherently Safer Chemical Processes

A Life Cycle Approach

2009

ISBN: 978-0-471-77892-9

Center for Chemical Process Safety (CCPS)

Guidelines for Hazard Evaluation Procedures

2008

ISBN: 978-0-471-97815-2

Center for Chemical Process Safety (CCPS)

Guidelines for Performing Effective Pre-Startup Safety Reviews

2007

ISBN: 978-0-470-13403-0

Bender, H. F., Eisenbarth, P.

Hazardous Chemicals

Control and Regulation in the European Market

2007

ISBN: 978-3-527-31541-3

Bhagwati, K.

Managing Safety

A Guide for Executives

2006

ISBN: 978-3-527-31583-3

Vogel, G. H.

Process Development

From the Initial Idea to the Chemical Production Plant

2005

ISBN: 978-3-527-31089-0

Edited by Jürgen Schmidt

Process and Plant Safety

Applying Computational Fluid Dynamics

WILEY-VCH Verlag GmbH & Co. KGaA

The Editor

Professor Dr.-Ing. Jürgen Schmidt
juergen.schmidt@onlinehome.de
Karlsruhe Institute of Technology
Faculty of Chemical and Process Engineering
Engler Bunte Institute
Engler Bunte Ring 1
76131 Karlsruhe,
Germany

and

BASF SE
Safety an Fluid Flow Technology
67056 Ludwigshafen
Germany

All books published by **Wiley-VCH** are carefully produced. Nevertheless, authors, editors, and publisher do not warrant the information contained in these books, including this book, to be free of errors. Readers are advised to keep in mind that statements, data, illustrations, procedural details or other items may inadvertently be inaccurate.

Library of Congress Card No.: applied for

British Library Cataloguing-in-Publication Data
A catalogue record for this book is available from the British Library.

Bibliographic information published by the Deutsche Nationalbibliothek
The Deutsche Nationalbibliothek lists this publication in the Deutsche Nationalbibliografie; detailed bibliographic data are available on the Internet at http://dnb.d-nb.de.

© 2012 Wiley-VCH Verlag & Co. KGaA,
Boschstr. 12, 69469 Weinheim, Germany

All rights reserved (including those of translation into other languages). No part of this book may be reproduced in any form – by photoprinting, microfilm, or any other means – nor transmitted or translated into a machine language without written permission from the publishers. Registered names, trademarks, etc. used in this book, even when not specifically marked as such, are not to be considered unprotected by law.

Composition   Thomson Digital, Noida, India
Printing and Binding   betz-druck GmbH, Darmstadt
Cover Design   Schulz Grafik-Design, Fußgönheim

Printed in the Federal Republic of Germany
Printed on acid-free paper

Print ISBN: 978-3-527-33027-0
ePDF ISBN: 978-3-527-64574-8
ePub ISBN: 978-3-527-64573-X
Mobi ISBN: 978-3-527-64575-6

This book was published with the generous support from the following companies

Linde AG, Pullach, Germany

BASF SE, Ludwigshafen, Germany

Merck KGaA, Darmstadt, Germany

Braunschweiger Flammenfilter
GmbH, Braunschweig,
Germany

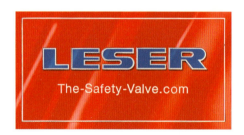

LESER GmbH & Co. KG,
Hamburg, Germany

ANSYS Germany GmbH,
Darmstadt, Germany

Germanischer Lloyd SE, Hamburg, Germany

Bopp & Reuther, Mannheim, Germany

consilab Gesellschaft für Anlagensicherheit mbH, Frankfurt, Germany

EPSC, Rugby, Warwickshire, Great Britain

REMBE GmbH, Brilon, Germany

Siemens AG, Frankfurt, Germany

Contents

Preface *XIX*
List of Contributors *XXI*

1 **Computational Fluid Dynamics: the future in safety technology!** *1*
Jürgen Schmidt

2 **Organized by ProcessNet: Tutzing Symposion 2011 CFD – its Future in Safety Technology'** *5*
Norbert Pfeil
2.1 ProcessNet – an Initiative of DECHEMA and VDI-GVC *5*
2.1.1 The ProcessNet Safety Engineering Section *6*
2.2 A Long Discussed Question: Can Safety Engineers Rely on Numerical Methods? *7*

3 **CFD and Holistic Methods for Explosive Safety and Risk Analysis** *9*
Arno Klomfass and Klaus Thoma
3.1 Introduction *9*
3.2 Deterministic and Probabilistic Design Tasks *11*
3.3 CFD Applications on Explosions and Blast Waves *12*
3.4 Engineering Methods: The TNT Equivalent *22*
3.5 QRA for Explosive Safety *25*
3.6 Summary and Outlook *27*
References *28*

Part One CFD Today – Opportunities and Limits if Applied to Safety Techology *31*

4 **Status and Potentials of CFD in Safety Analyses Using the Example of Nuclear Power** *33*
Horst-Michael Prasser
4.1 Introduction *33*
4.2 Safety and Safety Analysis of Light Water Reactors *33*
4.3 Role and Status of Fluid Dynamics Modeling *36*

4.4	Expected Benefits of CFD in Nuclear Reactor Safety 37
4.5	Challenges 40
4.6	Examples of Applications 42
4.6.1	Deboration Transients in Pressurized Water Reactors 42
4.6.2	Thermal Fatigue Due to Turbulent Mixing 47
4.6.3	Pressurized Thermal Shock 49
4.7	Beyond-Design-Based Accidents 53
4.7.1	Hydrogen Transport, Accumulation, and Removal 53
4.7.2	Aerosol Behavior 59
4.7.3	Core Melting Behavior 60
	References 66

Part Two Computer or Experimental Design? 69

5	**Sizing and Operation of High-Pressure Safety Valves** 71
	Jürgen Schmidt and Wolfgang Peschel
5.1	Introduction 71
5.2	Phenomenological Description of the Flow through a Safety Valve 71
5.3	Nozzle/Discharge Coefficient Sizing Procedure 72
5.3.1	Valve Sizing According to ISO 4126-1 73
5.3.2	Limits of the Standard Valve Sizing Procedure 74
5.3.3	Valve Sizing Method for Real Gas Applications 74
5.3.4	Numerical Sizing of Safety Valves for Real Gas Flow 77
5.3.5	Equation of State, Real Gas Factor, and Isentropic Coefficient for Real Gases 78
5.3.6	Comparison of the Nozzle Flow/Discharge Coefficient Models 80
5.4	Sizing of Safety Valves Applying CFD 82
5.4.1	High Pressure Test Facility and Experimental Results 82
5.4.2	Numerical Model and Discretization 86
5.4.3	Numerical Results 87
5.5	Summary 90
	References 93

6	**Water Hammer Induced by Fast-Acting Valves – Experimental Studies, 1D Modeling, and Demands for Possible Future CFX Calculations** 95
	Andreas Dudlik and Robert Fröhlich
6.1	Introduction 95
6.2	Multi-Phase Flow Test Facility 97
6.3	Extension of Pilot Plant Pipework PPP for Software Validation 99
6.4	Experimental Set-Up 99
6.5	Experimental Results 100
6.5.1	Experimental Results – Thermohydraulics 100

6.6	2 Case Studies of Possible Future Application of CFX	103
6.6.1	1D Modeling of Kaplan Turbine Failure	105
6.6.2	Simulation Results – Closing Time 10 s, Linear	105
6.7	Possible Chances and Difficulties in the Use of CFX for Water Hammer Calculations	106
6.7.1	Benchmark Test for Influence of Numerical Diffusion in Water Hammer Calculations	107
6.8	CFD – The Future of Safety Technology?	109
	References	110
7	**CFD-Modeling for Optimizing the Function of Low-Pressure Valves**	**113**
	Frank Helmsen and Tobias Kirchner	
	References	119

Part Three Fire and Explosions – are CFD Simulations Really Profitable? 121

8	**Consequences of Pool Fires to LNG Ship Cargo tanks**	**123**
	Benjamin Scholz and Gerd-Michael Wuersig	
8.1	Introduction	123
8.2	Evaluation of Heat Transfer	125
8.2.1	Simplified Steady-State Model (One-Dimensional)	125
8.2.2	Different Phases of Deterioration	126
8.2.3	Possibility of Film Boiling	127
8.2.4	Burning Insulation	128
8.3	CFD-Calculations	128
8.3.1	Buckling Check of the Weather Cover	129
8.3.2	Checking the CFD Model	129
8.3.3	Temperature Evaluation of Weather Cover/Insulation	131
8.3.3.1	Temperature Distribution inside the Insulation	131
8.3.3.2	Hold Space Temperature Distribution During Incident	132
8.3.4	Results of CFD Calculation in Relation to Duration of Pool Fire Burning According to the Sandia Report	133
8.3.5	CFD – the Future in Safety Technology?	136
8.4	Conclusions	136
	References	137
9	**CFD Simulation of Large Hydrocarbon and Peroxide Pool Fires**	**139**
	Axel Schönbucher, Stefan Schälike, Iris Vela, and Klaus-Dieter Wehrstedt	
9.1	Introduction	139
9.2	Governing Equations	139
9.3	Turbulence Modeling	140
9.4	Combustion Modeling	141
9.5	Radiation Modeling	142
9.6	CFD Simulation	144

9.7	Results and Discussion 145
9.7.1	Flame Temperature 145
9.7.2	Surface Emissive Power (SEP) 147
9.7.3	Irradiance 149
9.7.4	Critical Thermal Distances 150
9.8	Conclusions 154
9.9	CFD – The Future of Safety Technology? 154
	References 155
10	**Modeling Fire Scenarios and Smoke Migration in Structures 159**
	Ulrich Krause, Frederik Rabe, and Christian Knaust
10.1	Introduction 159
10.2	Hierarchy of Fire Models 161
10.3	Balance Equations for Mass, Momentum, and Heat Transfer (CFD Models) 162
10.4	Zone Models 164
10.5	Plume Models 164
10.6	Computational Examples 166
10.6.1	Isothermal Turbulent Flow through a Room with Three Openings 166
10.6.2	Buoyant Non-Reacting Flow over a Heated Surface 168
10.6.3	Simulation of an Incipient Fire in a Trailer House 170
10.6.4	Simulation of Smoke Migration 174
10.7	Conclusions 175
10.8	CFD – The Future of Safety Technology? 175
	References 177

Part Four CFD Tomorrow – The Way to CFD as a Standard Tool in Safety Technology 179

11	**The ERCOFTAC Knowledge Base Wiki – An Aid for Validating CFD Models 181**
	Wolfgang Rodi
11.1	Introduction 181
11.2	Structure of the Knowledge Base Wiki 182
11.2.1	Application Challenges (AC) 182
11.2.2	Underlying Flow Regimes (UFR) 183
11.3	Content of the Knowledge Base 184
11.4	Interaction with Users 185
11.5	Concluding Remarks 185
12	**CFD at its Limits: Scaling Issues, Uncertain Data, and the User's Role 189**
	Matthias Münch and Rupert Klein
12.1	Numerics and Under-Resolved Simulations 190

12.1.1	Numerical Discretizations and Under-Resolution *190*	
12.1.2	Turbulence Modeling *191*	
12.1.2.1	Reynolds-Averaged Navier–Stokes (RANS) Models *192*	
12.1.2.2	Large Eddy Simulation (LES) Models *194*	
12.2	Uncertainties *196*	
12.2.1	Dependency of Flow Simulations on Uncertain Parameters: Basic Remarks *196*	
12.2.2	Polynomial Chaos and Other Spectral Expansion Techniques *198*	
12.3	Theory and Practice *199*	
12.3.1	Reliability of CFD Program Results *200*	
12.3.1.1	Verification and Validation *200*	
12.3.1.2	The User's Influence *200*	
12.3.2	Examples *201*	
12.3.2.1	User's Choice of Submodels *201*	
12.3.2.2	Influence of a Model's Limits of Applicability *202*	
12.3.2.3	The Influence of Grid Dependency *206*	
12.3.2.4	Influence of Boundary Conditions *207*	
12.4	Conclusions *208*	
	References *210*	
13	**Validation of CFD Models for the Prediction of Gas Dispersion in Urban and Industrial Environments** *213*	
	Michael Schatzmann and Bernd Leitl	
13.1	Introduction *213*	
13.2	Types of CFD Models *214*	
13.3	Validation Data *215*	
13.3.1	Validation Data Requirements *215*	
13.3.2	Analysis of Data from an Urban Monitoring Station *218*	
13.4	Wind Tunnel Experiments *227*	
13.5	Summary *229*	
	References *231*	
14	**CFD Methods in Safety Technology – Useful Tools or Useless Toys?** *233*	
	Henning Bockhorn	
14.1	Introduction *233*	
14.2	Characteristic Properties of Combustion Systems *234*	
14.2.1	Ignition of Flammable Mixtures *234*	
14.2.2	Ignition Delay Times *237*	
14.2.3	Laminar Flame Velocities *239*	
14.2.4	Turbulent Flame Velocities *242*	
14.3	Practical Problems *247*	
14.3.1	Mixing of Fuels with Air in Jet-In-Cross-Flow Set-ups *247*	
14.3.2	Chemical Reactors for High-Temperature Reactions *251*	
14.4	Outlook *256*	
	References *257*	

Part Five Dynamic Systems – Are 1D Models Sufficient? 259

15 Dynamic Modeling of Disturbances in Distillation Columns 261
Daniel Staak, Aristides Morillo, and Günter Wozny
15.1 Introduction 261
15.2 Dynamic Simulation Model 262
15.2.1 Column Stage 263
15.2.1.1 Balance Equations 264
15.2.1.2 Phase Equilibrium 265
15.2.1.3 Incoming Vapor Flow 265
15.2.1.4 Outgoing Liquid Flow 265
15.2.1.5 Additional Equations 266
15.2.2 Relief Device 266
15.3 Case Study 268
15.4 CFD- The Future of Safety Technology? 269
15.5 Nomenclature 272
References 274

16 Dynamic Process Simulation for the Evaluation of Upset Conditions in Chemical Plants in the Process Industry 275

16.1 Introduction 275
16.1.1 Dynamic Process Simulation for Process Safety 276
16.2 Application of Dynamic Process Simulation 277
16.2.1 Rectification Systems 277
16.2.1.1 General 277
16.2.1.2 Verification of the Dynamic Process Simulation 278
16.2.1.3 Process Safety-Related Application of a Dynamic Process Simulator 284
16.2.2 Hydrogen Plant 288
16.2.2.1 General 288
16.2.2.2 Model Building and Verification of the Dynamic Process Simulation 289
16.3 Conclusion 293
16.4 Dynamic Process Simulation – The Future of Safety Technology? 293

17 The Process Safety Toolbox – The Importance of Method Selection for Safety-Relevant Calculations 295
Andy Jones
17.1 Introduction – The Process Safety Toolbox 295
17.2 Flow through Nitrogen Piping During Distillation Column Pressurization 296
17.2.1 Initial Design Based on Steady-State Assumptions 296
17.2.2 Damage to Column Internals 297

17.2.3	Dynamic Model of Nitrogen Flow Rates and Column Pressurization *297*
17.3	Tube Failure in a Wiped-Film Evaporator *301*
17.3.1	Tube Failure – A Potentially Dangerous Overpressurization Scenario *301*
17.3.2	Required Relieving Rate Based on Steam Flow – An Unsafe Assumption *303*
17.3.3	Required Relieving Rate Based on Water Flow – An Expensive Assumption *303*
17.3.4	Dynamic Simulation of Wiped-Film Evaporator – An Optimal Solution *304*
17.4	Phenol-Formaldehyde Uncontrolled Exothermic Reaction *306*
17.4.1	Assumptions Regarding Single-Phase Venting *306*
17.4.2	Will Two-Phase Venting Occur? *306*
17.4.3	Effect of Disengagement Behavior on Required Relieving Rate and Area *307*
17.5	Computational Fluid Dynamics – Is It Ever Necessary? *308*
17.5.1	Design of Storage Tanks for Thermally Sensitive Liquids *308*
17.5.2	Dispersion of Sprayed Droplets during Application of a Surface Coating *308*
17.5.3	Dispersion of Heat and Chemical Substances *309*
17.6	Computational Fluid Dynamics – The Future of Safety Technology? *309*
	References *311*
18	**CFD for Reconstruction of the Buncefield Incident** *313*
	Simon E. Gant and G.T. Atkinson
18.1	Introduction *313*
18.2	Observations from the CCTV Records *314*
18.2.1	Progress of the Mist *314*
18.2.2	Wind Speed *317*
18.2.3	Final Extent of the Mist *317*
18.2.4	What Was the Visible Mist? *318*
18.3	CFD Modeling of the Vapor Cloud Dispersion *318*
18.3.1	Initial Model Tests *318*
18.3.2	Vapor Source Term *319*
18.3.3	CFD Model Description *320*
18.3.4	Sensitivity Tests *320*
18.3.4.1	Grid Resolution *321*
18.3.4.2	Turbulence *321*
18.3.4.3	Ground Topology *322*
18.3.4.4	Hedges and Obstacles *323*
18.3.4.5	Ground Surface Roughness *324*
18.3.4.6	Summary of Sensitivity Tests *325*
18.3.5	Final Dispersion Simulations *325*

18.4	Conclusions 328
18.5	CFD: The Future of Safety Technology? 328
	References 329

Part Six   Contributions for Discussion 331

19   Do We Really Want to Calculate the Wrong Problem as Exactly as Possible? The Relevance of Initial and Boundary Conditions in Treating the Consequences of Accidents 333
Ulrich Hauptmanns

19.1	Introduction 333
19.2	Models 334
19.2.1	Leaks 334
19.2.1.1	Leak Size 334
19.2.1.2	Geometry of the Aperture 335
19.2.2	Discharge of a Gas 335
19.2.2.1	Filling Ratio 336
19.2.2.2	Duration of Release 336
19.2.2.3	Ambient Temperature and Pressure 336
19.2.3	Atmospheric Dispersion 337
19.2.3.1	Wind Speed 337
19.2.3.2	Eddy Coefficient 337
19.2.4	Health Effects 338
19.3	Case Study 339
19.3.1	Deterministic Calculations 339
19.3.2	Sensitivity Studies 339
19.3.3	Probabilistic Calculations 342
19.4	Conclusions 345
	References 346

20   Can Software Ever be Safe? 349
Frank Schiller and Tina Mattes

20.1	Introduction 349
20.2	Basics 350
20.2.1	Definitions 350
20.2.2	General Strategies 351
20.2.2.1	Perfect Systems 351
20.2.2.2	Fault-Tolerant Systems 352
20.2.2.3	Error-Tolerant Systems 353
20.2.2.4	Fail-Safe Systems 354
20.3	Software Errors and Error Handling 354
20.3.1	Software Development Errors 355
20.3.1.1	Errors in Software Development 355
20.3.1.2	Process Models for Software Development 355
20.3.2	Errors and Methods concerning Errors in Source Code 358

20.3.2.1	Errors in Source Code	*358*
20.3.2.2	Methods for Preventing Software Errors	*362*
20.4	Potential Future Approaches	*366*
20.5	CFD - The Future of Safety Technology?	*367*
	References	*367*

21 CFD Modeling: Are Experiments Superfluous? *369*
B. Jörgensen and D. Moncalvo
References *371*

Index *373*

Preface

Nice toy or reliable toolbox? There is no precise opinion about Computational Fluid Dynamic (CFD) results. Today, Computational Fluid Dynamics (CFD) calculations are standard in many applications with the exception of the conservative subject area of safety engineering. This was the focus of discussions between experts in the fields of safety engineering and CFD at a symposium in 2011 entitled "CFD - the future in safety technology?" (50th Tutzing symposium, 2011). When human lives or comparable commodities are at risk, then typically only conventional calculation methods are employed. As demonstrated by the eruption of the Eyjafjallajökull volcano on Iceland in 2010, it did not make sense to simulate the trajectory of the ash cloud with CFD tools when the effects of the ash particles on the jet engines of an airplane could not be satisfactorily modeled. On the other hand, CFD is already being used successfully in many areas of safety engineering, for example, for the investigation of an incident in Hemel Hempstead, England, during which an explosion led to a devastating fire at the Buncefield oil-products storage depot (the fifth largest of its type in the United Kingdom). So, when should CFD simulations be used in safety engineering?

The possible applications and limitations of CFD modeling in the area of safety engineering are discussed in this book, which covers a variety of topics, including the simulation of flow-through fittings, the consequences of fire and the dispersion of gas clouds, pressure relief of reactors, and the forensic analysis of incidents.

The contributions to this book propose the establishment of CFD programs as reliable standard tools in safety engineering. In order to accomplish this goal in the future, experts from the fields of safety engineering and CFD simulation must regularly exchange their knowledge and especially their different methodologies for dealing with certain topics. Hopefully this book will act as a catalyst for the development of deeper synergy between the two groups.

2011 *Jürgen Schmidt*

List of Contributors

G.T. Atkinson
Health & Safety Laboratory
Mathematical Sciences Unit
Fluid Dynamics Team
Harpur Hill
Buxton SK17 9JN
UK

Henning Bockhorn
Karlsruhe Institute of Technology
Engler-Bunte-Institute Division of
Combustion Technology (EBI/VBT)
Engler-Bunte-Ring 1
76131 Karlsruhe
Germany

Andreas Dudlik
Fraunhofer-Institut für Umwelt-,
Sicherheits- und Energietechnik
(Fraunhofer UMSICHT)
Osterfelder Straße 3
46047 Oberhausen
Germany

Robert Fröhlich
Fraunhofer-Institut für Umwelt-,
Sicherheits- und Energietechnik
(Fraunhofer UMSICHT)
Osterfelder Straße 3
46047 Oberhausen
Germany

Simon E. Gant
Health & Safety Laboratory
Mathematical Sciences Unit
Fluid Dynamics Team
Harpur Hill
Buxton SK17 9JN
UK

Ulrich Hauptmanns
Otto-von-Guericke-Universität
Magdeburg
Fakultät für Verfahrens- und
Systemtechnik
Institut für Apparate- und
Umwelttechnik (IAUT)
Abteilung Anlagentechnik und
Anlagensicherheit
Universitätsplatz 2
39106 Magdeburg
Germany

Frank Helmsen
Braunschweiger Flammenfilter GmbH
Industriestr. 11
38110 Braunschweig
Germany

W. Henk
Linde AG
Dr.-Carl-von-Linde-Str. 6–14
82049 Pullach
Germany

Tobias Kirchner
Braunschweiger Flammenfilter GmbH
Industriestr. 11
38110 Braunschweig
Germany

Rupert Klein
Freie Universität Berlin
FB Mathematik und Informatik
Institut für Mathematik
Arnimallee 6
14195 Berlin-Dahlem
Germany

Arno Klomfass
Fraunhofer Institute for High Speed Dynamics,
Ernst-Mach-Institute
Freiburg
Germany

Christian Knaust
BAM Bundesanstalt für
Materialforschung und -prüfung
Division 7.3 'Fire Engineering'
Unter den Eichen 87
12205 Berlin
Germany

M. Koch
Linde AG
Dr.-Carl-von-Linde-Str. 6–14
82049 Pullach
Germany

Ulrich Krause
BAM Bundesanstalt für
Materialforschung und -prüfung
Division 7.3 'Fire Engineering'
Unter den Eichen 87
12205 Berlin
Germany

Andy Jones
Evonik Degussa Corporation
Process Technology and Engineering
4301 Degussa Rd.
Theodore, AL 36590-0606
USA

B. Jörgensen
LESER GmbH & Co. KG
Wendenstrasse 133–135
20537 Hamburg
Germany

Bernd Leitl
University of Hamburg
Meteorological Institute
KlimaCampus
Bundesstr. 55
20146 Hamburg
Germany

Tina Mattes
Technische Universität München
Institute of Automation and
Information Systems
Automation Group
Boltzmannstr. 15
85748 Garching
Germany

D. Moncalvo
LESER GmbH & Co. KG
Wendenstrasse 133–135
20537 Hamburg
Germany

Aristides Morillo
BASF SE
Carl-Bosch-Straße 38
67063 Ludwigshafen
Germany

Matthais Münch
Freie Universität Berlin
FB Mathematik und Informatik
Institut für Mathematik
Arnimallee 6
14195 Berlin-Dahlem
Germany

Karl Niesser
Linde AG
Dr.-Carl-von-Linde-Str. 6–14
82049 Pullach
Germany

Wolfgang Peschel
BASF SE
Carl-Bosch-Straße 38
67063 Ludwigshafen
Germany

Norbert Pfeil
BAM Federal Institute for Materials
Research and Testing
ProcessNet Safety Engineering Section
12200 Berlin
Germany

Horst-Michael Prasser
ETH Zürich
Institut für Energietechnik
ML K 13
Sonneggstr. 3
8092 Zürich
Switzerland

Frederik Rabe
BAM Bundesanstalt für
Materialforschung und -prüfung
Division 7.3 'Fire Engineering'
Unter den Eichen 87
12205 Berlin
Germany

Wolfgang Rodi
Karlsruhe Institute of Technology
Institute for Hydromechanics
Kaiserstr.12
76128 Karlsruhe
Germany

Stefan Schälike
University of Duisburg-Essen
Institute of Chemical Engineering I
Universitätsstr. 5–7
45141 Essen
Germany

and

BAM Federal Institute for Material
Research and Testing
Division 2.2
'Reactive Substances and Systems'
Unter den Eichen 87
12205 Berlin
Germany

Michael Schatzmann
University of Hamburg
Meteorological Institute
KlimaCampus
Bundesstr. 55
20146 Hamburg
Germany

Frank Schiller
Technische Universität München
Institute of Automation and
Information Systems
Automation Group
Boltzmannstr. 15
85748 Garching
Germany

Jürgen Schmidt
juergen.schmidt@onlinehome.de
Karlsruhe Institute of Technology
Faculty of Chemical and Process Engineering
Engler Bunte Institute
Engler Bunte Ring 1
76131 Karlsruhe
Germany

and

BASF SE
Safety an Fluid Flow Technology
67056 Ludwigshafen
Germany

Benjamin Scholz
Germanischer Lloyd SE
Department of Environmental Research
Brooktorkai 18
20457 Hamburg
Germany

Axel Schönbucher
University of Duisburg-Essen
Institute of Chemical Engineering I
Universitätsstr. 5–7
45141 Essen
Germany

Daniel Staak
Technische Universität Berlin
Berlin Institute of Technology
Sekretariat KWT 9
Straße des 17. Juni 135
10623 Berlin
Germany

Klaus Thoma
Fraunhofer Institute for High Speed Dynamics, Ernst-Mach-Institute
Freiburg
Germany

Iris Vela
University of Duisburg-Essen
Institute of Chemical Engineering I
Universitätsstr. 5–7
45141 Essen
Germany

Klaus-Dieter Wehrstedt
BAM Federal Institute for Material Research and Testing
Division 2.2 'Reactive Substances and Systems'
Unter den Eichen 87
12205 Berlin
Germany

Anton Wellenhofer
Linde AG
Process & Environmental Safety – TS
Dr.-Carl-von-Linde-Str. 6–14
82049 Pullach
Germany

Günter Wozny
Technische Universität Berlin
Berlin Institute of Technology
Sekretariat KWT 9
Straße des 17. Juni 135
10623 Berlin
Germany

Gerd-Michael Wuersig
Germanischer Lloyd SE
Department of Environmental Research
Brooktorkai 18
20457 Hamburg
Germany

1
Computational Fluid Dynamics: the future in safety technology!
Jürgen Schmidt

Safety engineering is based on reliable and conservative calculations. With Computational Fluid Dynamics (CFD) tools, the knowledge of certain physical processes is deepened significantly. However, such programs are currently not standard. In safety engineering more stringent demands for accuracy must be set, for example, as compared to methods for the optimization of plants. The methods must, among other things, be sufficiently validated by experiences or experimental data and fully documented (method transparency). In addition, they must be comprehensible, reproducible, and economical to apply. The necessary demands on precision can usually only be met by model developers, program suppliers, and users of the CFD codes (common sense application).

The developers of models must document their models, and the assumptions under which the models were derived must be fully understandable. Only if the application range is carefully described can a responsible transfer to other fluids and parameter rages take place at some later time. Unlike simple empirical correlations, CFD models, with their many sub-models, often appear complex and not transparent. The validation is usually done only on certain individual data points or by measuring global parameters such as pressures and mass flows. This makes it difficult to assess whether a method is more generally applicable in practice. Margins of error cannot be estimated, or only very roughly. There are relatively for model validations for typical questions in the field of safety engineering. However, even there only models and methods with sufficiently well-known uncertainties should be applied.

It is still not enough if only the model application ranges are transparent. In addition it should be possible to review the CFD program codes. Most codes are not currently open source. Moreover, frequent version changes and changes in the program codes complicate any review. Generally accepted example calculations which can be used for revalidation (safety-relevant test cases) are usually lacking. There are often demands for open-source programs among the safety experts. This certainly facilitates the testing of models. On the other hand, in practice it is then only barely comprehensible what changes were made in a program in any particular case.

CFD calculations are reasonably possible in safety technology only with a good education and disciplined documentation of the results.

A university education should provide any students with:

- A fundamental knowledge of numerical modeling, including an understanding of the mathematical solution procedures, their use and application boundaries, and the influence of initial and boundary conditions.
- Experience in the application of safety-related models and methods.
- Analytical skills to be able to evaluate safety-related calculation results for abnormal operation conditions on the basis of experimental studies performed with other fluids and under normal conditions.
- A training in how to assess the self-evident plausibility of calculation results with the help of shortcut methods.
- A technical safety mindset and approach in dealing with computational methods and the evaluation of results.

These requirements are currently being taught in their entirety in hardly any of the major universities. Students often lack the mathematical skills of numerical modeling, a deeper understanding of turbulence models, or simply the experimental experience to assess calculation results. At some institutions, CFD codes are used as black boxes. Student training needs to be adjusted. A major effort to teach these necessary skills is essential.

Particularly in safety engineering, CFD programs are currently (still) used by a relatively small circle of experts. Careful documentation of results in this area is particularly important. In addition to input and output data, the initial and boundary values as well as the chosen solution method and model combinations must be recorded. These data are often very extensive. It may therefore be useful to keep all programs and necessary files long-term on appropriate computers. Again, it would be helpful if certain practices were well established and documented as standard – this is lacking in safety engineering.

In addition to the required computational results, sensitivity analysis of individual parameters is desirable. With CFD programs a deeper understanding of the physical processes can arise from that analysis. Alternatively it may turn out that the chosen combination of models is not suitable to solve a specific problem. Even with sensitivity analysis, the user has the duty to responsibly perform and document them as an additional part of the actual calculations.

For a third party, the CFD calculation results can in principle only be evaluated and understood from a safety point of view with much more effort. Even the inspection authorities must have sufficient expertise. For the industrial application of CFD programs in the field of safety technology, the exchange between learners and experts, and training specifically with experts from both safety engineering and CFD, are necessary. At the symposium in Tutzing, a "CFD's license" was proposed. The ensuing discussion revealed the following applications for CFD calculations in safety engineering:

1) To gain additional in-depth knowledge and understanding of physical processes. This is especially true if the effect of individual parameters to be investigated or detailed information about the spatial and temporal distribution of individual parameters is required.
2) To visualize process operational work flows.
3) To use as the sole source of information in areas where no experiments are possible (hazardous materials, very high pressures).
4) To examine boundary conditions as specified for conventional models.
5) To interpolate experimental results.

6) To question conventional methods and standards. This includes the improvement of these methods and the reduction of safety margins due to higher accuracy of the models.

CFD programs are already used in the field of safety technology for the optimization of valve operations, the investigation of fires and explosions, the examination of single-phase fluid flows, the propagation of liquid pools from leaks, and generally for the investigation of incidents. In contrast, there are also some areas where the CFD computer codes should not be used, namely:

- when simple models are adequate,
- if they are the only source of information to design safety devices,
- for unknown or fluctuating initial and boundary conditions,
- for extrapolations beyond a range with experimentally validated data,
- if only insufficient property data are available, or
- to solve very complex problems with many parameters.

Typically, established and conventional methods are applied to design safety measures and to size safety devices. With increasing risk, these standards are more important. For most of the safety experts it is currently not viable to size safety devices solely on the basis of CFD simulations. It is however expected that this will change in the future.

According to the information of the participants, 48% of the participants of the Tutzing symposium in 2011 with a safety-related background and 68% of the numerically trained participants trust in CFD simulations applied for safety engineering tasks. Training and experience, experimental validation of models, and the definition of standards (Best Practice) are the relevant criteria in order to further increase confidence.

In summary, the discussion in the Tutzing symposium has shown that CFD computer codes are used in safety technology with different intensity according to specific tasks. CFD has arrived in safety technology! The advantages of these tools show up in all areas of technology. But the dangers in the application of safety technology have also been impressively demonstrated, for example:

1) The extrapolation of validated results from highly non-linear CFD models can lead to extreme errors.
2) In safety technology, initial and boundary value problems often cannot be defined with the necessary accuracy. This may result in large errors or large uncertainties in the results of a CFD simulation. In many cases these uncertainties cannot be quantified.
3) The most often used eddy viscosity turbulence models dampen smaller fluctuations and in principle do not allow for the adequate resolution of a problem in some cases. In contrast, Large Eddy Simulation (LES) or Direct Numerical Simulation (DNS) are typically more precise but increase the computation time enormously.

BASF SE, Ludwigshafen

Only with sufficient training and experience in dealing with the CFD models and their solution methods can questions in the field of safety technology be answered responsibly. Any 'black box' CFD application mentality in which results are firstly obtained by systematic variation of models and adaptation of internal model parameters to very few experimental data and secondly presented as validated results and subsequently used for extrapolations must be strictly avoided. Of course, this applies to any type of modeling in safety technology. An extended study program at German and international universities is needed to inculcate the necessary safety skills and mindset in the next generation of students. At the same time, interdisciplinary numerical, experimental, and safety skills must be taught – just a new kind of computational Safety Engineering. For practical application in industry the idea of a CFD license or quality labels should be pursued.

The CFD computer codes should be supplemented as a standard tool by best practice guidelines in the field of safety technology and by many test cases from the professional safety community. The research and development work on the way to such standard tools (and common sense) can only enhance the training of safety engineers in the field of CFD, the acceptability of the methods, and ultimately the current state of safety technology.

With CFD tools, the demand for necessary safety measures and economic operation of plants can be merged. The knowledge so gained is considerable, and the trend toward making increasing use of these tools is already equally considerable. As long as the results are physically based on a meaningful theory and are responsibly weighted by safety considerations, this is the right way into the future of safety technology.

The 50th Tutzing Symposium 2011, organized by the community of safety technology of the Dechemás ProcessNet initiative, has brought experts from the fields of safety engineering and numerical modeling together for a first major exchange of views. Only when these two disciplines grow closer together will CFD be able to establish itself as a standard tool in all areas of safety technology.

Members of the safety community in Germany who participated in the 50th Tutzing Symposium in 2011.

2
Organized by ProcessNet: Tutzing Symposion 2011 'CFD – its Future in Safety Technology'

Norbert Pfeil

This contribution to the present book is intended to act as a bridge between the continuous work of ProcessNet's Safety Engineering Section and the 50th Tutzing Symposion entitled 'CFD – its Future in Safety Technology?' held in May 2010 at the Evangelische Akademie Tutzing at Lake Starnberg, Bavaria. It may also hopefully make both ProcessNet and its Safety Engineering Section better known, particularly in the international process and plant safety community.

2.1
ProcessNet – an Initiative of DECHEMA and VDI-GVC

In December 2006 DECHEMA (the German society for chemical engineering and biotechnology) and VDI-GVC (the German society of engineers and society for chemical and process engineering) united all their chemical and process engineering activities under a common umbrella: ProcessNet. Nearly a hundred committees of both societies with a wide variety of scientific and technical tasks of various kinds organized their work within the eight ProcessNet sections

    Chemical Reaction Technology
    Fluid Dynamics and Separation
    Particle Technology and Product Design
    Process, Apparatus, and Plant Technology
    Safety Engineering
    SuPER (Sustainable Production, Energy and Resources)
    Materials, Construction, Lifetime
    Education and Innovation.

Since then, ProcessNet has acted successfully as the one German platform on chemical engineering with more than 5000 members from the sciences, industry, and administration, exchanging ideas and experiences within ProcessNet's committees and at the numerous events organized year by year. Papers dealing with strategic topics have been published to support ongoing scientific and societal discussions and

---

Process and Plant Safety: Applying Computational Fluid Dynamics, First Edition. Edited by J. Schmidt.
© 2012 Wiley-VCH Verlag GmbH & Co. KGaA. Published 2012 by Wiley-VCH Verlag GmbH & Co. KGaA.

also to trigger funding policy initiatives. Further details on ProcessNet are available on its website: www.procesnet.org.

2.1.1
The ProcessNet Safety Engineering Section

Safety engineering was first treated as a distinct topic within DECHEMA and GVC in 1978, by establishing a joint Research Committee. In the inaugural meeting of this Committee on 5 July 1978, its tasks were characterized as

- development of safety concepts for chemical engineering
- public relations by conferences, publications, and discussion
- initiation and coordination of working items including research work

aiming to tackle the increasingly demanding safety requirements in chemical plants by taking on direct responsibility for these matters instead of reacting to requirements imposed from outside.

The first two working parties 'Safe Designing' and 'Risk, Damage Analysis, and Reliability' were established in 1978 as a first step toward today's structure of the section with working parties covering all relevant areas of safety engineering, as follows:

- Safety Parameters
- Chemical Process Safety
- Safe Design of Chemical Plants
- Preventive Industrial Fire Safety
- Safeguarding of Industrial Process Plants by Means of Process Control Engineering
- Releases and Impacts of Hazardous Materials
- Electrostatics
- Lessons from Process Safety Incidents
- Risk Management

The working parties listed above, with about 130 appointed members all in all, observe recent scientific, technical and legal developments and identify topics which need further consideration, for instance, by research, by more intensive exchange of ideas, or by making specific knowledge available to everybody responsible in the field of process and plant safety.

However, the working parties are only one part of the Safety Engineering Section. Another no less important part is the group of people who have subscribed to the Section, currently including 550 members, from industry, science, and administration.

A steering board of elected members plus the chairpersons of the working parties or temporary working groups takes care of both the safety engineering community – for instance by initiating events like the Tutzing Symposion 2011 – and the strategic development of the working structure of the Section. Aims and tasks of the currently existing working parties/groups are to be found on the Section's website, together with much relevant information for and from the community concerning institutions, events, publications, training courses and so on – most of it in the German language.

2.2
A Long Discussed Question: Can Safety Engineers Rely on Numerical Methods?

Safety engineers are conservative, for sure. Their recommendations need to err on the safe side. Therefore, they tend to suspect the results of numerical simulations of insufficiently known accuracy carried out with complex software the performance of which they do not understand completely. This subject has been been discussed repeatedly within the Safety Engineering Section and its working parties.

However, methods like CFD (Computational Fluid Dynamics) are successfully used in the chemical industry for optimizing process, apparatus, and plant design. Therefore, it is not surprising that several working parties regard numerical modeling or CFD as current working items. The question which has to be answered today seems no longer to be whether or not CFD could be used in process and plant safety, but how it is to be used most appropriately. And when the steering board of the Safety Engineering Section was required once again to organize DECHEMA's traditional Tutzing Symposium and to decide what the general topic would be, the unanimous vote was CFD and safety engineering. Tutzing Symposia are very popular because Tutzing Castle is a perfect venue for about 100 experts to gather together over a couple of days for a meeting which is more a conclave than a normal conference. It was very much hoped that the special atmosphere in the Evangelische Akademie Tutzing would lead to substantial answers to the question 'CFD – its Future in Safety Technology?'. The contributions in this book will show that this hope was well justified.

3
CFD and Holistic Methods for Explosive Safety and Risk Analysis

Arno Klomfass and Klaus Thoma

3.1
Introduction

After more than 30 years of intense development Computational Fluid Dynamics (CFD) has become an indispensable tool in many scientific and engineering disciplines. Early CFD development was strongly promoted by the aerospace and defense industries. In the 1980s, numerical solutions to the three-dimensional Euler equations of inviscid flow became possible, and in the 1990s the full Navier–Stokes equations became solvable for an entire aircraft [1]. While CFD in the aeronautics community was mainly concerned with external flows of gases and combustion processes in rocket motors and jet engines, the defense industry developed so-called hydrocodes or wave propagation codes, which enabled the simulation of high speed processes in compressible fluids for the analysis of explosion and penetration processes [2]. The characteristic features of these codes are the explicit time integration and the capability to treat multiple materials within the same computational grid and to model materials with strength, that is, solids. The first computational methods for the simulation of large-scale gas explosions for the oil and gas industry also appeared in the mid-1980s [3, 4]. Since the early days, CFD has developed into a versatile tool for many industrial branches. Today, an evolution into multi-physics simulation can be recognized, as modern software concepts permit the coupling of simulation codes from different disciplines: fluid–structure coupling is the most prominent example [5, 6]. This successful evolution was enabled through three factors: the development of advanced software architectures, the invention of versatile numerical discretization and integration techniques, and the incredible and ongoing development of computer performance. While a majority of early CFD codes only worked on Cartesian grids, most modern codes are based on flexible unstructured grid concepts, which enable the usage of body-fitted grids even for complex geometries. Further to this, mesh-free methods for particular applications such as flows through complex geometries with changing topologies have been developed. Such methods are used, for example, in the simulation of airbag inflation in automotive crash simulation [7].

Today's CFD tools are embedded into an overall computer-aided engineering (CAE) environment, which enables almost continuous work flows from CAD to grid generation to numerical solution, visualization, and design optimization. The integration of several tools into a common working environment with graphical user interfaces has been driven by the commercial software developers and has led to a high level of acceptance and application in practical engineering. The availability of almost intuitively operated user interfaces has also opened up the world of CFD for users with limited background knowledge in numerical mathematics and fluid dynamics. From this point of view, convenient and efficient usability might be an issue when it comes to application of CFD tools for safety-critical tasks.

Typical applications of CFD codes in the field of explosive safety and risk analysis fall into the following categories:

(A) Design of devices: efficient and safe functioning of process components, such as heat exchangers, mixers, valves, pipes, containments, and so on.
(B) Assessment of the consequences of device failures: leakage, dispersion and mixing of agents, explosions and fires
(C) Derivation and calibration of simplified quantitative device models for usage in system modeling and risk analysis.

Applications of type A or B provide estimates of mechanical and thermal loads exerted by the working medium on the device itself or of loads transmitted by the ambient environment onto other objects, buildings, or persons. Based on these load estimates a device can be dimensioned and constructed or, as in case B, the consequences of a failure can be assessed. In case C, the performance characteristics of a device are analyzed under all operational configurations with the aim of deriving a fast-running algebraic or differential equation model which can then be embedded into a full system simulation.

Before CFD is applied in safety and risk analysis there are preliminary tasks to be accomplished. These concern the identification of potential failure sources and failure effects. This stage of a safety analysis, the hazard identification study, requires a qualitative approach rather than a quantitative one. Structured approaches for hazard identification are the HAZOP (Hazard and Operability Study) or PAAG (in German: *Prognose, Auffinden, Abschätzen, Gegenmaßnahmen*) techniques, where a multi-disciplinary expert team investigates potential causes and consequences of failure on the basis of an existing facility design plan [8]. The findings obtained from these procedures will define the demand for detailed investigations of critical system components, which might then be performed with CFD or other suitable methods.

A further element of explosive safety analysis is the Quantitative Risk Analysis (QRA), which – in contrast to the hazard identification process – leads to a quantitative assessment of risk [9, 10]. A short survey on the principles of QRA is given in Chapter 5. Here it shall suffice to mention that a QRA demands quantitative knowledge about effects of failure events. CFD or other suitable methods must be used to accomplish this task (Case B in the above list).

There are two alternatives in the application of CFD methods. These are conducting experiments (typically on a scale model) or using (typically simplifying) engineering methods – in this book also referred to as 'short cut' methods. Before the advent of powerful CFD tools reliable predictions could in many cases only be made by means of experiments. These were carried out with prototypes or scale models if scaling laws permitted the transfer of the results to the original scale. Today model scale experiments have been widely replaced by CFD simulations; however, they still serve for validation of new CFD methods or models. So-called engineering methods are typically based on combinations of theoretical models, scaling laws, and empirical correlations derived from experimentally or numerically obtained data bases. These engineering methods have the basic property of providing – in comparison with CFD – less accurate predictions with less effort and fewer input parameters at a limited range of applicability. In Chapter 4 an example is discussed.

In the remainder of this article we illustrate the state of the art in CFD for explosions and blast waves and show how engineering methods and CFD can be combined with QRA into holistic approaches to explosive and safety analysis.

3.2
Deterministic and Probabilistic Design Tasks

Many design tasks are deterministic by definition. This is the case when an exact solution is sought on the basis of a model with precisely given parameters and boundary and initial conditions. The dimensioning of a vessel for static internal overpressure can be considered as an example. Given the vessel dimensions and the internal overpressure a unique value for the required wall thickness can be reliably calculated with validated and established procedures.

Other tasks are probabilistic by definition. An example for this type is the crash worthiness of an automobile. Not only must a range of impact velocities, relative positions and different crash barriers be considered. A reliable design assessment requires also that manufacturing tolerances and statistical fluctuations are taken into account as, due to the non-linear nature of strongly deforming structures, significant bifurcations can occur in their response as a result of only small causes. A robust design must eliminate such bifurcations in order to guarantee a predictable behavior. The overall design will then be driven by a scenarios derived from QRA.

As a rule it can be stated that the more elementary a device is the more is its design governed by deterministic approaches, and the more complex a system is, the more is its design driven by probabilistic approaches. This also affects the choice of methods and models employed for the different design tasks in safety and risk analysis. To a large extent the selection of methods and models is based on the accuracy of definition of the scenario, the required accuracy of the results, and the effort required for the application of the particular method.

The deterministic design tasks concerning the functioning and efficiency of devices will typically be based on the most accurate methods available, as the normal operation mode of a facility can be well defined and optimized. This justifies the effort required for accurate 3D CFD simulations.

On the other hand, a very large number of evaluations with hypothetical or uncertain scenario definitions can be necessary for a probabilistic evaluation in

QRA. In this case the requirement for accuracy is clearly reduced, and fast-running engineering methods can be employed for the analysis, c.f. Figure 3.1.

The challenge in the design of a safe complex system therefore lies in the complete identification and suitable modeling of the multiple interactions of numerous components. These interactions can be established on different levels, for instance, by a working medium, through electronic control loops, or just through spatial proximity of components. The typically non-linear behavior of devices under extreme (out of the operational envelope) conditions raises an additional challenge for the selection of suitable models for safety and risk analysis. In order to accomplish this task, CAE methods and system simulation must be applied in a holistic approach.

3.3
CFD Applications on Explosions and Blast Waves

The fact that most CFD codes reliably provide accurate predictions for a wide range of flow problems rests upon four premises:

- The codes are based on *first principles*: these are the conservation equations of continuum mechanics, where the conservation of mass, momentum, and energy are formulated as partial differential equations. The applicability of these laws of physics is almost universal. Further specification and closure of the equations just requires initial and boundary conditions and suitable constitutive models for the considered materials. Parameters of these models are typically measurable properties. A good CFD code should require very few non-physical parameters for the fine tuning of the numerical solution method.
- The numerical method produces approximate solutions in a consistent and stable manner, and therefore the results converge with increasing grid resolution (decreasing grid cell sizes) against the exact solution of the mathematical model. However, even with powerful computers it is in many cases impossible to achieve converged solutions for large 3D problems. Nevertheless, it is at least possible to assess the attainable accuracy by means of *grid convergence studies*.
- Code developers follow the principles of *verification and validation*. Here, verification means that the equations are correctly programmed and correctly solved; validation means that a physical model (e.g., an equation of state or a certain turbulence model) provides predictive results compared to experiments. Both issues can in practice only be achieved in an incomplete manner. The code development is a continuous process and errors cannot be generally excluded; the number and diversity of validation experiments is limited. The responsibility for the validity of produced results is therefore on the user. A certain level of expertise in fluid dynamics and numerical simulation is thus required.
- Useful *options and interfaces*: Sophisticated codes offer a variety of grid concepts and solution methodologies, such that the user can choose the best possible approach for the problem at hand. Valuable options include the choice between stationary and moving or deforming grids (Euler- or ALE-grids, ALE = Arbitrary

Lagrange Euler), body-fitted or Cartesian grids with embedded geometries, implicit or explicit time integration, compressible and incompressible fluid models, remapping of solutions between different grids at different stages of a simulation, and interfaces which allow the implementation of user-specific models. Only with such options can the full potential and efficiency of CFD be utilized.

In the following we will illustrate some state-of-the-art examples of CFD applications for explosion and blast wave problems. The results presented were obtained with the APOLLO code, which has been developed over the past 15 years at the Fraunhofer CMI for the simulation of explosions, blast waves, and high speed compressible fluid flows [11–16].

APOLLO is based on a 3D finite-volume approximation of the conservation equations for transient flows of compressible fluids. It uses a Godunov type method for the flux calculation and explicit time integration to solve the equations on block-structured grids, which can be either stationary or moving, and deforming, body-fitted or Cartesian. A non-diffusive transport of multiple materials is achieved by a Volume-of-Fluid method. The code is entirely written in *Fortran* and is parallelized with MPI. The parallelization utilizes the block structure of the grid such that the blocks are distributed onto several processors in such a way that a reasonable load balancing can be realized. Typical 3D applications are executed with 10–100 processors and grid sizes of up to around 10^8 cells.

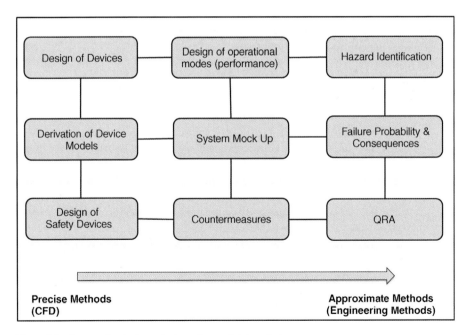

Figure 3.1 Survey of tasks in explosive safety and risk analysis.

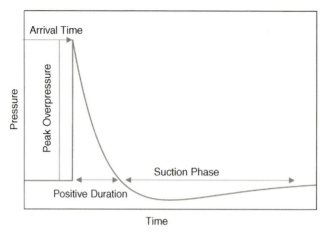

Figure 3.2 Friedlander function as approximation for pressure transients in blast waves.

For the simulation of the detonation of solid explosives, such as TNT, the JWL equation of state is used for the gaseous detonation products. The detonation process itself is modeled by the Chapman-Jouget model [17]. Ambient air is treated as a thermally perfect gas with temperature-dependent specific heat. Viscosity, heat conduction, and diffusion are negligible on the considered temporal and spatial scales and are thus not modeled in the examples shown. For the initial validation of the detonation and blast propagation model we conducted 1D-simulations of the detonation of a spherical charge of 1 kg TNT in free field. The pressure transients generated in the far field of such a detonation can be approximated by the so-called Friedlander function [18], c.f. Figure 3.2.

$$p(t) = p_0 + P\left(1 - \frac{t^*}{T}\right)\exp\left(-\alpha\frac{t^*}{T}\right); \quad t^* = t - t_{arr}$$

The Friedlander function features the parameters peak overpressure P, arrival time t_{arr}, positive duration T and a wave shape coefficient α. The dependences of these parameters on the distance and the charge mass have been intensively investigated and empirical correlations have been derived by Graham and Kinney from the measured values [18]. Figure 3.3 shows a comparison between the correlation of the peak overpressure and the values obtained with the APOLLO code. The figure also includes the results obtained with the commercial code ANSYS AUTODYN, which uses essentially the same physical model.

It can be seen that the peak overpressure is fairly well predicted by the simulations, while a small systematic deviation is observed for the positive duration value.

This discrepancy is diminished if the combustion process between air and detonation gases is taken into account, which follows the primary detonation with a certain time delay. The combustion between these gases (mainly the reaction of CO in the detonation products with atmospheric oxygen) is controlled by their mixing at the surface of the detonation fireball (the cloud of detonation gases). This mixing is

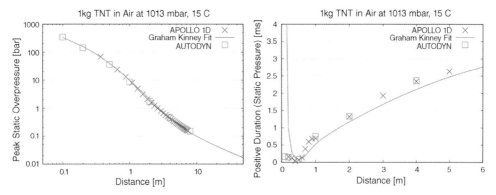

Figure 3.3 Comparison of blast parameters computed with APOLLO and AUTODYN with empirical correlations; (a) peak overpressure, (b) positive duration.

caused by the Rayleigh–Taylor and the Richtmyer–Meshkov instabilities, which lead to the formation of a turbulent mixing layer from an initially smooth contact surface. This phenomenon is responsible for the typical cauliflower shape of a detonation cloud (see Figure 3.4).

The fluid dynamic instabilities and the subsequent mixing and combustion can only be resolved in 3D simulations with sufficiently high grid resolutions. Some stills taken from an APOLLO simulation are shown in Figure 3.5. The upper picture in Figure 3.5 shows the surfaces of the fireball and the leading shock wave front at a selected time, and the lower picture shows, for the same time instant, the location of the combustion zone in a symmetry plane. The flame can be recognized as a thin curled-up sheet between the ambient air and the detonation products.

The instabilities are produced in the numerical simulation without artificial triggering and grow naturally without additional modeling. As a suitable technique to capture this phenomenon with reasonable computational effort we chose a time-dependent grid, which stretches with the expanding fireball (i.e., the cloud of detonation gases). This grid has an initial resolution of 0.5 mm and initially covers

Figure 3.4 Surface of the fireball from the detonation of 50 kg TNT showing the typical cauliflower shape caused by Richtmyer–Meshkov and Rayleigh–Taylor instabilities; still taken from a high-speed video.

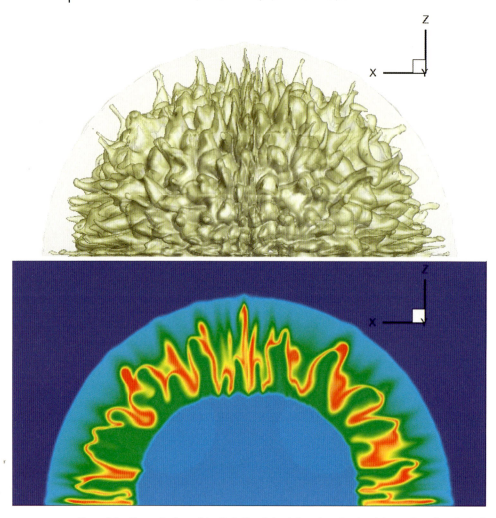

Figure 3.5 Above: Surface of the fireball and shock front in a simulated detonation of a spherical, centrally ignited charge of 1 kg TNT at $t = 0.5$ ms after initiation. Below: Temperature distribution in a symmetry plane at $t = 0.5$ ms showing the hot combustion zone of air and detonation gases in red.

a domain between $x, y \in [-10$ cm: 10 cm$]$, $z \in [0: 10$ cm$]$, which contains the spherical charge with a radius of about 5 cm. From this initial state the grid is continuously and uniformly stretched according to a prescribed, time-dependent boundary velocity. At about 1 ms after initiation the fireball attains its maximum radius of about 80 cm. The blast wave continuously separates from the fireball and propagates into the environment with a decreasing velocity which asymptotically reaches the ambient sound speed. The grid had about 64 million grid cells and the simulation was performed on 64 processors within about 4 h.

The effects of the combustion which follows the detonation are best observed in a closed chamber, as the reflections of the shock waves from the chamber walls lead to

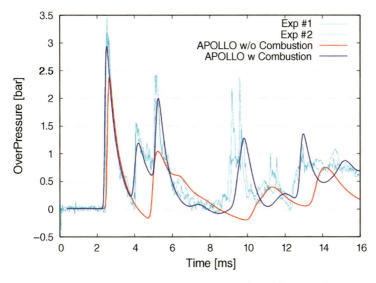

Figure 3.6 Comparison of simulation results obtained from different models with experimentally measured pressure transients: detonation of 1 kg TNT in an air-filled closed room with dimension 4 × 5 × 2.5 m; the significant effect of the combustion of detonation gases in air is recognized.

an enhancement of the mixing and combustion. An example of an experimental validation for such a case is given in Figure 3.6.

The next step in a typical CFD application in the context of explosive safety analysis may concern the interaction of the blast wave with an object, for instance, a building or a critical facility component. At that stage fluid–structure interaction takes place.

In many cases the characteristic response time of the loaded object is large compared to the characteristic loading time, which can be represented by the positive duration of the blast wave. In these cases loading and response are decoupled and the loading of the object can be calculated with CFD alone under the assumption of a rigid object. However, if this is not the case, fluid–structure coupling must be accounted for in a coupled and simultaneous solution of both domains, CFD and CSD (Computational Structural Dynamics). Such a case may exist for very large explosions at large distances which produce large positive durations or if the explosion takes place under water. In underwater explosions a very strong coupling exists between the nearly incompressible medium water and the immersed solid structures due to the similarity of the impedances of water and for example, concrete or steel [16]. It should be mentioned that coupled simulations can also be useful for cases without significant interdependency between loading and response. This is the case when strongly non-uniform, time-dependent pressure distributions exist, which must be transferred from the CFD code to a CSD code. Instead of transferring hundreds of loading curves manually between the codes, one can conveniently use a coupled code to achieve this task.

3 CFD and Holistic Methods for Explosive Safety and Risk Analysis

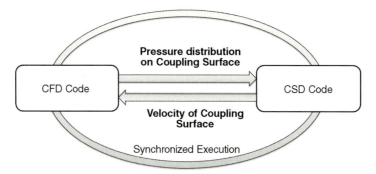

Figure 3.7 Schematic of data exchange in fluid–structure coupling for explosively loaded structures.

Coupled fluid–structure simulations for explosive loadings are mostly realized by loose coupling of existing CFD and CSD codes. The CFD and CSD codes run in parallel and exchange the relevant loading and response data on the coupling surface, which is the loaded surface of the object. In explosive loading scenarios the CFD code transfers the surface pressure distributions to the CSD code, which in turn transfers the surface velocity distributions to the CFD code, c.f. Figure 3.7.

The main software, technical, and algorithmic tasks required for the code coupling are:

- Discrete data on the coupling surfaces must be transferred between codes
- The time stepping must be synchronized
- Discrete data must be interpolated between CFD and CSD grids.

For the communication between the codes, the MPI (Message Passing Interface) environment can be utilized, which is used for code parallelization anyway. There is also software specifically developed for code coupling, such as MpCCI [19]. Today, many commercial codes are already equipped with MpCCI interfaces for code coupling. These technical and algorithmic aspects of code coupling are taken care of by the code developers. It is, however, left to the user to select an appropriate grid concept for the coupling. The available choices are schematically shown in Figure 3.8.

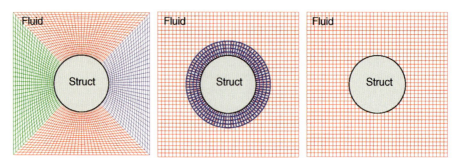

Figure 3.8 Alternative concepts for fluid–structure coupling. From left to right: ALE coupling with multi-block grid, coupling with overset grid (blue), embedded structure.

In the so-called ALE coupling a body-fitted grid is used for the fluid domain. The fluid grid stays attached to the moving or deforming structure and thus requires additional operations for grid adaption. The coupling surface is always well defined and its localization in the grid is logically constant. The ALE option is therefore often the best choice in terms of efficiency and accuracy. However, it is limited to small deformations and small displacements of the loaded objects, as severe grid deformations can impair the efficiency and accuracy significantly. If large deformations and large displacements or even topological changes (e.g., fragmenting objects) occur, a direct embedding of the structure into a stationary fluid grid is an appropriate choice. In this case the coupling surface is not fixed in the grid and must be re-localized when deformations or displacements occur. This localization can be most efficiently performed on equidistant Cartesian grids. A special treatment of masked and cut fluid cells is required. The existence of cut cells impairs the maximum possible time step size and also negatively affects the local accuracy. However, the direct embedding offers the widest range and flexibility in applications and eliminates the need for the generation of body-fitted grids. For cases with large displacements but small deformations, a hybrid concept with overset grids can be used. Here, a thin zonal grid stays attached to the coupling surface (as in ALE coupling) and communicates with a stationary background grid. In this approach the interpolation between the background and the overset grid costs additional computing time and negatively affects the accuracy. However, it also offers a high degree of flexibility, as multiple objects can be easily placed into and moved through a single background grid.

A different type of fluid–structure coupling is required when entire facilities are to be modeled and simulated, such as complete offshore platforms. In this case the size of rigging components (pipes etc.) is often comparable to or less than the grid cell size, which can be realized in the simulation. In this case, a sub-cell modeling of structural components has to be applied. Similarly to multi-phase flows, the effect of the small structures on the fluid is modeled with sink or source terms in the momentum and energy equations using drag coefficient and heat transfer models. Such concepts are used in commercial simulation codes such as FLACS or AUTOREAGAS.

A small-scale validation experiment for a coupled fluid–structure simulation is illustrated in Figure 3.9. In this example a shock wave is generated in an air-filled shock tube. The shock wave propagates through the tube (from left to right in the pictures, which show only the transparent measurement section of the tube) and refracts around a thin, vertically mounted aluminum sheet which fills the entire width of the rectangular cross section. This configuration was simulated with a coupled version of APOLLO and our CSD code SOPHIA. An excellent match with respect to both the shock wave refraction and the aluminum sheet deformation can be seen.

A three-dimensional large-scale validation experiment was conducted in collaboration with the Bundeswehr Technical Center WTD 52, which operates a large blast simulator facility (LBS). This LBS is a tunnel about 100 m long with a cross section of 75 m^2 (see Figure 3.10). At its closed end, a blast wave generator system, which consists of 100 identical pressure reservoirs, is mounted. These can be selectively filled with compressed air (up to 200 bar). The reservoirs are initially sealed with

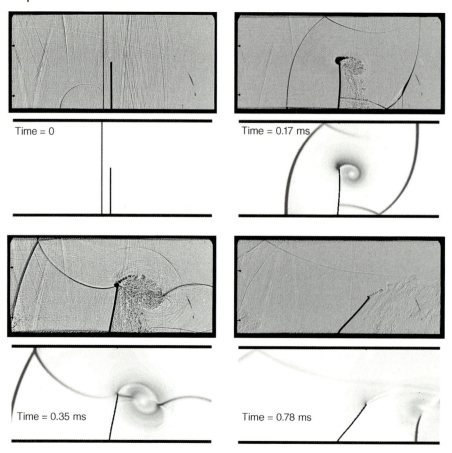

Figure 3.9 Results from a shock tube experiment and a fluid–structure coupled simulation. The height of the aluminum sheet is about 4 cm; the overpressure behind the shock wave is about 1.5 bar.

plastic membranes and are opened simultaneously with small amounts of explosive cord. The supersonic jets which then emerge from the nozzles into the tunnel produce a long duration blast wave in the tunnel air. Adjustment of the blast parameters is achieved through reservoir pressures and number of used reservoirs. Military equipment, vehicles, or field camp structures can be placed in the tunnel to test their resistance against long-duration blast waves.

For code validation purposes, a simple container structure was mounted in the tunnel (Figure 3.11). Its loading, deformation, and internal pressure build-up were measured. As the container was air-tight, the pressure transients inside the container are caused solely by acceleration and deformation of the container walls. Thus, the predictive capability of the fluid dynamic simulation and the fluid–structure coupling could be assessed. Figure 3.12 shows a representative comparison of simulation results with measured pressure transients. We used on overset grid to place the container model in the already existing tunnel grid. The overset grid itself was an ALE

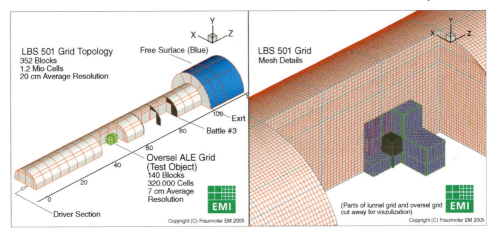

Figure 3.10 (a) 3D simulation model of the LBS; cut-away zones enable the view onto the overset ALE grid (green), which is attached to the test object. (b) Sectional view of the tunnel background grid (red), the overset grid (blue), and the surface of the test object (gray).

grid adapted to the container motion. The simulation was run on 16 processors and took a few hours CPU time (on 2005 hardware).

Given the successful validation cases shown here and many others to be found in the literature the question arises under what conditions could a CFD prediction fail? There are four potential reasons for erroneous results:

- Non-expert user selects inappropriate computational strategy
- The selected physical model is inappropriate for the specific application
- Results are not grid converged
- Side effects of the numerical method influence the result (e.g., numerical diffusion)
- Selected boundary and initial conditions do not represent the real scenario.

Figure 3.11 Picture of the container in the tunnel after blast wave loading.

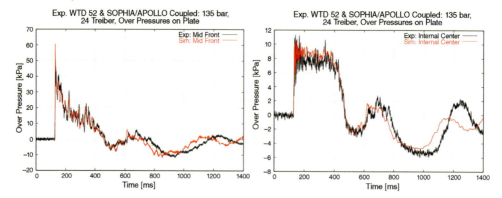

Figure 3.12 Comparison between measured and computed pressure–time histories. (a) Measurement point on the center of the container front side facing the incoming blast wave (propagating from left to right in Figure 3.10). (b) Measurement location at the central point inside the container.

In most cases, combinations of the above reasons will be found to have been responsible for reducing the predictive accuracy. The most important precondition for a successful CFD simulation must be that it is accomplished by an engineer who conceives a useful abstraction of a real life scenario into an appropriate model description. This holds for both the application of CFD codes and engineering methods.

3.4
Engineering Methods: The TNT Equivalent

Most so-called engineering methods can be speedily evaluated, are widely applicable, require few input parameters, and provide a suitable degree of accuracy. Most methods or procedures incorporate to a certain degree theoretical models, mainly similarity laws, which allow the reduction of the number of independent model parameters. Model parameters are typically calibrated against experimental measurements. However, parameters for engineering methods can also be deduced with the help of reliable first-principle simulation tools, such as CFD.

There are numerous standard methods in each field of engineering. A prominent example in the field of explosive safety is the TNT equivalent, which we now briefly describe. The success of the concept of TNT equivalence is based on its simplicity and wide-ranging applicability. The latter stems from the scaling property of blast waves, which states that for spherical or hemispherical waves all blast parameters depend on a scaled distance. According to Sachs the scaled distance can be defined as [20]:

$$z_{\text{Sachs}} = \frac{R}{(E/p_0)^{1/3}}$$

where R is the true distance from the center of explosion and E and p_0 are the explosive energy and the ambient pressure. A more widely used scaling law has been

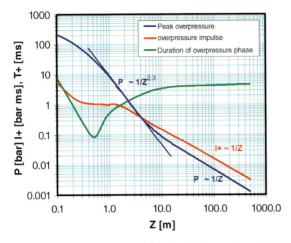

Figure 3.13 Blast parameters of a spherical detonation of 1 kg TNT in Air: Peak Overpressure P, positive impulse I+ and positive duration T+.

suggested by Hopkinson and Cranz [18]. They normalize the distance with the cube root of the explosive mass M:

$$z_{\text{Hopkinson}} = \frac{R}{M^{1/3}}$$

Figure 3.13 shows the important blast parameters, peak overpressure, positive duration, and positive impulse, as a function of distance for a 1 kg charge of TNT. With the help of the scaling laws, these parameters can be easily transferred to other charge sizes [18].

The scaling laws have been experimentally validated over a wide parameter range. The scaling property allows the investigation of blast waves in model scale. The scaling laws also express the fact that aside from the total energy release, the nature of the explosive source does not (significantly) affect the blast wave. Thus there exists a similarity of blast waves generated by different energy sources, and the effectiveness of different explosives can be quantified in terms of their TNT equivalence.

There are, however, limitations to this. The similarity law only holds for distances which are large compared to the charge size and for explosive sources with negligible energy release times. These conditions are well satisfied for most solid explosives but not so for gas explosions. The volumetric energy density and the detonation velocities of explosive gas mixtures are significantly less than those for solid explosives; furthermore, rarely are explosive gas clouds spherical or of uniform composition, as they are mostly produced by an accidental leakage. The application of TNT equivalence for the assessment of blast waves generated by gas explosions is therefore questionable.

To assess the TNT equivalence for gas explosions we have performed an APOLLO simulation of the detonation of a spherical methane–air cloud with uniform stoichiometric composition. In this simulation it was assumed that the exploding

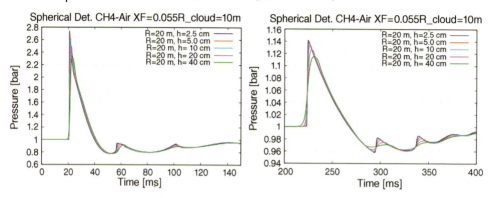

Figure 3.14 Pressure transients at R = 20 m and 100 m calculated with APOLLO for a spherical detonation of a uniform stoichiometric methane–air cloud with radius 10 m.

gas is initially at rest and in a state of ambient pressure and temperature (1 bar, 288 K). Detonation of the unconfined cloud with radius 10 m was initiated at its center. The cloud was surrounded by pure air at the same pressure and temperature.

Although it is known that methane–air mixtures in free field are not prone to deflagration–detonation transitions, we assumed that the reaction is completely detonative. The simulation thus provides a conservative estimate for the peak overpressures. Figure 3.14 shows, as an example, pressure–time curves at distances of 20 m and 100 m from the cloud center computed with different grid resolutions. In Figure 3.15 we have compared the peak overpressures and positive durations calculated with the highest grid resolution with the Graham–Kinney correlations for spherical TNT detonations for several TNT masses.

It is found that close to the cloud surface the TNT equivalent with respect to the overpressure is about 1000 kg, while at larger distances the equivalent mass increases beyond 2000 kg. At small distances the positive duration is more than double the value of any considered TNT masses and at the largest considered distance the value approaches a TNT equivalent of 4000 kg. While the estimates for the peak overpres-

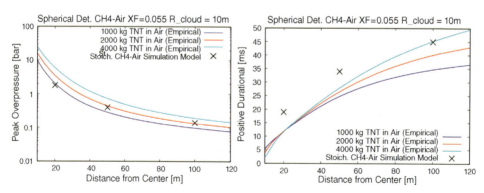

Figure 3.15 Peak overpressures and positive durations produced by the detonation of the uniform stoichiometric methane–air cloud with R = 10 m compared to values generated by TNT detonations.

sure maybe acceptable, the predicted impulse – being a function of both overpressure and duration – will be off by a significant factor.

For comparison we can consider the TNT equivalence rules for hydrocarbon–air mixtures as stated in the Gas Explosion Handbook [21]:

$$M_{TNT} = \eta\, 10\, M_{HC}; \quad \eta \in [0.03 : 0.05]$$
$$M_{TNT} = 0.16\, V_{cloud}\, [kg/m^3]$$

In our example, the cloud volume V_{cloud} is about $4200\,m^3$ and it contains about $M_{HC} = 270\,kg$ Methane, such that the above formulae provide equivalent TNT masses between about 135 kg and 670 kg. These values are clearly below our evaluations. The reason is that the formulae were derived from experiments and true events, where mostly deflagrations occur instead of detonations.

This leads back to a crucial aspect in gas explosion modeling with CFD. In principle, the deflagration-to-detonation transition (DDT) can be simulated on the basis of suitable turbulent combustion models. However, the appearance of DDT depends on several influencing factors. These are

- The strength of the ignition source
- Generation of turbulence through obstacles and blockage in the exploding cloud
- Ambient turbulence fields in the cloud resulting from leakage with high velocities (e.g., following the rupture of a pressure vessel).

In general we cannot attribute exact conditions to a hypothetical event in an explosive safety study. It is then obligatory to identify and consider the worst case, which often leads to the mandatory assumption of a detonative explosion, although this might be unlikely to occur. Again, it is left to the expert to conceive a credible conservative model of a real scenario.

3.5
QRA for Explosive Safety

Quantitative risk analysis is a holistic approach to assess risks in complex systems. The quantitative measure of risk is based on the definition:

$$Risk = \sum Probability * Effect * Exposure$$

where probability, effect, and exposure each refer to a specific event, that is, a cause of hazard (e.g., breakage of a certain pressure vessel in an array of such vessels), and the sum indicates the cumulative consideration of all possible events. The event probability is expressed in terms of the statistically expected number of such events per time interval (e.g., events per year), and the probability of exposure refers, for example, to persons or movable objects being close to the hazard source at the respective time of failure. The effects of a single event can be numerous and distinct with respect to miscellaneous types of risks (fatality of persons, interruption of production, etc.). The process of a QRA is schematically illustrated in Figure 3.16.

Figure 3.16 Schematic of QRA process.

For each particular manifestation of a hazard, a quantitative statement on effects and consequences must be provided by suitable models and methods. This typically needs consideration of multiple aspects in consecutive stages, an example being depicted in Figure 3.17.

Obviously the modeling of an event requires contributions from numerous engineering disciplines. At that stage the adequacy of accuracy and effort plays a role in model selection, as the weakest (most inaccurate) component in the overall model makes the effort for high accuracy in another component ineffectual. This is the reason why most QRA studies make use of very simplified models and do not employ CFD directly. In many cases it would indeed be impossible to conduct such studies with CFD tools for two reasons:

- It is hardly possible to incorporate a general purpose CFD tool into an existing QRA software framework. In general, the conduct of a CFD simulation (consisting of several steps from grid generation to boundary and initial value assignments) is still too complex for such an incorporation, which would require a full control through some easy-to-handle sets of parameters.
- The required computational effort (in terms of run time) would be excessive for the large number of simulations typically required in probabilistic evaluations.

The current role of CFD in QRA is therefore to support the derivation of simplified engineering models, which can be incorporated into the QRA framework. A good example of such an application is a parametric model for the effectiveness of blast protection walls, which has been derived from CFD calculations [23]. The same holds for the application of CSD methods in QRA.

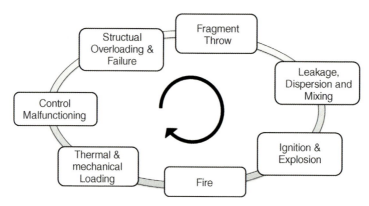

Figure 3.17 Example of consecutive events following some original cause.

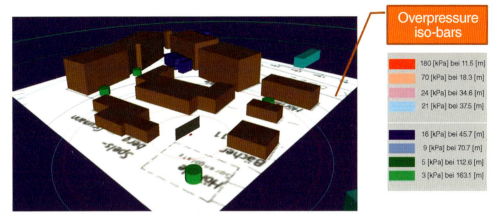

Figure 3.18 Example application of the German Explosive Safety QRA Software (ESQRA).

QRA models which are currently being developed at the Fraunhofer EMI mostly use TNT equivalence models to assess blast wave parameters and employ single degree of freedom models to estimate structural response and damage [22, 24–26], c.f. Figure 3.18. With such simple models, fast running tools can be designed, which offer a high degree of application flexibility at a balanced ratio of accuracy and computational effort.

3.6
Summary and Outlook

The predictive capabilities of today's CFD codes in general permit their application in explosive safety design and analysis tasks. In fact, CFD has the largest potential for accurate predictions from all available theoretical methods. Reference to suitable validation experiments should, however, always be made when new physical models are used or novel or extreme scenarios are considered. It is up to the experienced engineer to ensure that the simulation model is conservative from the point of view of safety.

A direct incorporation of CFD and CSD codes into holistic modeling and simulation frameworks such as QRA is today rarely possible because of the large user interaction and computer effort required for these codes. Instead CFD and CSD can be used for the derivation of simplified engineering models, which are then included in such frameworks.

The further evolution of computer performance will bring holistic simulations and CAE tools closer together, as run times will be significantly reduced. Likewise, the further development of application-specific CFD and CSD codes will foster this trend, as they can become very efficient and reliable standard tools for certain areas of applications. The current trend to Multi-Physics simulation already indicates that system simulation and CAE will come closer together. A further indication of this trend is the noticeable demand for customized simulation software seen by many Fraunhofer institutes.

References

1 Vos, J.B., Rizzi, A., Darracq, D., and Hirschel, E.H. (2002) Navier–Stokes solvers in European aircraft design. *Progress in Aerospace Sciences*, **38**, 601–697.

2 Benson, D.J. (1992) Computational methods in Lagrangian and Eulerian hydrocodes. *Computer Methods in Applied Mechanics and Engineering*, **99**, 2–3.

3 van Wingerden, C.J.M., Van Den Berg, A.C., and Zeeuwen, J.P. (1985) Validation of numerical codes for the simulation of blast generated by vapor cloud explosions. Tenth international colloquium on dynamics of explosions and reactive systems, Berkeley, California.

4 Bakke, J.R., Bjerketvedt, D., Bjørkhaug, M., and Sørheim, H.R. (1989) Practical applications of advanced gas explosion research: FLACS – a predictive tool. International Symposium on Loss prevention and safety promotion in the process industries, 1989, Oslo, Norway.

5 Wolf, K. (ed.) (2010) Multiphysics Simulation – Advanced Methods for Industrial Engineering, 1st International Conference, Bonn, Germany, June.

6 N.N. Revolutionizing Engineering Science through Simulation – Report of the National Science Foundation Blue Ribbon Panel on Simulation-Based Engineering Science (2006)

7 Hiermaier, S. (2007) *Structures under Crash and Impact*, Springer.

8 Crawley, F. and Preston, M. (2002) HAZOP: Guide to best practice for the process and chemical industries. *Institute of Chemical Engineers*.

9 N.N. Anwendung von Quantitativen Risikoanalysen (QRAs) – Positionspapier der Fachgemeinschaft Sicherheitstechnik, Arbeitsausschuss Risikomanagement (www.process.net). (last accessed May 2011)

10 Antonioni, G., Bonvicini, S., Spadoni, G., and Cozzani, V. (2009) Development of a framework for the risk assessment of Na-Tech accidental events. *Reliability Engineering & System Safety*, **94**, 9. (last accessed May 2011)

11 Klomfass, A. (2009) Numerical investigation of fluid dynamic instabilities and pressure fluctuations in the near field of a detonation, in *Predictive Modeling of Dynamic Processes* (ed. S. Hiermaier), Springer.

12 Herzog, O. and Klomfass, A. (2010) A new AMR-Solver for Precise and Fast Simulations of Blast Events in Urban Scenarios. Proceedings of the 21st Symposium on Military Aspects of Blast and Shock, Jerusalem, Israel.

13 Klomfass, A. (2007) A coupled Euler-Lagrange Method for Fluid–Structure Interaction Problems involving complex Geometries and large Deformations, Eccomas Thematic Conference: Coupled Problems 2007, Ibiza.

14 Klomfass, A. (2007) A Cartesian Grid Finite-Volume Method for the Simulation of Gasdynamic Flows about Geometrically Complex Objects. Proc. 26th International Symposium on Shock Waves, Göttingen, Germany.

15 Mangerig, I., Gebbeken, N., Döge, T., and Klomfass, A. (2006) Von der Explosion zur Bemessungslast und zur Brandeinwirkung, 2. Workshop BAU-PROTECT, Berichte aus dem Konstruktiven Ingenieurbau, Neubiberg.

16 Klomfass, A., Heilig, G., and Thoma, K. (2005) Analysis of blast loaded structures by numerical simulation, in *Fluid Structure Interaction and Moving Boundary Problems* (eds S.K. Chakrabarty, S. Hernandez, and C.A. Brebbia), WIT Press.

17 Davis, W.C. and Fickett, W. (1979) *Detonation*, University of California Press.

18 Kinney, G.F. and Graham, K.J. (1985) *Explosive Shocks in Air*, Springer.

19 http://www.mpcci.de (last accessed May 2011)

20 Baker, W.E., Cox, P.A., and Westine, P. (1983) *Explosion hazards and evaluation*, Elsevier.

21 http://www.gexcon.com//handbook/GEXHBcontents.htm (last accessed May 2011)

22 Radtke, F.K.F., Stacke, I., Rizzuti, C., Brombacher, B., Voss, M., and Häring, I. (2010) Newest Developments in the

German Explosive Safety Quantitative Risk Analysis Software (ESQRA-GE). 34th DoD Explosives Safety Seminar 2010, Portland, Oregon.

23 Borgers, J. and Vantomme, J. (2010) van der Stoel A. Blast Walls Reviewed. Proceedings of the 21st Symposium on Military Aspects of Blast and Shock, Jerusalem, Israel.

24 Riedel, W., Thoma, K., Mayrhofer, Chr., and Stolz, A. (2010) Engineering and numerical tools for explosion protection of reinforced concrete. Multi-Science Publishing, *International Journal of Protective Structures*, **1**, 85–101.

25 Riedel, W., Fischer, K., Erskine, J., Cleave, R., Hadden, D., and Romani, M., Modeling and validation of a wall-window retrofit system under blast loading, for submission to. *Int J. Comp & Struct.*

26 Fischer, K., and Häring, I., (2009) SDOF response model parameters from dynamic blast loading experiments. *Engineering Structures*, **31** (8), 1677–1686.

27 http://www.processnet.org/ereignisdb.html.

Part One
CFD Today – Opportunities and Limits if Applied to Safety Techology

4
Status and Potentials of CFD in Safety Analyses Using the Example of Nuclear Power

Horst-Michael Prasser

4.1
Introduction

The present paper gives a brief overview of nuclear reactor safety, the state-of-art in modeling safety-relevant thermal–hydraulic processes, and the role CFD (computational fluid dynamics) plays or may play in the future in this regard. Examples selected for illustration were taken, as far as possible, from work in which the author was involved in one way or another (HZDR, PSI), supplemented by selected papers by groups from all over the world. The reader is kindly asked to accept the incompleteness of this selection. In the space available, it was impossible to present a comprehensive review of the vast literature existing in the numerous fields of research that are linked to nuclear reactor safety. We apologize to the many groups and authors who significantly contributed to today's state-of-art but who have not been mentioned in the present paper, which is rather intended to combine brief explanations of the safety issues with an assessment of the perspectives and difficulties of the introduction of CFD into the safety analyses.

4.2
Safety and Safety Analysis of Light Water Reactors

Nuclear safety is based on a defense-in-depth concept aiming at a reliable enclosure of the highly radioactive material present in the core of the nuclear reactor. A system of nested barriers is constructed around the irradiated fuel, which is material of the highest specific radioactivity (Figure 4.1). At the same time, nuclear fuel produces a considerable amount of thermal energy, which originates from the fission reaction during operation and from decay of radioactive fission products, persisting even after the termination of the fission reaction. Both, but in the long term especially the decay heat, can potentially threaten and damage the above-mentioned barriers. Safety systems and other safety measures are implemented in order to protect the barriers. All together, this makes up the defense-in-depth.

Process and Plant Safety: Applying Computational Fluid Dynamics, First Edition. Edited by J. Schmidt.
© 2012 Wiley-VCH Verlag GmbH & Co. KGaA. Published 2012 by Wiley-VCH Verlag GmbH & Co. KGaA.

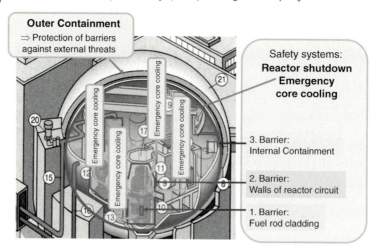

Figure 4.1 Barriers against the release of radioactive substances from the reactor as most important element of the Defense-in-depth concept.

The protection of barriers is first of all a problem of reliable heat removal, mostly by a moving fluid, but also in some cases (e.g., in high-temperature gas-cooled reactors) by thermal radiation. In the case of nuclear reactors, the conditions for the heat removal may be quite unfavorable: high heat flux densities close to values typical of different mechanisms of heat transfer crisis may be combined with a loss of coolant due to the failure of a pipe or another part of the equipment. Heat transfer crisis, or even worse, the dryout of the complete reactor core with its fuel elements, would lead to an overheat of the fuel rods and rapid oxidation of the shells of these fuel rods (the so-called fuel rod cladding), consisting of zirconium alloys in today's Light Water Reactors. Finally, the overheating can even lead to a point where the fuel itself can melt. Such a progression of the accident is called core damage. The consequence is a fall of the first barrier against a release of highly radioactive material - the fuel rod cladding and the consequent loss of the fuel integrity.

To avoid core damage, emergency core cooling systems are activated in an early stage of an accident. These are designed to maintain or recover the heat removal in case of a failure of systems for normal operation, especially in the case of Loss-Of-Coolant Accidents (LOCA). The initiated processes are nearly always accompanied by the occurrence of a highly dynamic gas–liquid two-phase flow, whereas the flow structure varies within the full range from bubbly flow to annular and finally droplet flow. The reliable functioning of the emergency core cooling has to be demonstrated for a comprehensive list of so-called Design-Based Accidents (DBA), combined with the assumption of the unavailability of parts of the safety system or, in particular, the emergency core cooling (ECC) system and the station power supply. In view of the recent nuclear disaster in Fukushima, it is worth noting that external events initiating an accident, such as airplane crashes, earthquakes, and floods are on the list of DBAs, too. The goal is to handle all these accident scenarios (caused by many different initiating events, the biggest of which is the instantaneous failure of a main circulation line in the case of a Pressurized Water Reactor plant) in such a way that the fuel rod cladding stays intact and continues to act as first barrier, unless it can be proven that those scenarios are extremely unlikely or impossible. The fluid dynamics to be modeled in this stage is characterized by a two-phase flow, partially in strong non-equilibrium, coupled with rather complex heat transfer processes at the fuel rod surfaces and in components of the safety system.

Today, nuclear safety does not stop at the threshold to core damage. Nuclear power plants of the third generation (Generation III) are equipped with autonomous engineered safety systems aimed at enclosing the radioactive inventory released in case of a core damage inside the hermetic reactor building, the so-called containment. Running reactors (Gen II) have been back-fitted with systems that provide the operators with sufficient means to achieve the same or at least to significantly reduce the releases in case of such a severe accident. In turn, operators are provided with elaborated Severe Accident Management Guidelines (SAMG), representing an addition, for severe accidents, to the operational handbook. The difference between Gen II and Gen III lies in the fact that back-fitted existing plants still rely on the right human action, to be performed a relatively short period after the accident becomes severe. Those types of reactors that are available for new power plant constructions can instead handle the core damage without human interference for very long periods of time, typically between 24 and 72 h, which makes human errors very unlikely.

Accidents during which the first barrier of the fuel rod cladding fails are called Beyond-Design Based Accidents (BDBA). This is a certain linguistic paradox, because a safe enclosure of the radioactive inventory even in case of a core damage is in fact 'design against beyond-design-based accidents'. The difference is in the treatment of both classes of accidents in the licensing process: Design-Based Accidents must be governed in a deterministic way for a strictly prescribed list of scenarios, including postulated failures of safety systems, whereas for Beyond-Design-Based Accidents the frequency of occurrence of a large release of radioactive material and the corresponding exposure of the population, calculated on the basis of a Probabilistic Safety Analysis (PSA), has to be lower than a legal limit.

The fluid dynamics phenomena occurring during postulated severe accident scenarios are much more complex, because starting from the oxidation of the cladding, a number of difficult to model chemical processes start to become relevant. The fluid dynamics modeling has to be coupled to chemical reaction kinetics, as far as this is necessary for applications in chemical process engineering. Typical reactions, besides the above-mentioned exothermal zirconium–water reaction, include the interaction of molten uranium and plutonium oxide with metallic elements, reactions of aerosols of radioactive iodine and cesium with organic substances (e.g., in the paint of the walls of the reactor building), the combustion of hydrogen produced in the zirconium-water reaction, the interaction of different metals resulting in the formation of new alloys, and the associated degradation of mechanical properties and the change in the melting point, as well as finally the interaction of the extremely hot core melt with the concrete of the reactor building floor and the accompanying production of non-condensable gases.

Furthermore, the flow becomes essentially multi-phase and multi-component. Hydrogen adds to the gaseous phase. When released from the reactor to the containment, a three-component mixture of air, vapor, and hydrogen is formed. Measures have to be designed in order to avoid the formation of explosive mixtures, which requires three-dimensional simulations in very large domains such as the containment compartments. Another liquid phase is formed by the core melt, which by itself can divide into two non-miscible phases, a metallic and an oxidic one. Melting and solidification processes add the challenge of a second phase transition, which has to be modeled. Gas-solid flows are significant when aerosols of iodine, cesium, and other volatile components are formed during the core degradation. Very often, it is necessary to couple fluid dynamics with mechanical models of the solid structures, in order to predict reduction in structural integrity and eventually failure. This may include erosion by melting of solid material, plastic deformation, creep, and brittle fracture.

Last but not least, in a nuclear reactor there is a quite strong coupling between the kinetics of the chain reaction and the density and temperature of the coolant, which very often also acts as a moderator, that is, a substance slowing down neutrons. Light water reactors are designed such that a negative feedback exists between cooling and reactor power, that is, an increase of the coolant temperature leads to a damping of the chain reaction, and thus to a decrease of the reactor power. Conversely, a sudden decrease of the coolant temperature will lead to a reactivity increase and, in extreme cases, to a power excursion. Since the moderator properties are the result of the power density distribution in the reactor core imposed on a certain three-dimensional distribution of the coolant mass flow density, a correct prediction of the kinetic behavior of the reactor can only be made by means of a 3D neutronics modeling of the reactor core coupled to a 3D CFD simulation for the flow in the core or at least at the core inlet. This is important for the modeling of so-called Reactivity-Induced Accidents (RIA).

In the majority of accident sequences, the fission reaction is stopped and neutron kinetics does not play a role, except for the need to correctly reflect the decay heat production. Additionally, fluid dynamics has to be coupled to models of the mechanical behavior of the fuel, of the reactor pressure vessel and its internals, when the consequences of power excursions and/or thermal–hydraulic shocks have to be analyzed. The same is true for processes initiated by seismic loads or a station blackout.

4.3
Role and Status of Fluid Dynamics Modeling

Workhorses for all fluid dynamic issues connected to safety analyses are so-called thermal–hydraulic system codes (TH codes, such as RELAP, TRACE, ATHLET, CATHARE), which are based on the solution of mass, momentum, and energy conservation equations for the gas and liquid phase on a one-dimensional domain. The complex system of the power plant is represented by a network of prototypal elements (pipes, valves, pumps, etc.). For each of these, the set of governing equations is solved in the time domain. The most fundamental fluid element is a pipe consisting of a serial connection of nodes. More complex nodes, describing, for example, pumps or valves, are based on pipe elements complemented by additional terms reflecting the momentum input by the pump or the variable cross-section in case of a valve. State-of-art is set by so-called best-estimate codes that solve a two-fluid model of the two-phase mixture of vapor and water, consisting of six conservation equations for each node, completed by a large set of constitutive laws describing, for example, the interaction of the phases at the gas-liquid interface, the heat transfer with the walls, and the wall friction, as well as the physical properties of the fluid.

While the discretization of the plant is comparatively coarse and one-dimensional, the codes have reached perfection in their ability to reflect the transient behavior of the system as a whole. They are equipped with elements for the modeling of the whole control system, as well as the heat source caused by the nuclear reactions, or, alternatively, are coupled to 3D neutron kinetics codes that provide the nuclear heat production as a function of space and time.

Also, for the analysis of severe accidents, lumped parameter codes similar to the already mentioned thermal–hydraulic codes have reached an advanced state of art. Again, non-resolved phenomena are modeled on an empirical or phenomenological basis, using experimental data, which, of course, becomes more and more difficult to obtain with growing severity of the studied accident scenario. Often, original fluids, like molten core material or hydrogen, have to be replaced by less dangerous model fluids in the experiments. The transfer of results to the plant conditions becomes more and more challenging. The motivations for the stepwise introduction of CFD methods are similar to those in case of the analysis of Design-Based Accidents.

It is clear that fluid dynamics phenomena that manifest as a consequence of three-dimensional flow fields cannot be reflected by one-dimensional approximations. They have to be modeled by adding empirical correlations. The information needed to develop and tune these empirical constitutive equations is taken from experiments. Here, the main weakness of thermal–hydraulic system codes is manifest: the need for empirical correlations limits the applicability of the entire model to geometries and parameter ranges covered by the experiments used for model development and validation. Extrapolation to the plant scale is therefore accompanied by uncertainties and requires a conservative approach, that is, the introduction of large safety margins. The direct transfer to new designs and parameter ranges is not possible without new experiments.

4.4
Expected Benefits of CFD in Nuclear Reactor Safety

Fluid dynamics research in the field of nuclear safety in the past two decades is characterized by strong efforts to employ methods of computational fluid dynamics (CFD). Without exception, power plants that are in operation nowadays have been designed and licensed without the aid of CFD.

For the building license of a nuclear power plant, the licensee has to prove that the plant meets the safety requirements defined in the nuclear legislation, complemented by the guidelines of the national regulator. There is a strong tendency toward international harmonization, which is promoted by an intensive exchange of information between the regulators worldwide. Harmonized recommendations are issued by the International Atomic Energy Agency (IAEA), the International Nuclear Regulators Association (INRA) and, for Europe, by the Western European Nuclear Regulators' Association (WENRA). In Switzerland, the Federal Nuclear Safety Inspectorate (ENSI) is in charge of performing an independent assessment of all safety-related analyses of the licensee. This comprises deterministic and a probabilistic safety analysis. Modeling of the dynamic behavior of the entire plant system during accidents is part of the deterministic safety analysis.

The worldwide established practice is based on thermal–hydraulic modeling, fluid dynamic processes being involved in the given accident scenario. To address uncertainties that analyses necessarily contain, models and boundary conditions are selected in a conservative way, that is, lack of knowledge and accuracy is replaced by unfavorable assumptions in order to avoid results showing unrealistically high safety margins. Some regulators allow the application of a so-called best-estimate approach, where a full system model based on a state-of-art thermal–hydraulic representation of the plant is used together with realistic boundary conditions. This approach can be chosen only on condition that the licensee provides a full uncertainty analysis of the performed modeling, which requires comparatively high effort. Still, the benefit lies in the reduction of unnecessary conservatism, and thus in the possibility of coming to a more economic design of the plant.

CFD is still not accepted as evidence to prove the compliance of the safety requirements in a licensing process. It is a common opinion that CFD is not yet mature from several points of view. On the other hand, there are a number if issues which require the application of CFD due to their pronounced three-dimensional character. Scenarios such as boron dilution, main steam line breaks, thermal loads to structures due to turbulent mixing, and pressurized thermal shock, discussed below, cannot be analyzed without addressing very non-uniform fields of velocity, temperature and other passive scalars within the flow domain. In a transition period, CFD and dedicated experiments will have to be applied together to satisfy the requirements of the regulator. It is the clear goal of research to achieve a degree of maturity that will allow tackling issues of this kind by CFD only.

Significant progress in terms of the assurance of quality of CFD calculation has been made by the introduction of Best Practice Guidelines (BPG) by The European Research Community on Flow, Turbulence and Combustion (ERCOFTAC), at least concerning the issues of adequate modeling of turbulence, which requires considerable attention to an appropriate application of models, convergence and round-off errors, user and code errors, the search for a mesh-independent solution, as well as issues of validation and robustness addressed by sensibility tests.

Both licensees and regulators believe that CFD possesses great potential for the future development of nuclear safety as well as for the enhancement of efficiency and economy, the latter being in the interest of the utilities operating the plants.

In its purest form, the Direct Numerical Simulation, CFD is capable of predicting a flow field solely relying on the fundamental fluid dynamics equation, the Navier–Stokes equation. When coupled with an energy conservation equation, temperature fields and heat transfer can be calculated. If the energy equation for the fluid is coupled to a heat conductance equation in the solid structures forming the flow domain, conjugate heat transfer problems can be solved. All this is in principle true not only for single-phase flows, but also for any kind of a multiphase flow, as long as the interface between the participating phases is fully resolved in space and time. If numerical issues are properly addressed, such modeling should be able to provide correct answers on all scales and for all possible fluid conditions. Unfortunately, limited computational power restricts the applicability of such a fundamental modeling to generic problems in very small flow domains under simplified boundary conditions; thus, applicability to the industrial scale is excluded.

Industrial-scale plant and equipment, where the flow domain is frequently complex and where Reynolds numbers are high, can only be handled using models to describe turbulence as well as the structure of the interface between different phases on a small scale in order to avoid the necessity to resolve them. Only by doing so is an application to most of the practical problems feasible. A detailed classification of different modeling approaches is given by Bestion [1]. With regard to the complexity of the modeling and the possibility of losing the ability to fully resolve turbulence scales, turbulence models range from Large Eddy Simulations (LES) to transient and steady-state Reynolds Averaged Navier–Stokes (RANS) simulations, one of the simplest examples of which is the k-eps model.[a] Concerning the interface, modeling can range from a full resolution by surface tracing methods (level set, volume-of-fluid, etc.) via the two-fluid model, down to the homogeneous model coupled to transport equations for quantities describing the local interfacial area density or bubble number densities.

Independently of the model, the constitutive equations needed to close the set of conservation equations, which represent the given model, are significantly different from those closure laws used in state-of-art TH codes. The coarse discretization employed in 1D TH codes, mainly in the form of a serial connection of calculation cells, requires empirical correlations reflecting the behavior of more complex components, that is, determined by a three-dimensional flow field. Correlations of this type are geometry dependent and refer to the integral characteristics of the given element. A flow obstruction may serve as a simple example: this has to be represented by a pressure drop correlation, which is strongly geometry dependent. CFD models also need closure laws that in most cases are based on empirical information, but they are defined on the level of the calculation grid, that is, they are by nature independent of the global geometry of the component to be modeled. An equivalent to the mentioned pressure drop correlation would be Newton's hypothesis concerning the proportionality of velocity gradient and shear stress, together with no-slip boundary conditions at the wall. With this 'microscopic' closure law, a CFD model is in principle able to predict

---

a) k-eps models are based on a representation of the turbulence field by two scalar field variables, a turbulent kinetic energy (k) and a turbulent dissipation rate (eps). Such so-called two-equation models are able to predict three-dimensional isotropic turbulence.

the pressure drop caused by any kind of flow obstruction, whereas the specific geometry of each case is reflected by the meshing, providing the latter is fine enough.

This argument justifies the expectation that CFD models will be independent from specific empirical correlations used in the TH approach. This is considerable progress, since the correlations have to be updated each time qualitative changes in the geometrical boundary conditions, in the participating fluids or in the thermodynamic parameter range, take place. This is always associated with considerable experimental effort, which can be reduced when CFD is applied. It is furthermore expected that CFD will pose less need for justification of the scale-up to plant size and the range of operational parameters due to the fundamental character of the implemented models.

There are, furthermore, a considerable number of tasks where the performance of a component of an installation depends on establishing a three-dimensional flow field, to such an extent that simplified empirical models fitted to experiments are not able to cover the full range of phenomena at all. In reactor technology, this problem became more and more relevant with the introduction of passive safety systems. These often rely on the thermal and hydraulic capacity of large pools or tanks filled with liquid and/ or gas, which are brought into motion by comparatively small driving forces. This requires analyzing rather complex, three-dimensional flow fields, which are often driven or obstructed by gravity. Here, three-dimensional modeling is inevitable for a correct reflection of the functioning of a safety system. This is the domain of CFD and coarse-mesh three-dimensional models using elements of CFD modeling, the latter when the size of the flow domain would lead to too large meshes in the case of 'full' CFD, for example, in the field of containment thermal hydraulics.

A third aspect is the interpretation of experimental data. CFD is sometimes used to better understand experimental results. There is a two-way exchange of benefit here: experimentalists appreciate the support of CFD modeling for planning and interpreting their tests as much as theoreticians seek for appropriate experimental data for code validation.

Today, CFD is more often applied to particular problems occurring in the course of the accident scenario than to a full complex safety analysis, which will very likely still remain a domain of 1D TH codes for a long time. For example, the injection of cold ECC water into the hot, pressurized reactor pressure vessel can pose new problems associated with the thermal loads that occur. This phenomenon is called Pressurized Thermal Shock and constitutes an important application for CFD modeling, as shown below. Sometimes, three-dimensional flow phenomena determine the severity of events initiating accident scenarios, as, for example, coolant mixing in the case of deboration transients and steam line breaks in Pressurized Water Reactors (PWRs). This matter will be discussed in more detail in the later chapters. Finally, thermal fatigue can be the initiating cause of Loss-of-Coolant Accidents (LOCA). Here, turbulent three-dimensional mixing phenomena cause the damaging temperature fluctuations.

4.5
Challenges

In reactor technology, CFD faces a number of specific challenges, whereas some challenges are also characteristic of process engineering applications in general. It has already been mentioned that modeling of turbulence is challenging due to the extremely large scale of the components of nuclear power plant equipment. A mesh resolution coherent with the needs even of quite 'macroscopic' turbulence models, like the well-known k-eps model, can easily reach a size of several million cells, if, for example, a component of the size of the nuclear reactor pressure vessel has to be resolved in such a way that the resolution of boundary layers stays acceptable. This does not yet include a full CFD model of the reactor core, which would easily arrive at mesh sizes in the region of a billion cells. In 3D studies of the flow in the reactor pressure vessel, the core is today very often represented as a porous body, which allows reducing the size of the mesh. On the other hand, a full representation of the core is pursued by a number of groups, an approach which is part of the efforts to create the so-called 'numerical reactor'. Similar challenges arise in the case of containment flows, where we speak about domains of several tens of meters in size and it is impossible to resolve the geometry by a mesh fine enough for CFD simulations. Here, coarse-mesh models are needed, which constitute the basis of so-called containment codes. They contain empirical correlations, for example, for the wall friction instead of fundamental wall functions, and require more experimental validation.

The next challenge is posed by strong gradients in the physical properties of the fluid, which are often present in abnormal situations. This concerns mainly density and viscosity, whereas non-Newtonian behavior of fluids is practically not an issue. Density gradients may affect especially flow fields established by natural circulation or otherwise weak driving forces. Here, turbulence models still need to be perfected to reflect all occurring relevant phenomena.

A huge challenge results from the fact that the majority of accident scenarios in both Pressurized and Boiling Water Reactors are accompanied by the occurrence of a steam–water two-phase flow, which has to be modeled appropriately. It is commonly accepted that CFD has reached a considerable degree of maturity when applied to single phase flow problems, whereas two-phase or even multiphase CFD still requires huge research efforts to reach a comparable state. The reason is the complexity of models needed to describe mass, momentum, and energy transfer at the interface between the phases. As pointed out earlier, a full resolution of the dynamic behavior of the interface requires DNS (direct numerical simulation) techniques, which are not suitable for the solution of most of the practical problems. Two-phase flows in large, complex domains are accessible by two-fluid models, which have to be equipped with closure laws describing both the dynamics of the interface and the exchange processes at the interface (for mass, momentum, and energy). These closure laws are empirical and have to be deduced from experiments or more detailed simulations (e.g., DNS). The good news is that this kind of empirical modeling is geometry independent, as discussed above. Complications arise from the fact that there are interactions between the dynamics of the interface and the turbulence model to be considered. As a matter of fact, it is well known that the presence of a second phase can promote or damp turbulence. The wall functions used in combination with RANS models are also influenced by the presence of the second phase.

The phenomena of both types of heat transfer crises, namely the departure from nucleate boiling (DNB), also called film boiling, and the dryout, are not yet accessible by modeling based on fundamental principles alone. Here, one has to resort to empirical correlations. At present, there are ongoing attempts to transfer experimental correlations found for integral parameters in the cooling channel (such as mass flow, relative enthalpy, and average heat flux density) into correlations suitable for a local prediction of DNB and dryout on the basis of the local velocity and the velocity gradient at the wall, turbulent kinetic energy, and local subcooling or void fractions, which promises to remove a part of the geometry-dependent character of the empirical correlations. DNS of DNB and dryout will still require considerable research efforts. As a matter of fact, some researchers point out the necessity to couple fluid dynamics with molecular dynamics.

Another challenge is posed by the coupling to reaction kinetics, be it the neutronic model of the reactor or a set of chemical reactions, and by the coupling to dynamic models of the solid structures. The latter two issues will have to receive increasing attention the more this type of high-fidelity modeling penetrates into the field of beyond-design-based accidents. This is accompanied by the need to model the transport of an increasing number of species, such as boric acid, non-condensable gases (in particular hydrogen and aerosols), different liquid phases (like molten metals and molten fuel), and so on. Besides evaporation and condensation, melting and solidification processes have also to be included.

4.6
Examples of Applications

A summary of the application and a discussion of the state-of-art of CFD modeling as well as their validation against experimental data are given in a report issued by OECD-NEA/CSNI [2]. In the following subsections, selected examples are highlighted.

4.6.1
Deboration Transients in Pressurized Water Reactors

In a PWR, boric acid is used as a dissolved neutron absorber and is added to the coolant to compensate for the excess reactivity in the core, which is highest after loading fresh fuel. Consequently, the boric acid concentration has to be high after refueling and it is reduced by the operators continuously during the reactor cycle to counteract the decrease of reactivity due to the increasing fuel usage. When it arrives at zero, the next refueling is due. A deboration transient is a power excursion of the reactor caused by an insertion of a certain amount of insufficiently borated water at a stage when the boric acid concentration should still be high. It is assumed that a boron-free slug can accumulate in one of the loops of the primary circuit by an operator failure during a plant outage, for example, by feeding boron-free water into one of the loops in the absence of circulation – a so-called inherent deboration. The latter is possible during a small-break Loss-of-Coolant Accident (LOCA), when the coolant inventory in the primary circuit is reduced such that the U-tubes of the steam generators (primary side) are filled with steam and the circulation in the loops is interrupted. From then on, steam generated in the reactor can enter the U-tubes and condense there, which could lead, in specific conditions, to the accumulation of deborated water in the reactor loop seal, that is, at the lowest point of the cold leg of the primary circuit. This is because boric acid is not soluble in steam, so the primary circuit in this so-called boiler-condenser mode acts as a distillation process generating an increasing amount of boron-free coolant. In both cases, a reestablishment of circulation, either by a startup of the main coolant pumps, or by onset of natural circulation, can relocate the boron-free slug into the reactor and finally into the core, where it can cause a sudden growth of the multiplication factor and, consequently, a power excursion.

In such a case, there is always still a quite considerable inventory of correctly borated water in the non-affected loops as well as in the reactor pressure vessel. Therefore, when the deborated slug arrives in the reactor pressure vessel (RPV), it will mix with the borated water already present in the RPV and in the other loops, yielding a partial increase in the boric acid concentration in the deborated slug and consequently a reduction of the impact on the core reactivity. The response of the chain reaction is very sensitive to the actual boric acid concentration profile at the core inlet. The case of ideal mixing (i.e., uniform concentration distribution at the core inlet) will result in a much less pronounced power excursion than the case of absence of mixing, where a large perturbation affects a small part of the reactor core. In the latter case the boron-free slug arrives unmixed at the sector of the core inlet corresponding to the location of the inlet nozzle of the affected coolant loop, resulting in a large local power peak. Without a detailed, three-dimensional analysis, conclusions about the consequences of such an incident have to be drawn under very conservative assumptions.

The analysis of the coolant mixing during deboration transients was one of the first useful applications of CFD to a safety issue in Pressurized Water Reactors (PWR). The boric acid behaves as a passive scalar that is transported by the three-dimensional velocity field. The basis of the modeling is therefore a solution of the Navier–Stokes equation in the quite extensive and complex geometry of the reactor pressure vessel with all its internals. The main workhorses are transient RANS calculations based on a k-eps turbulence model or similar two-equation turbulence models, like the Shear Stress Transport (SST) model of CFX, which were found to reproduce experiments with the necessary accuracy. Anisotropy of the turbulent shear stresses, which calls for an application of Reynolds Stress Models (RSM), seems to play a minor role, since the mixing processes are mainly occurring in the bulk of the flow. Often, especially in scenarios of inherent boron dilution, temperature differences between boron-free condensate, well-borated reactor coolant inventory, and streams of injected cold water from the ECC system affect the flow field by gravitationally driven density effects (buoyancy). The fluid density is in general a function of both local boric acid concentration and temperature. For this reason, the addition of an energy conservation equation is often needed to obtain the 3D temperature distribution. The feedback of the density to the flow field is established by a buoyancy term in the momentum equation.

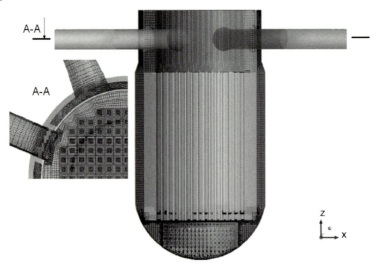

Figure 4.2 Nodalization of the ROCOM reactor pressure vessel model by Höhne et al. [3].

First modeling attempts tried to reproduce mixing experiments in scaled reactor models. It was successfully demonstrated that with CFD codes it is possible to reliably predict the boric acid distribution at the core inlet as a function of time. Höhne et al. [3] performed a detailed comparison between experimental results from the ROCOM test facility (see Ref. [4]) with CFX simulations. The scenario, an inherent boron dilution occurring in the course of a small-break LOCA, was predicted by GRS on the basis of thermal–hydraulic system codes calculations [5]. The scenario consisted in the accumulation of deborated condensate in the cold legs of two out of four loops of the primary circuit due to the establishment of a boiler–condenser mode. The slugs had a 2% lower density than the reactor inventory due to their higher temperature. They were brought into motion by a startup of the natural circulation in the affected loops reaching 5% of the nominal flow rate, while the circulation in the non-affected loops was all the time around 6% of the nominal value. ROCOM, a 1:5 scaled RPV of a German KONVOI reactor, was represented by a structured mesh of 6.5 million cells (Figure 4.2). It included all relevant parts of the cold legs of the four loops, the inlet nozzles, the flow blockage in the downcomer due to the connecting pieces between riser and outlet nozzles, the conical downcomer extension, the lower plenum, the core support plate, the perforated drum in the lower plenum with a full representation of all orifices, as well as the core simulator with 193 fuel element dummies, the upper plenum, and the outlet nozzles.

Calculations and experimental results for both cases (with and without density difference) are presented. In order to illustrate the agreement, time series of circumferential distributions of the measured and calculated transport scalar are reported as a function of time in Figure 4.3. In the figure, the results for the lower downcomer sensor are shown, which is located close to the lower edge of the core barrel (the figure was created by combining Figures 4.11 and 4.12 of the original

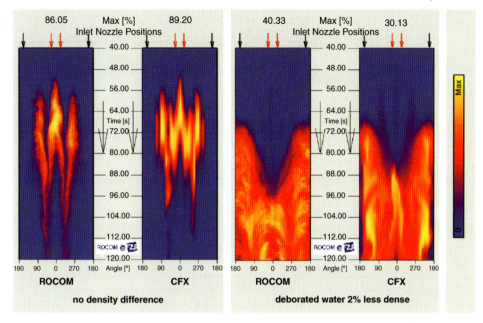

Figure 4.3 Circumferential profiles of the deboration as function of time at the lower end of the downcomer of a PWR [6].

paper). In the case without density difference, the deborated water arrives first at the position below the midpoint between both affected inlet nozzles, followed by two streams left and right from it. If the deborated water has a lower density than the reactor inventory, the pattern changes completely. The deborated water (yellow-red) arrives later, at the end of the downcomer, because it spreads around the circumference due to buoyancy. Maxima are shifted to the side opposite to the affected loops. Except for small deviations, the CFD calculation is found to be in satisfactory agreement with the experimental results qualitatively as well as quantitatively. Höhne and coauthors show that time histories at selected positions of the core inlet are also well captured.

A very illustrative presentation of the influence of density on a velocity field in the downcomer can be found in Höhne et al. [6], who calculated the transport of a plug of water with higher density compared to the reactor inventory, which is initially at rest (Figure 4.4). The velocity field which establishes under these conditions is characterized by a spread of the streamlines over the entire circumference of the downcomer, whereas directly below the nozzle of the injection a large vortex is formed as long as the plug of higher density has not yet arrived at the reactor inlet. Afterwards, the velocity field restructures due to the action of gravity. A plume of dense water is formed and the previous streamline pattern is considerably deformed.

Later the numerical models were used to model the real plant scale, which requires comparatively large meshes. Petrov and Manera [7] used a 10 million cell mesh to represent the pressure vessel of the AREVA EPR, currently being built in Finland.

Figure 4.4 Effect of higher density coolant injected into the reactor on the streamlines in the downcomer [6].

The authors report about the superiority of unstructured polyhedral cells (Figure 4.5) over structured grids, both in terms of accuracy and reduced meshing effort. This latter is quite important, since it offers the possibility to automatically mesh the reactor pressure vessel with all relevant internal structures in very great detail with CAD data as input.

Mixing processes similar to the one just described are relevant during main steam line break (MSLB) accidents as well. In this case, the pressure on the secondary side of the affected steam generator decreases rapidly, leading to an increased heat transfer

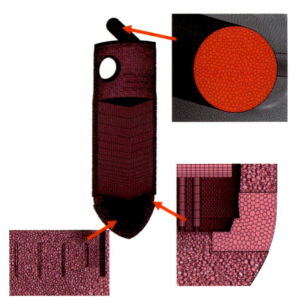

Figure 4.5 Some details of a 10 million unstructured polyhedral cell mesh (STAR-CCM +) of an EPR [7].

in the corresponding loop. Consequently, the temperature of the coolant in the affected loop decreases. The colder water will eventually reach the core, where it can cause a reactivity insertion (and thus an acceleration of the chain reaction) due to the increase in moderator density. Also in this case, a correct prediction of the core power response requires reliable information on the mixing with the coolant streams returning from the unaffected steam generators. A specific difference from boron dilution scenarios is the high flow rate in the main loops during an MSLB, since all main circulation pumps are in operation.

4.6.2
Thermal Fatigue Due to Turbulent Mixing

Cyclic thermal loads caused by the mixing of fluid streams at different temperatures can contribute to plant aging and pose safety-relevant problems for the integrity of the reactor circuit. A prediction of the behavior of a material requires, as an input, information on the amplitude and frequency of the temperature fluctuations induced in the solid structure by the contact with the mixing fluids. Flow conditions in known relevant cases are mainly single-phase. The challenge consists in the need to resolve turbulent mixing phenomena on the scale resulting from the relevant frequencies of the fluctuations.

Research in the field of thermal stripping and the related fatigue problem of piping was initiated in the 1980s in relation to the liquid metal fast breeder reactors (LMFBRs), where it is more significant than in systems with water as coolant due to the high thermal conductivity of the liquid metal coolant. Thermal fatigue phenomena were observed in the secondary loop of the French Phénix reactor and in a T-junction of the Superphénix reactor. Recently, benchmark studies to analyze these events were organized by IAEA [8]. The issue received strong renewed attention when in 2001 thermally induced cracks were found in a T-junction of the reactor circuit of the French NPP in Civaux 1 [9, 10]. Subsequently, a large number of experiments and simulations were carried out which led to the finding that LES coupled with conjugate heat transfer is able to provide the time histories of the temperature, which are needed for the structural analyses. The limited complexity and size of the flow domain of interest, as well as the fact that the problem is a single-phase flow phenomenon, played in favor of the application of LES; that is, the calculation costs stay within a feasible range despite the large grid resolutions needed for such a type of simulation.

In order to obtain the temperature fluctuations in the material of the pipe wall, an energy conservation equation is coupled to the momentum equation solved in LES. At the boundary, the local instantaneous fluid temperature is coupled to a thermal conduction model of the wall. A solution of this kind of conjugate heat transfer modeling is presented by Kuhn *et al.* [11], who took advantage of the fact that CEA had performed mixing experiments in a T-junction built from a thin brass pipe. This was observed from the outside by an infrared camera, the so-called FATHERINO experiment [12], providing experimental data for the wall temperature evolution.

As shown in Figure 4.6, LES is able to predict both mixing profiles and fluctuation amplitudes (RMS) of the mixing scalar (dimensionless temperature) after time

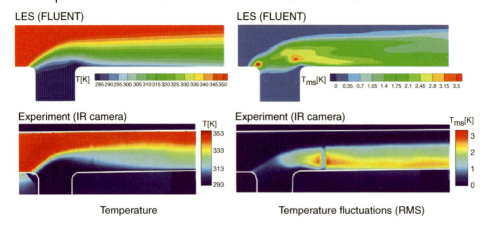

Figure 4.6 Measured (FATHERINO, [12]) and calculated (LES, FLUENT, [11]) temperature and fluctuations (RMS) at the outer surface of a 1 mm brass wall.

averaging of the time-resolved simulation results. The solution of the conjugate heat transfer task allows the prediction of the fully resolved time history of the temperature in the wall. Some questions remain open concerning the correct wall functions as well as the prediction of flow separation at curved surfaces when this has a relevant influence on the temperature field. It is believed that the problem of thermal stripping in T-junctions is basically solved by applying time-resolved LES. The large computational effort, though, drives further research in the direction of less time-consuming RANS methods, since clearly cyclic thermal loads may also occur in more complex and larger geometries than T-junctions, such as the downcomers of Pressurized and Boiling Water Reactors, the upper and lower reactor heads, and in different parts of the reactor installation of several potential candidates for nuclear power plants of Generation IV. It is therefore of considerable interest to develop and validate alternative less computationally expensive methods for the prediction of thermal fluctuations. These methods may be based on RANS and especially URANS models, which in their current state do not perform well enough to be applicable. An example of alternative methods an application of the Scale Adaptive Simulation (SAS) technique available in CFX [13], which combines the advantages of URANS and LES but requires further development to overcome its present poor accuracy. Fluctuations of the temperature can potentially be reflected by temperature fluctuation transport equations, which are a pendant to a RANS turbulence model derived from a Reynolds averaging of the energy equation [14]. Applied in URANS, this technique allows taking into account fluctuations that are not resolved on the grid scale, while those fluctuations arising from large-scale flow instabilities are directly predicted by the unsteady calculation [15]. Alternatives to transport equations of turbulent heat fluxes, similar to the Reynolds Stresses Model for the momentum equation, are higher-order algebraic models of the turbulent heat transport similar to higher-order eddy viscosity models proposed by CD-Adapco.

4.6.3
Pressurized Thermal Shock

A much more complex case of safety-relevant thermally induced stresses is so-called Pressurized Thermal Shock (PTS). The phenomenon is connected to the aging of the reactor pressure vessel caused by neutron irradiation of the steel, which leads to neutron-induced embrittlement of the RPV material. Neutrons leaking from the core during the operation of the reactor are absorbed by the surrounding material. Depending on the construction of the reactor, a more or less intense neutron flux impacts on the inner wall of the RPV, which is the most important pressure-carrying component of the entire plant. The phenomenon is mostly relevant for smaller Pressurized Water Reactors, because large PWRs and BWRs, especially those belonging to Generation III, have enough shielding capacity in the downcomer surrounding the reactor core to appropriately attenuate the neutron flux directed toward the RPV inner wall.

A failure of the RPV would have devastating consequences and has to be excluded by a full periodic surveillance, including ultrasonic and vortex current testing, to identify the formation of cracks at an early stage. On the other hand, the neutron irradiation causes a continuous increase in the transition temperature to brittle fracture. Non-irradiated reactor steel has a transition near or below ambient temperature. In the course of the accumulation of neutron fluence during the lifetime of the plant, this temperature can increase to more than $180\,°C$. In normal operation, the temperature of the vessel is always above this critical temperature. Also, with the low temperature reached during outages, there is no danger, because then the vessel is not pressurized. The potential risk of the occurrence of coolant at temperatures below the brittle fracture limit combined with high mechanical loads in the reactor walls only exists in the case of small-break LOCAs, when cold emergency core cooling water is injected into a still-pressurized reactor. In periodic safety assessments, it has to be demonstrated that there are enough margins against a brittle fracture for these cases. This requires both a thorough monitoring of the aging of the reactor steel, a forecast of the neutron fluence on the RPV wall (by means of Montecarlo codes), as well as a reliable prediction of the wall temperatures during all potential accident scenarios that may lead to fast cool-down without a concomitant fast depressurization of the reactor pressure vessel.

Clearly, the cold water from the emergency core cooling injection will mix with the hot inventory of the primary cooling circuit of the reactor before reaching the pressure vessel wall. Neglecting this process, when estimating the wall temperature, would lead to a much too conservative result. As a complication, the emergency core cooling water is often injected into a partially filled primary circuit. The contact of the injected water with steam formed during the transient may therefore further decrease the subcooling and thus contribute to a mitigation of the consequences. The modeling task is therefore essentially a two-phase CFD simulation, which must be able to reflect all relevant phenomena, as illustrated in Figure 4.7. Due to the presence of steam, the flow regime is expected to be stratified in the main circulation line of the primary circuit. The injected water jet comes into contact with the steam

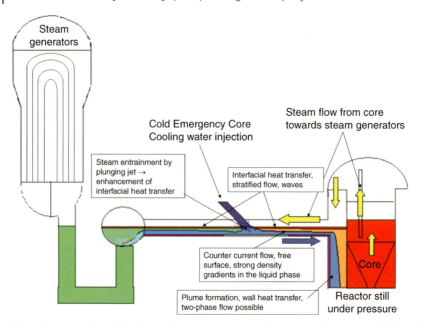

Figure 4.7 Main thermal hydraulic phenomena causing and mitigating the pressurized thermal shock (PTS) to the reactor pressure vessel wall.

directly after leaving the injection nozzle. While plunging into the stratified layer of water, steam bubbles are entrained. In both cases, heat exchange by condensation will increase the temperature of the injected water. Both phenomena have therefore to be addressed by the model.

Also, further on, the water masses flowing toward the reactor inlet stay in contact with steam. Inside the water layer, stratification stabilized by a density gradient has to be expected, which may delay the heat-up. At the entrance into the downcomer of the reactor, a water plume can be formed which could intensify the contact between water and steam. Finally, the heat exchange with the reactor wall has to be reflected correctly. During such an accident scenario, the walls of the reactor have usually a higher temperature than the coolant, and heat accumulated in the wall is released into the fluid, heating it up further. A robust and validated prediction of the real temperature histories in the reactor wall are of extreme interest, because the removal of conservative assumptions can significantly contribute to the prolongation of the lifetime of the power plant.

The current status of CFD modeling at CEA of the mentioned PTS phenomena is given by Martin [16]. The investigations are focused on the many 900 MW PWRs operated in France. Depending on the size of the break and the course of the accident scenario, either the cold leg of the primary circuit can remain completely filled with water or a stratified steam-water flow can be formed, especially when a larger break is assumed. The authors demonstrate that their in-house CFD code NEPTUNE-CFD is able to handle both cases. NEPTUNE-CFD is a classical six-equation two-fluid model

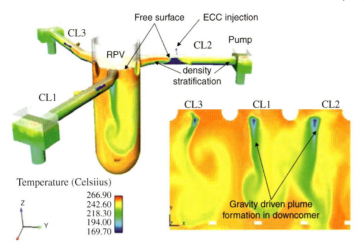

Figure 4.8 Free surface flow and stratification in the cold legs of the primary circuit in a 900 MW French PWR during ECC injection in the case of a small-break LOCA, code: NEPTUNE-CFD [16].

equipped with a k-eps model for the continuous phases. Heat transfer between fluid and wall as well as interfacial heat and mass transfer are reflected by the model. The initial and boundary conditions for the CFD simulation have to be taken from a 1D thermal–hydraulics system code (CATHARE), since the CFD simulation can reflect neither the entire reactor installation (i.e., it has to focus on the part where the relevant mixing processes happen) nor the entire process time of the small-break LOCA sequences, that often have a duration of more than an hour.

Presented results predict in a plausible way phenomena like stratified flow formation (the density stratification within the part of the pipe cross-section filled by the liquid phase), a counter-current flow of cold and hot layers of liquid in the pipe segment between ECC injection nozzle and pump model and the mixing between these layers, as well as the formation of oscillating density-driven plumes in the downcomer of the reactor (see Figure 4.8). Histories of fluid and RPV wall temperatures, needed for the analysis of the brittle fracture criterion, are obtained. Benchmarks between different codes suggest the robustness of the results, whereas their validation still awaits appropriate experimental data.

High-end two-phase flow simulations are based on the LES approach, which has to be coupled to a surface-tracking technique in order to reflect the dynamics of the gas–liquid interface. Labois and Lakehal [17] presented such a type of modeling for the French COSI experiment, where cold water was injected into a pipe filled with steam, similarly to ECC injection in a PWR. Turbulence is modeled by a modified V-LES approach, which takes a position between URANS and LES by adopting elements of a k-eps model in the scales unresolved by the grid, which allows a saving in computational effort. An important achievement is the progress in the modeling of the interfacial heat transfer and hence the condensation rate. 3D simulation results are thus in good agreement with the COSI experiment. An instantaneous snapshot of the calculated free surface shape and the predicted condensation rates is shown in Figure 4.9.

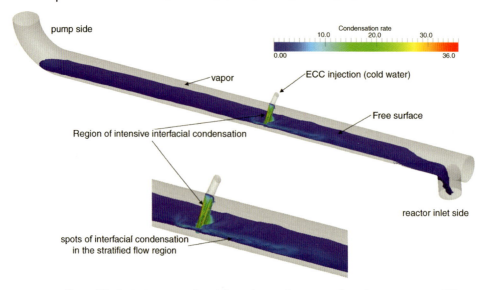

Figure 4.9 Instantaneous surface deformations and steam condensation rate contours [17].

As a result of the interfacial heat transfer, in combination with the turbulent single-phase mixing of the injected ECC water with the hot water resident in the pipe, temperature distributions as shown in Figure 4.10 are generated, which represent the input for the calculation of the wall temperatures relevant from the point of view of thermally induced stresses. It can be seen that the mixing processes in the lines of the primary circuit lead to a significant increase in the temperature of the injected water, and thus to a mitigation of the consequences of PTS. This and similar work clearly demonstrates the tremendous progress on the way to tackling the fluid dynamic phenomena relevant for PTS by means of CFD.

The theoretical studies are accompanied by a number of past and ongoing experimental programs needed for model and code validation, as well as for a demonstration of a reliable transfer of the results to the reference plant. As pointed out before, the regulator would not accept proofs built on evidence obtained by CFD simulations only. The scope ranges from cold, non-pressurized tests in transparent test channels (e.g., HYBISCUS) to a pressurized steam–water experiment carried out

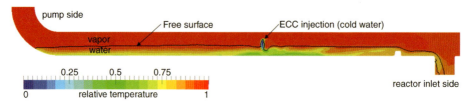

Figure 4.10 Temperature distribution as a result of mixing in liquid and interfacial heat transfer and condensation.

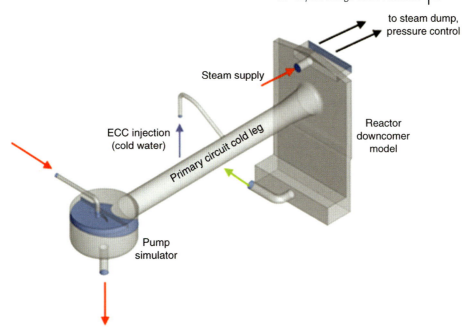

Figure 4.11 TOPFLOW-PTS experimental setup, taken from Apanasevich et al. [18].

in a special pressure vessel accommodating a medium-scale test rig in order to perform the experiments in pressure equilibrium (TOPFLOW-PTS, Figure 4.11).

Apanasevich et al. [18] presented results of steady-state pretest calculations for TOPFLOW-PTS, which were preformed using both a single fluid and a two-fluid model available in CFX (Figure 4.12). Interfacial heat transfer was considered using surface renewal theory according to Hughes and Duffey. This example shows that there is more than one candidate for the simulation of the complex phenomena expected during ECC injection and related to PTS.

4.7
Beyond-Design-Based Accidents

4.7.1
Hydrogen Transport, Accumulation, and Removal

As pointed out earlier, the number of modeling issues increases significantly when BDBA analyses have to be performed because numerous chemical species and reactions emerge that have to be taken into account, while the flow becomes essentially multiphase. BDBAs are characterized by damage to the fuel rods, which starts from an oxidation of the fuel rod cladding given by the reaction:

$$Zr + 2H_2O \rightarrow ZrO_2 + 2H_2 \tag{4.1}$$

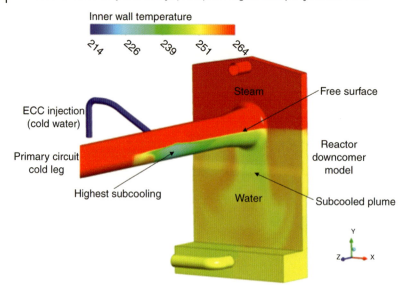

Figure 4.12 Temperatures at the inner wall of the structure of the TOPFLOW-PTS test facility during emergency core cooling injection into a semi-filled cold leg of the primary circuit [18].

The oxide is not mechanically stable, and the fuel rods disintegrate, forming a debris bed. The reaction in Eq. (4.1) progresses in a steam atmosphere and is exothermic. At a temperature of about 1200 °C it becomes self-sustaining. In the case of oxidation in air, a process that can start in the spent fuel basin in the case of a loss of cooling and water inventory, the reaction is even more energetic:

$$Zr + O_2 \rightarrow ZrO_2 \tag{4.2}$$

The production of hydrogen by the zirconium-water reaction (Eq. (4.1)) is an issue important for the integrity of the containment, since hydrogen can form an explosive mixture with air. The destruction of the reactor buildings (secondary containments) of units 1 and 3 in Fukushima Daiichi was caused by the detonation of hydrogen that escaped from the reactor and the primary containment caused by operations aimed at relieving overpressure.

The threat of such a reaction arises from the potential loss of the last barrier against radioactive releases. There are two counter-measures employed in today's nuclear power plants: either (a) the containment is inertialized by filling it with nitrogen during normal operation or (b) hydrogen recombiners from the type of catalytic devices or so-called spark igniters provide a slow and controlled oxidation of the hydrogen.

Passive recombiners (PAR) consist of a stack of plates or foils covered by a layer of noble metal, at the surface of which the oxidation of hydrogen takes place. The plates are placed in a casing with flame barriers at the inlet and the outlet. Heitsch [19] performed a detailed CFD simulation of the flow around one of the catalyst plates using the commercial code CFX 4.1. The model was extended by transport equations for the most important chemical species, including reaction kinetics as well as

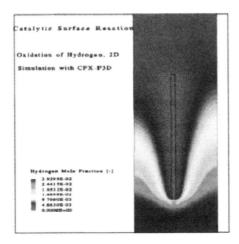 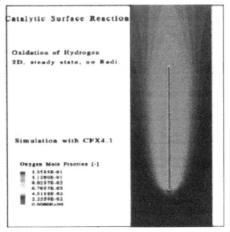

Figure 4.13 Hydrogen (a) and oxygen (b) mole fraction around a catalyst plate, calculated with CFX 4.1 [19].

models for adsorption and surface reactions. The overall performance is controlled by the kinetics of adsorption, desorption, and surface reactions on the catalytic surface. Since the kinetics of the surface reactions depend on the temperature of the catalyst plate, which required taking into account thermal conductivity of the catalyst plate and conjugate heat transfer. The geometry was simplified by modeling only one single catalyst plate assuming periodic boundary conditions within the stack. In the result, the performance was reproduced with good agreement with experiments (Figure 4.13).

Catalytic recombiners have to be arranged at different locations in the containment in order to avoid a dangerous local accumulation of hydrogen and to make sure that they start operating in time. Furthermore, their efficiency not only depends on the flow field established inside them, but also on the flow field in their surroundings, which is driven by buoyancy due to thermal and molar effects, since the oxidation is exothermal and the molar weight of the gas changes due to the reaction. A summary of open issues concerning the modeling of recombiners and their function in the containment atmosphere during severe accidents is given by Reinecke *et al.* [20]. To reflect the effect of recombiners on the global flow field and the integral efficiency of hydrogen removal, manufacturer correlations of the characteristics are implemented as source terms into coarse mesh containment codes as well as into CFD codes. The flow is single-phase but essentially multi-component (hydrogen, steam, oxygen, nitrogen, gaseous components that can poison the catalyst). Multiphase issues may play a role when the influence of aerosols or water spray has to be considered.

A detailed modeling of the flow field in a rectangular 2D domain with a pair of symmetrically arranged PARs performed with the commercial CFD code CFX revealed that density stratification can deteriorate the performance. As shown in Figure 4.14, natural circulation bringing the hydrogen-rich gas mixture to the catalytic surfaces can die out after the establishment of a thermal stratification. The

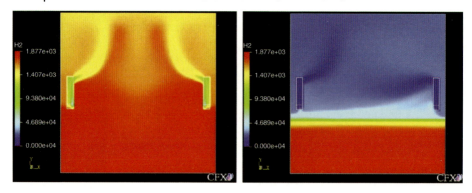

Figure 4.14 Hydrogen concentration in the space around a pair of catalytic recombiners showing the phenomenon of blockage due to the formation of a stable density stratification [20].

operation changes into a slow, diffusion-driven regime, which must be considered when designing the distribution of PARs within the containment. The real extent of deterioration of the efficiency could remain undiscovered if coarse mesh codes are used, since they predict unrealistically high rates of hydrogen ingress due to unphysical numerical diffusion.

The challenge of CFD modeling consists here in (a) the rather complex reaction kinetics and (b) the effect of strong density gradients on the turbulence quantities. In the latter respect, there is some similarity to thermal stratification issues discussed earlier in connection with cyclic thermal loads and PTS. The final step is the simulation of the overall distribution of hydrogen in the whole containment, which poses the additional complication of (c) very large meshes. Nonetheless, CFD codes have begun to acquire a position in this business already. Numerous efforts are undertaken to validate both coarse-mesh and CFD codes. A number of experiments on large-scale test facilities are available for code validation. Examples are the former Battelle Model Containment, PANDA (PSI, Switzerland) MISTRA (CEA, France), and THAI (Becker Technologies, Eschborn). Several international activities have been launched in the past, or are currently on-going, aimed at addressing the validation and further development of models for issues specific to severe accidents. The EU project ECORA [21], for example, dealt with turbulent transport in a complex, multi-compartment system. The activity took benefit from the fact that individual components of the model can be studied individually. The experiments taken as a basis for the benchmark were performed at the PANDA facility, where steam was injected into the previously air-filled compartments of the test facility. Figure 4.15 shows a comparison of calculated and measured temperature distributions within a steam plume caused by a lateral injection, as well as steam contents as a function of height. Codes applied by the participants were CFX, FLUENT and TONUS (CEA). Two-equation turbulence models were used. The agreement concerning the position of the plume was good for all calculations except the one performed with the finest mesh. Axial concentration profiles were reproduced with acceptable accuracy (Figure 4.15, right side).

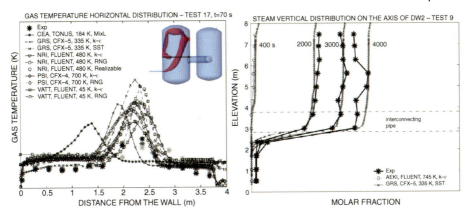

Figure 4.15 Horizontal profiles of the calculated and measured gas temperature across the injected steam plume in the left PANDA drywell vessel and selected results for the axial steam content profiles [21].

In such large-scale containment tests, hydrogen is usually replaced by helium. Another example of such types of experiments and post-test calculations was published by Heitsch *et al.* [22], who performed post-test calculations of the HYJET experiments at the former Battelle model containment (Figure 4.16). The facility was rather large-scale for the times in which the experiments were carried out, with a volume of 600 m^3 and a total height of 9 m. The computational model used by Heitsch included the effect of steam condensation, which has a great effect on the estab-

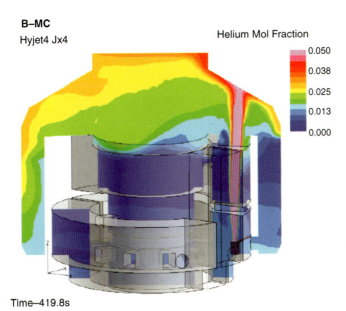

Figure 4.16 Evolution of helium molar fraction during an experiment at the Battelle Model Containment (HYJET experiments) calculated with CFX [22].

lishment of regions with high hydrogen (represented by helium) concentrations. The authors found that the RNG k-eps model provided the best agreement, which confirms that the simplifying assumption of isotropic turbulence in containment flows leads to acceptable results. However, as stated by Heitsch, the opportunities for a detailed comparison were limited due to the relatively sparse instrumentation of the quite early test facility.

Recently, numerous experiments have been carried out to study the interaction of jets and plumes of steam impacting on a stable stratification layer of helium. It was found that codes often fail to predict accurately the erosion of the light gas layer by a heavier jet, which is attributed to a wrong prediction of turbulent dispersion in cases of stable density stratifications. Here, work is in progress and results may be expected in the near future. Sometimes experiments are conducted in rather complicated scenarios involving all three important gases – air, steam, and hydrogen (helium); see for example, Paladino et al. [23]. There, the authors observed a rather good performance of the coarse-mesh code GOTHIC and state that deviations decrease when the mesh is refined. In Figure 4.17 it is shown that the injected steam jet is stopped when impacting the helium layer in the top region of the left vessel. Helium is eroded out of this layer and transported to the right vessel. In the interconnecting pipe, a single-phase counter-current flow is predicted. All modeled phenomena are in good agreement with the experiment.

For some particular issues, CFD codes were even applied to model the real containment scale. An example of the modeling of a hydrogen explosion was reported by Baraldi et al. [24] in a simplified EPR geometry. Both, CFX and a dedicated combustion code (REACFLOW) were applied. Special attention was given to the

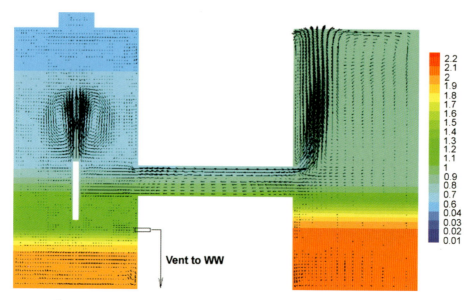

Figure 4.17 Density (color) and velocity field in the drywell vessels of PANDA, calculated with GOTHIC [23].

mixing effect induced by the walking grid platforms separating the floors in the steam generator and circulation pump compartments. Summarizing, it can be stated that coarse-mesh codes like GOTHIC and GASFLOW, as well as specialized combustion codes, are already being applied for containment safety analyses, whereas CFD codes are still in the phase of validation and some first attempts of applications for real containment geometries.

4.7.2
Aerosol Behavior

Overheated – even if not yet molten – nuclear fuel releases radioactive material, which consists of volatile fission products. These are divided into fission gases, that is, radioactive isotopes of krypton and xenon on the one hand side, and aerosol-forming radionuclides, the most prominent being iodine (mainly I-131) and cesium (mainly Cs-137). While the noble gases completely dissolve in the gaseous phase, aerosols have to be treated as individual phases with their own velocity field. Under the action of gravity and due to a number of other effects like thermophoresis and strong accelerations in curved streamlines, they can be deposited. High velocity gradients at the surface can mobilize them, and spray flows or bubbling though water reservoirs can scrub them.

The modeling of aerosol flows is essentially multiscale. In today's containment codes, they are advected by the vector field of the gaseous phase and dispersed by turbulence. Deposition and remobilization are modeled by empirical correlations. Often, chemical reactions take place that change the behavior of aerosols. This leads to the necessity to include separate individual mass conservation equations for the different species, for example, for inorganic and organic iodine compounds, respectively, completed by reaction kinetic models.

The application of DNS and LES methods is focused mainly on processes on the micro-scale. Researchers try to predict aerosol deposition as a function of the velocity gradient, the turbulent kinetic energy, dissipation rate, and temperature gradients. The goal is a fundamental access to closure laws needed for modeling the macro-scale.

On the plant scale, the goal is to couple RANS simulations of the global velocity field with a Lagrangean particle transport model based on discrete or continuous random walk approaches to reflect the turbulent dispersion of the aerosol particles [25]. Deposition is assumed when particles arrive in the first cell close to the boundary, completed by empirical efficiency coefficients. This method is in the state of validation. The author describes post-test calculations in order to study the interplay between thermophoresis forces and turbulent dispersion. The latter effect counteracts deposition, and neglecting it would lead to unrealistically high depositions. Among others, CFD calculations were performed on the small scale of the steam generator heat exchanger pipe of the PHEBUS facility ([26], see Figure 4.18), where strong temperature gradients are present. It was shown that turbulent dispersion significantly reduces the aerosol deposition in the steam generator tube, and brings the theoretical results into good agreement with the deposition measured

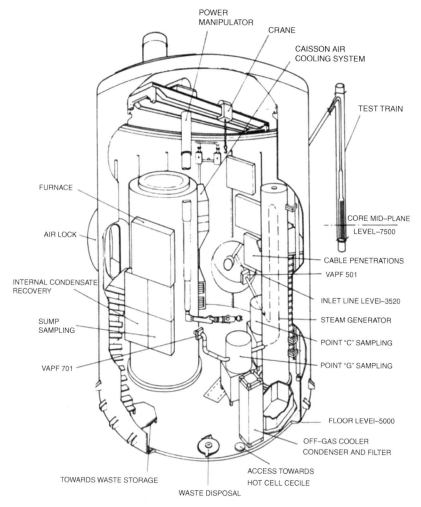

Figure 4.18 PHEBUS test facility at CEA in Cadarache, France [26].

at PHEBUS ([27], see Figure 4.19). It is worthwhile mentioning that this validation was carried out for real prototypical aerosols, which are highly radioactive. Nonetheless there is still a long way to go to an application of the described approach to large containment scales, which will remain for quite a while a domain of coarse mesh Navier–Stokes solvers equipped with empirical correlations.

4.7.3
Core Melting Behavior

Core melting occurs when the debris formed after fuel rod failure remains without adequate cooling and temperatures rise to values well above 2000 °C (pure UO_2 melts at 2800 °C – eutectic mixtures with other structural materials may melt at

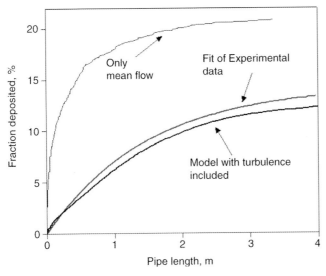

Figure 4.19 Integral aerosol deposition in the steam generator tube of PHEBUS [27].

lower temperatures). The melt is liquid with quite low viscosity, and can therefore relocate from the damaged core into the lower plenum of the reactor. There, it starts to attack the reactor pressure vessel. It is clear that the outcome of this attack depends on the distribution of the thermal load. A uniformly distributed heat flux over the entire surface of the lower reactor head coming in contact with the melt is less dangerous than the formation of hot regions, where the vessel is very likely to fail.

It deserves to be mentioned that all relevant physical phenomena connected with core damage have been studied experimentally on rather comprehensive scales and partially even with real nuclear material. One of the leading experiments on the damage of a real irradiated fuel element, including the integral plant behavior during BDBA, is PHEBUS, which was mentioned in the previous section [26]. Some examples of other key experiments are the test on melt-concrete interaction in Karlsruhe [28], the experiments on natural circulation, stratification and heat transport in core melts relocated into the lower head of the reactor pressure vessel, like the SIMECO tests carried out with model fluids [29], as well as experiments on reactor vessel integrity including creep-failure carried out at the KTH in Stockholm [30]. This list is far from exhaustive.

Modeling has to take into account processes of melting and solidification. The latter is observed as crust formation in those places where the molten material is sufficiently cooled. Once relocated into the lower head of the reactor pressure vessel, the melt starts attacking the steel. The typical situation (Figure 4.20) is characterized by an oxidic melt pool consisting mainly of UO_2 covered by molten metal originating from the melting of structural materials (cladding, control rods, RPV internals). The fact that the metallic melt has a much higher thermal conductivity, together with natural circulation flows in the melt pool, leads to a focusing of the

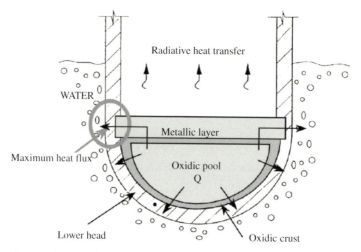

Figure 4.20 Typical situation after relocation of a core melt into the lower head of a reactor pressure vessel, causing the focus effect of the heat flux [29].

heat flux on a comparatively small part of the vessel surface. This leads to an aggravation of the attack. According to the experiments of Sehgal [29], the most probable location of a vessel failure is not at the lowest point of the lower plenum, but rather at the side where the focus of the thermal flux is observed. Additionally, there is the risk of local hot spots. The correct simulation of the melt behavior is therefore essential for the prediction of whether the reactor pressure vessel might fail, and especially for confirming the effectiveness of external cooling measures which aim at confining the melt inside the reactor pressure vessel. The latter is called 'in-vessel core retention' and has been developed for some reactor types of Generation III.

Computations of the heat flux focus effect are presented by Caroli et al. [31], who used the finite element code CASTEM2000 for a coupled fluid dynamic and structural modeling. The latter included ablation of the pressure vessel wall as well as melt solidification. Thermal radiation from the upper surface of the melt pool was taken into account. The mesh was two-dimensional assuming axial symmetry. The calculations aimed at a demonstration of an in-vessel core retention strategy; that is, the presence of water in the reactor pit was assumed and modeled by a heat transfer correlation for nucleate boiling, extended by a model for the boiling crisis, which was set at the outer surface of the vessel wall. On the inside, perfect contact between molten core material and vessel steel was assumed, which is conservative, since there is evidence that a gap between melt crust and vessel wall can contribute to the cooling. The flow field in the melt pool as well as vessel ablation and crust formation are shown in Figure 4.21 for a selected instant during the process. The mechanism of the heat flux focusing by the metallic melt layer on top of the pool is well reproduced, and causes the vessel ablation to progress fastest at the side of the reactor pressure vessel.

Tran and Dinh [32] use the concept of effective convectivity, which replaces the ordinary advective term in the energy transport equation and allows calculation of 3D

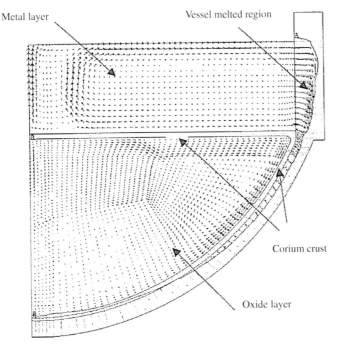

Figure 4.21 Flow field, focus effect, vessel ablation, and crust formation in the lower reactor plenum in the case of impact by a two-layer core-melt computed by Caroli et al. [31] using the finite element code CASTEM2000.

temperature fields and heat fluxes to the structures without solving the Navier–Stokes equations in large, complex geometries, which is especially computationally expensive in situations with high Rayleigh numbers, as in the case of melt pools. Real velocities are replaced by characteristic velocities calculated by algebraic correlations containing models for the heat transfer at the boundary of the flow domain, that is, at the reactor walls. The authors apply CFD to calibrate these models by comparison with small-scale experiments.

In Tran and Dinh [33], the same authors apply the model, which is essentially a coarse-mesh approach, to the lower head of the reactor pressure vessel of a BWR. The challenge consists here in the presence of the guide tubes of the control rod drives, which penetrate the vessel wall from below. These significantly contribute to the heat removal from the melt and therefore cannot be neglected (Figure 4.22).

Some types of Generation III Light Water Reactor concepts rely on an in-vessel core retention concept. Stosic et al. [34] describe the solution applied in the project of the SWR1000 (today called KERENA). In the article, the result of a full-scope CFD simulation of the core damage is presented, including the relocation of the molten material into the lower head of the RPV as well as the modeling of the external RPV cooling. The calculations were performed by Kolev using the in-house code iva4 (see Ref. [35]).

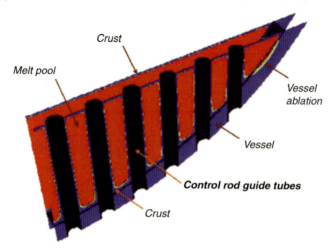

Figure 4.22 Temperature field in a core melt relocated into the lower head of a BWR pressure vessel, where the presence of control rod guide tubes leads to an enhancement of the cooling and a mitigation of the consequences [32].

4.8
Summary

In the last years, CFD has found its way into the field of nuclear reactor safety. Although not yet recognized as hard proof for licensing relevant safety analyses, it is used to deliver indications on the effect of three-dimensional phenomena which are not reflected by today's system codes. Nuclear regulators increasingly ask for CFD analyses in certain cases in order to check if conservative assumptions required in deterministic safety analyses are correct. The need for 3D analyses is increasing with the growing interest in passive safety systems, which gained additional momentum after the accident in Fukushima. Further urgency comes from the fact that with increasing design optimization and erosion of safety margins more accurate methods are needed for better estimation of safety margins. CFD is also used to achieve a better understanding of single phenomena as well as the performance of single components of the plant, and it is widely used to design and interpret experiments. On the small scale, CFD, and in particular high-resolution methods such as DNS, LES, and Lagrangean particle tracking are used to derive and validate global, empirical correlations for an application in coarse-mesh codes and 1D thermal–hydraulic system codes. In some cases, like the Pressurized Thermal Shock (PTS) problem and the further development of passive systems, there is a growing need to validate CFD until it achieves sufficient maturity for full acceptance in the licensing process.

Two fields have been completely excluded from the scope of the present paper. First of all, CFD is becoming the driving tool for the optimization of nuclear fuel elements. There is tremendous progress in 3D modeling of the flow field in fuel rod bundles with a full geometrical representation of the spacer geometry and its influence on heat transfer. Spacer grids are primarily installed to keep the rather long and thin fuel rods in the correct position, so that it is is very important to maintain the correct free cross-section for the coolant flow. In the past decades, spacers were increasingly used to enhance heat transfer by adding functional elements of flow control – so-called vanes. Today, both single-phase and two-phase flows under adiabatic and non-adiabatic conditions are successfully modeled by CFD codes, including convective and boiling heat transfer. The results are at least qualitatively correct, and quantitative agreement with available experiments is being continuously improved. The calculations already contribute significantly to the design optimization. This field would deserve individual reporting.

Other emerging fields of application are novel reactor concepts of Generation IV, especially with regard to gaseous and liquid metal coolants. The latter pose new challenges, above all due to the fact that low-Prandtl-number heat transfer issues have to be tackled. Here, some similarities to the described issues connected with the CFD modeling of the heat transfer in case of core melt accidents are evident. Gaseous coolants, on the other hand, show strong thermal stratification effects due to thermally induced large density differences. Here, benefit can be taken from recent studies on containment phenomena applied to LWR plants.

List of Acronyms

BDBA Beyond Design Based Accident
BPG Best Practice Guidelines
BWR Boiling Water Reactor
CEA Commissariat à l'énergie atomique
CFD Computational Fluid Dynamics
DBA Design Based Accident
DNB Departure from Nucleate Boiling
DNS Direct Numerical Simulation
ECC Emergency Core Cooling
ERCOFTAC European Research Community on Flow, Turbulence, and Combustion
HZDR Helmholtz-Zentrum Dresden-Rossendorf (former FZR)
IAEA International Atomic Energy Agency
LES Large Eddy Simulation
LOCA Loss-of-Coolant Accident
LWR Light Water Reactor (PWR or BWR)
PAR Passive Recombiners (catalytic recombiners for hydrogen)
PSA Probabilistic Safety Analysis
PSI Paul Scherrer Institute, Switzerland
PTS Pressurized Thermal Shock
PWR Pressurized Water Reactor
RANS Reynolds Averaged Navier Stokes
RPV Reactor Pressure Vessel

RSM Reynolds Stresses' Model
SAS Scale Adaptive Simulation
TH Thermal Hydraulics
URANS Unsteady RANS
INRA International Nuclear Regulators Association
WENRA Western European Nuclear Regulators' Association
ENSI Federal Nuclear Safety Inspectorate
1D One-Dimensional
2D Two-Dimensional
3D Three-Dimensional

References

1 Bestion, D. (2010) Extension of CFD codes application to two phase flow safety problems. *Nuclear Engineering and Technology*, **42** (4), 1–12.
2 OECD (January 2008) Assessment of Computational Fluid Dynamics (CFD) for Nuclear Reactor Safety Problems. Nuclear Safety, NEA/CSNI/R(2007)13.
3 Höhne, T., Kliem, S., Rohde, U., and Weiss, F.-P. (2008) Boron dilution transients during natural circulation flow in PWR – Experiments and CFD simulations. *Nuclear Engineering and Design*, **238**, 1987–1995.
4 Prasser, H.-M., Grunwald, G., Höhne, T., Kliem, S., Rohde, U., and Weiss, F.-P. (2003) Coolant mixing in a PWR – deboration transients, steam line breaks and emergency core cooling injection – experiments and analyses. *Nuclear Technology*, **143**, 37–56.
5 Pointner, W. and Wohlstein, R. (2003) Initiating events and scenarios of boron dilution transients (in German). Proceedings of the Topical session, Experimental and Theoretical Investigations on Boron Dilution Transients in PWRs, Annual meeting on Nuclear Technology, pp. 5–22.
6 Höhne, T., Kliem, S., and Bieder, U. (2006) Modeling of a buoyancy-driven flow experiment at the ROCOM test facility using the CFD codes CFX-5 and Trio U. *Nuclear Engineering and Design*, **236**, 1309–1325.
7 Petrov, V. and Manera, A. (June 13–17 2010) Development and validation of a CFD model for the EPR pressure vessel. Proc. Int. Congress on Advances in Nuclear Power plants (ICAPP'10), San Diego, California, USA (SB-RND-ACT-001-08.008).
8 Gelineau, O., Sperandio, M., Simoneau, J.P., Hamy, J.M., and Roubin, P.H.L. (2002) Thermomechanical and thermalhydraulic analyses of a T-junction using experimental data. In: IAEA-TECDOC-1318, pp. 31–62.
9 Shah, V.N. et al. (August 1999) Assessment of field experience related to pressurized water reactor primary system leaks. Proceedings of the ASME Pressure Vessel & Piping 1999, Boston, MA, USA.
10 Jungclaus, D. et al. (7–12 June 1998) Common IPSN/GRS safety assessment of primary coolant unisolable leak incidents caused by stress cycling. Proceeedings of the NEA/CSNI Specialists' Meeting on Experience with Thermal Fatigue in LWR Piping Caused by Mixing and Stratification, OECD Nuclear Energy Agency, Paris, France.
11 Kuhn, S., Braillard, O., Niceno, B., and Prasser, H.-M. (2010) Computational study of conjugate heat transfer in T-junctions. *Nuclear Engineering and Design*, **240**, 1548–1557.
12 Braillard, O. (5 June 2008) FATHERINO experiment for thermal fatigue studies in a mixing tee and nozzle – support of TH and mechanical analysis of thermal load determination. In: Älvkarleby, Sweden.

13 Frank, T., Lifante, C., Prasser, H.M., and Menter, F. (2010) Simulation of turbulent and thermal mixing in T-junctions using URANS and scale-resolving turbulence models in ANSYS CFX. *Nuclear Engineering and Design*, **240**, 2313–2328.

14 Manera, A., Prasser, H.-M., Lechner, R., and Frank, T. (2009) Prediction of Temperature Fluctuations by means of Steady RANS for the Estimation of Thermal Fatigue. NURETH-13, Kanazawa City, Japan, Sept. 27–Oct. 2, 2009, paper N13P1312.

15 Merzari, E., Khakim, A., and Ninokata, H. (2008) Toward an Accurate URANS Approach for the Prediction of the Flow in a T-junction. *NUTHOS*-7, Seoul, Korea, October 5–9, Paper 164.

16 Martin, A., Cornille, S., Lestang, F., Bellet, S., Barbier, A., Vit, C., and Huvelin, F. (2009) CFD use in PTS safety analysis state of art and challenges for industrial applications. The 13th International Topical Meeting on Nuclear Reactor Thermal Hydraulics (NURETH-13), Kanazawa City, Japan, Sept. 27–Oct. 2, 2009, Paper N13P1307.

17 Labois, M. and Lakehal, D. (14–16 September 2010) PTS Prediction using the CMFD code TransAT: The COSI Test Case. CFD4NRS-3, Workshop on Experimental Validation and Application of CFD and CMFD Codes to Nuclear Reactor Safety Issues, Washington D.C., USA, paper 11.4.

18 Apanasevich, P., Lucas, D., and Höhne, T. (14–16 September 2010) Pre-Test CFD Simulations on TOPFLOW-PTS Experiments with ANSYS CFX 12.0. CFD4NRS-3, Workshop on Experimental Validation and Application of CFD and CMFD Codes to Nuclear Reactor Safety Issues, Washington D.C., USA, paper 11.1.

19 Heitsch, M. (2000) Fluid dynamic analysis of a catalytic recombiner to remove hydrogen. *Engineering and Design*, **201**, 1–10.

20 Reinecke, E.-A., Bentaib, A., Kelm, S., Jahn, W., Meynet, N., and Caroli, C. (2010) Open issues in the applicability of recombiner experiments and modelling to reactor simulations. *Progress in Nuclear Energy*, **52**, 136–147.

21 Andreani, M., Haller, K., Heitsch, M., Hemström, B., Karppinen, I., Maceke, J., Schmid, J., Paillere, H., and Toth, I. (2008) A benchmark exercise on the use of CFD codes for containment issues using best practice guidelines: A computational challenge. *Nuclear Engineering and Design*, **238**, 502–513.

22 Heitsch, M., Baraldi, D., and Wilkening, H. (2010) Simulation of containment jet flows including condensation. *Nuclear Engineering and Design*, **240**, 2176–2184.

23 Paladino, D., Zboray, R., Benz, P., and Andreani, M. (2010) Three-gas mixture plume inducing mixing and stratification in a multi-compartment containment. *Nuclear Engineering and Design*, **240**, 210–220.

24 Baraldi, D., Heitsch, M., and Wilkening, H. (2007) CFD simulations of hydrogen combustion in a simplified EPR containment with CFX and REACFLOW. *Nuclear Engineering and Design*, **237**, 1668–1678.

25 Dehbi, A. (2009) A stochastic Langevin model of turbulent particle dispersion in the presence of thermophoresis. *International Journal of Multiphase Flow*, **35**, 219–226.

26 von der Hardt, P. and Tattegrain, A. (1992) The Phebus fission product project. *Journal of Nuclear Materials*, **188**, 115–130.

27 Dehbi, A.(17 October 2007) Thermophoretic Deposition in the Phébus Steam Generator Tube: a Revisit. CACIC Meeting, Aix-en-Provence, France.

28 Alsmeyer, H. (1987) Beta experiments in verification of the WECHSL Code: Experimental results on the melt-concrete interaction. *Nuclear Engineering and Design*, **103**, 115–125.

29 Sehgal, B.R. (2006) Stabilization and termination of severe accidents in LWRs. *Nuclear Engineering and Design*, **236**, 1941–1952.

30 Sehgal, B.R., Theerthan, A., Giri, A., Karbojian, A., Willschütz, H.G., Kymäläinen, O., Vandroux, S., Bonnet, J.M., Seiler, J.M., Ikkonen, K., Sairanen, R., Bhandari, S., Bürger, M., Buck, M., Widmann, W., Dienstbier, J., Techy, Z.,

Kostka, P., Taubner, R., Theofanous, T., and Dinh, T.N. (2003) Assessment of reactor vessel integrity (ARVI). *Nuclear Engineering and Design*, **221**, 23–53.

31 Caroli, C., Miliozzi, A., and Milillo, F. (1999) Thermo-mechanical behaviour of the reactor pressure vessel and corium molten pool in a severe accident with core melt down. SMiRT 15, Seoul, Korea, paper P02/2.

32 Tran, C.-T. and Dinh, T.-N. (2009) The effective convectivity model for simulation of melt pool heat transfer in a light water reactor pressure vessel lower head. Part I: Physical processes, modeling and model implementation. *Progress in Nuclear Energy*, **51** (2009), 849–859.

33 Tran, C.-T. and Dinh, T.-N. (2009a) The effective convectivity model for simulation of melt pool heat transfer in a light water reactor pressure vessel lower head. Part II: Model assessment and application. *Progress in Nuclear Energy*, **51**, 860–871.

34 Stosic, Z.V., Brettschuh, W., and Stoll, U. (2008) Boiling water reactor with innovative safety concept: The Generation III + SWR-1000. *Nuclear Engineering and Design*, **238**, 1863–1901.

35 Kolev, N.I. (2001) SWR 1000 severe accident control through in-vessel melt retention by external RPVcooling. 9th International Conference on Nuclear Engineering – ICONE-9, Nice, France, April 9–12, 2001.

Part Two
Computer or Experimental Design?

5
Sizing and Operation of High-Pressure Safety Valves

Jürgen Schmidt and Wolfgang Peschel

5.1
Introduction

Safety valves are presently sized according to ISO 4126-1. This standard is based on equations for ideal gases and is not valid if the gas properties are significantly affected by the system pressure, that is, if the gas behaves as a real gas. In practice this effect must be taken into consideration typically when sizing a safety valve for set pressures not far from the saturation line or above the thermodynamic critical point. Even for a real gas the valve sizing procedure can be based on a nozzle/discharge coefficient model, as recommended in ISO 4126-1. Then, the discharge coefficient has to be measured experimentally under conditions representative or equal to the sizing conditions. But most of the current test facilities allow only for valve tests up to about 25 MPa. As an alternative, the valve may be sized by computational fluid dynamics. This permits not only optimization of the valve capacity but also investigation of the valve characteristics. Any valve experiment would be unnecessary if CFD calculations are reliable and the sizing effort is comparable to current sizing procedures.

In the following, three sizing methods are compared with each other: the nozzle/discharge coefficient model for an ideal and a real gas and a CFD sizing procedure. Comparisons with experimental data are carried out to evaluate the methods.

5.2
Phenomenological Description of the Flow through a Safety Valve

The flow channel formed during the flow of a fluid through a safety valve has a very complex shape (Figure 5.1). The fluid entering the valve is at first strongly accelerated, and it is subsequently strongly deflected under the valve disc. The exact contour of the flow channel is determined by the inlet conditions, the type of gas, and the geometry of the valve. In the valve seat (Point 1 in Figure 5.1) or under the valve disc (Points 2 and 3 in Figure 5.1), as a rule, a critical pressure ratio is established, where the gas reaches the velocity of sound. Immediately thereafter, the pressure decreases very strongly, in some cases significantly below ambient pressure, and a region of

---

Process and Plant Safety: Applying Computational Fluid Dynamics, First Edition. Edited by J. Schmidt.
© 2012 Wiley-VCH Verlag GmbH & Co. KGaA. Published 2012 by Wiley-VCH Verlag GmbH & Co. KGaA.

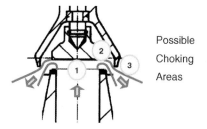

Figure 5.1 Example of the contour of flow in a safety valve with the places where the velocity of sound may be developed.

supersonic velocity is formed in the valve housing. At the periphery of this region, the flow is extremely decelerated (compression shock), and the pressure decreases to almost the valve outlet flange pressure. Typically, at very high set pressures, the pressure in the outlet flange is again decreased by a second compression shock down to almost the back pressure of the valve. Any change of conditions in the inlet may result in a change of the position and size of the narrowest cross section of flow and the location of pressure shocks in the valve and the outlet line.

5.3
Nozzle/Discharge Coefficient Sizing Procedure

Most safety valves in industry are used at set pressures below 5 MPa (50 bar). For gas service they are sized according to EN-ISO 4126-1 [1] (or according to national regulations derived from this).

In general, the seat cross sectional area of a safety valve A_{seat} must be at least large enough to discharge the minimum required mass flow rate $Q_{m,OUT}$ according to the sizing scenario for a pressurized system. For simplification, the dischargeable mass flow rate $Q_{m,SV}$ through the valve is calculated by means of an equivalent nozzle flow model for ideal gases corrected by an experimentally determined discharge coefficient $K_{d,g}$. With an appropriate nozzle flow model, all property data at the operating conditions are covered, that is, only the geometric effects are included in the discharge coefficient. The more accurately the nozzle flow model reproduces the actual mass flow rate, the less the discharge coefficient is influenced by property data. This leads to the sizing criteria for safety valves:

$$Q_{m,SV} = K_{d,g} \cdot Q_{m,nozzle} \geq Q_{m,OUT} \tag{5.1}$$

With the sizing coefficient C_g for a nozzle with a throat area equal to the seat area of the safety valve, the mass flow rate is defined in a dimensionless form:

$$C_g = \frac{Q_{m,nozzle}}{A_{seat} \cdot \sqrt{2 \cdot \frac{p_0}{v_0}}} \tag{5.2}$$

Eqs. (5.1) and (5.2) are definitions, which are valid for both ideal gas and real gas flow.

5.3.1
Valve Sizing According to ISO 4126-1

The sizing coefficient for a steady state discharge of an ideal gas through a nozzle without friction and heat exchange with the wall reads as (isentropic flow of an ideal gas):

$$C_{g,\text{ISO}} = \sqrt{\frac{\kappa_0}{\kappa_0 - 1} \cdot (\eta)^{2/\kappa_0} \cdot \left[1 - (\eta)^{(\kappa_0 - 1)/\kappa_0}\right]}; \quad \eta = \frac{p}{p_0} \tag{5.3}$$

Eq. (5.3) was developed for plenum flow with negligible inlet velocity. If the inlet velocity may not be neglected, a (fictitious) stagnation pressure – the actual inlet pressure increased by the pressure difference computed by an isentropic deceleration of flow to the velocity of zero – must be introduced in Eq. (5.3). This equation is not valid if gases condense or chemical reactions occur.

The critical pressure ratio is the lowest pressure ratio that can be established under these flow conditions in the narrowest flow cross section of the valve:

$$\eta_{\text{crit,ISO}} = \left(\frac{2}{\kappa_0 + 1}\right)^{\frac{\kappa_0}{\kappa_0 - 1}} \tag{5.4}$$

According to EN-ISO 4126-1, the isentropic exponent κ is to be related to the entry conditions ($\kappa = \kappa_0$). Since no additional information has been provided for the calculation, it is obvious to compute the isentropic exponent as well as the sizing coefficient for an ideal gas:

$$\kappa_0 = \frac{c_{p,0}}{c_{p,0} - R} \tag{5.5}$$

The specific volume at stagnation conditions in Eq. (5.2) should be determined not for ideal gases but rather for a real gas. No method is recommended in EN-ISO 4126-1 for computing the real gas factor.

The valve discharge coefficient is in general determined experimentally by the valve manufacturer for a fixed nominal lift. It corresponds to the average value of the ratio from mass flows measured and calculated with the nozzle flow model [1]. At least three measurements are performed at each of three set pressures on two or three sizes of the same valve type or three significantly different springs in the same valve.

$$K_{d,g} = \frac{1}{n} \cdot \sum_n \left(\frac{Q_{m,\text{exp}}}{Q_{m,\text{nozzle}}}\right) \tag{5.6}$$

This testing of safety valves at a fixed lift presumes that the valves will open at least to the nominal lift in case of an emergency relief.

Overall, the sizing method according to EN-ISO 4126-1 is based on very simple equations, mainly for ideal gases. In the literature the validity of this sizing method was shown by numerical calculations up to pressures of about 10 MPa (100 bar) [2, 3]. Although there are no limits of application formulated in the standards for high pressure valves, the method should not be applied if real gas effects are significant. Furthermore, the discharge coefficients for safety valves are most often measured at pressures between 0.1 and 5 MPa and assumed to be constant up to arbitrary sizing conditions. Measurements under increased sizing conditions are restricted by the application limit of current test facilities, which amounts to about 25 MPa (250 bar).

5.3.2
Limits of the Standard Valve Sizing Procedure

Despite the complex flow phenomena, the technical rules and standards give simple nozzle equations without application limits for the sizing of safety devices, and the valve manufacturers report only two discharge coefficients as constant values for all gases/vapors and liquids in their catalogs. Physically, this is only possible if the flow in the valves behaves similarly. The location where the velocity of sound is established must be essentially the same, and the narrowest flow cross section in the valve may not vary significantly (Reynolds and Mach number similarity).

Near the thermodynamic critical point, conversely, gases no longer behave ideally but as real gases. As a result of this, the stagnation temperature in the nozzle varies; therefore, the gas generally becomes cooler in the presence of moderate inlet pressures (negative Joule–Thomson effect); however, it heats up considerably if the inlet pressure is very high (positive Joule–Thomson effect). This real expansion behavior is displayed by many gases even at very low pressures (e.g., methane). It influences the location and size of the narrowest flow cross section and thereby the flow characteristics of the valve. It can also be changed by wall friction and the transport of heat to the wall if the gas densities are so high that these effects are no longer negligible. In addition, the similarity of the flow may also be nullified by geometrical factors. Safety valves of the same type are frequently not geometrically similar at different nominal widths for economic and engineering reasons. Since these effects are not accounted for at all in the simple nozzle flow model, they would have to be included in a new flow model or in the discharge coefficients of the valves. As a consequence, the coefficients according to ISO 4126-1 cannot be constant. Additionally, the discharge coefficients for gases and liquids at the thermodynamic critical point should even be the same for physical reasons, because the densities of both phases are identical there. The discharge coefficients, however, are not measured under the above-noted boundary conditions.

It is expected that the discharge coefficients for gases/vapors are influenced by a whole series of parameters (gas type, valve geometry and property data) which are presently not fully accounted for either in the nozzle model per EN-ISO 4126-1 or in the testing procedure.

5.3.3
Valve Sizing Method for Real Gas Applications

Compared to an ideal gas, the caloric and thermal properties of the gas depend not only on temperature but also on pressure. In the case of stationary flow of a real gas through an adiabatic nozzle, the stagnation enthalpy between the entrance and the narrowest flow cross section, the nozzle throat, remains constant (h_t = const). The change in enthalpy corresponds to the change in kinetic energy, and the irreversible frictional pressure loss is expressed by the resistance coefficient ς:

$$-\int_0^{\text{throat}} dh = \frac{w^2}{2}(1+\varsigma); \qquad w_0 \ll w \tag{5.7}$$

$$Q_{m,\text{nozzle}} = \frac{A_{\text{throat}}}{v_{\text{throat}}} \cdot \sqrt{\frac{-2}{1+\varsigma} \cdot \int_0^{\text{throat}} dh} = C_g \cdot A_{\text{throat}} \cdot \sqrt{2 \cdot \frac{p_0}{v_0}} \tag{5.8}$$

The mass flow rate is obtained by a progressive variation of the state conditions in the nozzle throat at a constant stagnation enthalpy until the back pressure or a maximum of the mass flow rate (critical pressure ratio) is reached.

This method is suitable when state diagrams of the gases are available or can be plotted (see for example, Thompson et al. [4]). It applies not only for pure substances but also for mixtures of gases. If graphical methods are not efficient, the mass flow rate through a nozzle can also be determined with an equation for real gases, which can be derived from the first law of thermodynamics [5, 6]:

$$dh = c_p \cdot dT + \left[v - T \cdot \left(\frac{\partial v}{\partial T} \right)_p \right] \cdot dp \tag{5.9}$$

The second term corresponds to the Joule–Thomson effect, which describes the variation of the stagnation temperature during the expansion of the gas. This term is equal to zero for ideal gases.

The temperature dependence of the specific volume at constant pressure may be derived from the second law of thermodynamics:

$$ds = c_p \cdot \frac{dT}{T} - \left(\frac{\partial v}{\partial T} \right)_p \cdot dp \tag{5.10}$$

By means of Eqs. (5.9) and (5.10) the derivation of an analytical equation for the mass flow rate through the nozzle is quite simple if the change of state between the entry and the nozzle throat is assumed to be isentropic ($ds = 0$) and the frictional losses neglected at this time ($\varsigma = 0$) as well as the heat transfer with the wall are accounted for by the velocity coefficient φ; otherwise the equations cannot be analytically integrated.

$$\varphi = \frac{w}{w_{is}} \tag{5.11}$$

5 Sizing and Operation of High-Pressure Safety Valves

Herewith, the solution of the integral Eq. (5.7) is traced back to the calculation of an isentropic change of state of a real gas:

$$C_g = \varphi \cdot \left(\frac{v_0}{v_{throat}}\right) \cdot \sqrt{\frac{-\int_0^{throat} dh_{is}}{v_0 \cdot p_0}} \quad \text{where} \quad dh_{is} = v \cdot dp \tag{5.12}$$

The change of enthalpy within the isentropic nozzle flow can be calculated by using the first and second laws of thermodynamic if the equation of state for real gases,

$$v = \frac{Z \cdot R \cdot T}{p}, \tag{5.13}$$

including the real gas factor Z and the isentropic exponent κ is introduced

$$dh_{is} = Z \cdot R \cdot \frac{\kappa \cdot [1 + K_T]}{\kappa \cdot [1 - K_P] - 1} \cdot dT = \frac{Z \cdot R}{\Pi} \cdot dT \tag{5.14}$$

$$\Pi = \frac{\kappa \cdot [1 - K_P] - 1}{\kappa \cdot [1 + K_T]} \tag{5.15}$$

K_T and K_p are the gradients of the real gas factor, which correspond to a value of zero for ideal gases:

$$K_T = \left(\frac{T}{Z} \cdot \frac{\partial Z}{\partial T}\right)_p ; \quad K_P = \left(\frac{p}{Z} \cdot \frac{\partial Z}{\partial p}\right)_T \tag{5.16}$$

The isentropic exponent and the specific heat capacity for ideal and real gases are given in Table 5.1.

Eq. (5.12) can be integrated if the average values of the parameters Z and Π, $\langle Z \rangle$ and $\langle \Pi \rangle$, in each case averaged between the stagnation conditions in the entry and the nozzle throat conditions, are introduced. The velocity in the nozzle throat in the case of an isentropic change of state is obtained from this:

$$C_g = \varphi \cdot \frac{v_0}{v_{throat}} \cdot \sqrt{\frac{1}{\langle \Pi \rangle} \cdot \frac{\langle Z \rangle}{Z_0} \cdot \left(1 - \frac{T_{throat}}{T_0}\right)} \tag{5.17}$$

$$\langle \Pi \rangle = \left(\frac{\kappa \cdot [1 - K_P] - 1}{\kappa \cdot [1 + K_T]}\right)\Big|_{T_0}^T = \frac{\langle \kappa \rangle \cdot [1 - \langle K_P \rangle] - 1}{\langle \kappa \rangle \cdot [1 + \langle K_T \rangle]} \tag{5.18}$$

With the isentropic relationships for real gases,

$$\frac{T}{T_0} = \left(\frac{p}{p_0}\right)^{\langle \Pi \rangle} = (\eta)^{\langle \Pi \rangle}, \tag{5.19}$$

$$\frac{v_0}{v} = \left(\frac{p}{p_0}\right)^{\frac{1}{\langle \kappa \rangle}} \tag{5.20}$$

and the velocity coefficient φ, the sizing coefficient for a nozzle in case of a frictional flow of a real gas is obtained,

$$C_g = \varphi \cdot \sqrt{\frac{1}{\langle \Pi \rangle} \cdot \frac{\langle Z \rangle}{Z_0} \cdot (\eta)^{2/\langle \kappa \rangle} \cdot \left(1 - (\eta)^{\langle \Pi \rangle}\right)} \tag{5.21}$$

For an ideal gas and a frictionless flow through an adiabatic nozzle, where $\langle Z \rangle = Z = 1 \rightarrow K_T = K_p = 0$, $\varphi = 1$ is valid, Eq. (5.21) merges into Eq. (5.3).

Typical values for the velocity coefficient in a frictional nozzle flow are 0.9 to 1; Sigloch (Ref. [7], p. 392) gives an average value of 0.97. Numerical calculations on a high pressure safety valve lead to the conclusion that the velocity coefficient is close to 1 [8]. The real gas factor is generally determined with a suitable equation of state.

The critical pressure ratio η for calculating the sizing coefficient [see Eq. (5.3) (ideal gas) or Eq. (5.21) (real gas)] corresponds to the maximum of either the back pressure ratio $\eta = \eta_b = p_b/p_0$ or the critical pressure ratio $\eta = \eta_{crit} = p_{crit}/p_0$.

Care must be taken applying Eq. (5.21) near the critical point. Here, mean averaging of the property data is not allowed due to the highly non-linear changes of the real gas factor and its gradients. In this case only the numerical solution of the first and second laws of thermodynamics seems meaningful.

5.3.4
Numerical Sizing of Safety Valves for Real Gas Flow

The most precise 1 D method for calculating the critical mass flux through a safety valve accounting for real gas effects is a numerical integration of Eq. (5.12), defining the sizing coefficient for a nozzle for real gas service. For the integration along the nozzle path, from the inlet to the throat, the specific heat capacity $c_p(p, T)$ and the specific volume $v(p, T)$ of the real gas are needed depending on temperature *and* pressure. For most gases in practice the specific heat capacity is only available for ideal gases $c_{p_{id}}(T)$ – it depends only on temperature, whereas the specific volume can be deduced from an equation of state (see next chapter). Consequently, Eq. (5.12) should be integrated via the following thermodynamic path:

1) Isothermal change of state at inlet stagnation temperature T_0 between the inlet stagnation pressure p_0 and zero pressure, where the gas behaves ideally
2) Isobaric change of state of an ideal gas at zero pressure from inlet stagnation temperature T_0 to the nozzle throat temperature T_{throat}
3) Another isothermal change of state at nozzle throat temperature T_{throat} between the zero pressure and the pressure in the nozzle throat p_{throat}.

Herewith, the first law of thermodynamic, Eq. (5.9), reads:

$$\int_0^{throat} dh_{is} = \int_{T_0}^{T_{throat}} cp_{id}(T) \cdot dT$$

$$- \int_{p_0}^{0} Z(p, T_0) \cdot R \cdot T_0 \cdot K_T(p, T_0) \cdot \frac{dp}{p} \tag{5.22}$$

$$- \int_0^{p_{throat}} Z(p, T_{throat}) \cdot R \cdot T_{throat} \cdot K_T(p, T_{throat}) \cdot \frac{dp}{p}$$

Because the throat conditions p_{throat} and T_{throat} are not known, the second law of thermodynamic, Eq. (5.10), must be solved simultaneously. Generally, an isentropic change of state can be assumed (ds = 0) as long as heat transfer and friction to the wall does not significantly affect the result [3]:

$$0 = \ln\left(\frac{p_0}{p_{throat}}\right)_s + \int_{T_0}^{T_{throat}} \frac{c_{p_{id}}(T)}{R \cdot T} dT$$

$$\int_{p_0}^{0} (Z(p, T_0) \cdot (1 + K_I(p, T_0)) - 1)\frac{dp}{p} \quad (5.23)$$

$$- \int_{0}^{p_{throat}} (Z(p, T_{throat}) \cdot (1 + K_T(p, T_{throat})) - 1)\frac{dp}{p}$$

Applying the averaging procedure, both equations will merge into the analytical form for real gases as derived in the previous chapter. Following the ideal gas assumptions leads to the procedure given in ISO 4126-1.

Generally, the sizing coefficient of Eq. 5.12 in combination with Eqs. 5.22 and 5.23 would be lowered due to frictional effects and increased due to cooling heat transfer through the wall, but these effects are seldom dominant [3].

Based on the definition of enthalpy, the specific heat capacity of a real gas may be derived to calculate the real gas isentropic coefficient (see Table 5.1):

$$c_p(T, p) = c_{p_{id}}(T) - \frac{d}{dT}\left[\int_0^p Z(p, T) \cdot R \cdot T \cdot K_T(p, T) \cdot \frac{dp}{p}\right] \quad (5.24)$$

Table 5.1 Property data of ideal and real gases.

Ideal gas	Real gas
$\kappa = \dfrac{c_{p_{id}}}{c_{p_{id}} - R}$	$\kappa = \dfrac{c_p}{c_p \cdot [1 - K_P] - Z \cdot R \cdot [1 + K_T]^2}$
$c_{p_{id}} = \dfrac{\kappa \cdot R}{\kappa - 1}$	$c_p = Z \cdot R \cdot \dfrac{\kappa \cdot [1 + K_T]^2}{\kappa \cdot [1 - K_P] - 1}$
$c_{v_{id}} = \dfrac{R}{\kappa - 1}$	$c_v = Z \cdot R \cdot \dfrac{[1 + K_T]^2}{\kappa \cdot [1 - K_P]^2 - [1 - K_P]}$

5.3.5
Equation of State, Real Gas Factor, and Isentropic Coefficient for Real Gases

In industry, predominantly cubic equations of state (EoS) are used, because very many coefficients are tabulated. The most widely used EoS can be expressed as, see table 5.2:

$$p = \frac{R \cdot T}{v-b} - \frac{\Theta}{v^2 + \delta \cdot v + \varepsilon} \cdot \frac{v}{R \cdot T} \tag{5.25}$$

Table 5.2 Coefficients of the cubic equations of state.

Equation of state	Θ	δ	ε
Redlich Kwong [9]	$a/\sqrt{T}$	b	0
Soave Redlich Kwong [10]	$a \cdot \alpha(T)$	b	0
Peng Robinson [11]	$a \cdot \alpha(T)$	$2 \cdot b$	$-b^2$

Table 5.3 Coefficients of the cubic equations of state.

Equation	Ω_1	Ω_2	S
Redlich Kwong	0.42748	0.08664	0
Soave Redlich Kwong	0.42748	0.08664	$0.48 + 1.574 \cdot \omega - 0.176 \cdot \omega^2$
Peng Robinson	0.45724	0.0778	$0.37474 + 1.54226 \cdot \omega - 0.26992 \cdot \omega^2$

Here, a and b are component-specific parameters that are adapted to measured data or can be calculated from the thermodynamic critical data of the gas, see table 5.3:

$$a = \Omega_1 \cdot \frac{R^2 \cdot T_c^2}{p_c}, \quad b = \Omega_2 \cdot \frac{R \cdot T_c}{p_c} \tag{5.26}$$

$$\alpha(T) = \left[1 + S \cdot \left(1 - \left(\frac{T}{T_c}\right)^{0.5}\right)\right]^2 \tag{5.27}$$

$$\omega = -\log\left(\frac{p}{p_c}\right)_{T/T_c = 0.7} - 1 \tag{5.28}$$

The equations have not been developed for very high pressures (100 MPa and above). Here there are special equations of state available for individual gases [12]. In this pressure range a comparison with measured data is always recommended. In the case of gas mixtures, the coefficients a and b can be determined with the aid of mixing rules [13–15].

The real gas factor Z is calculated from the equation of state by adjustment of Eq. (5.25) with the specific volume $v = Z \cdot R \cdot T/p$. The equations for computing the real gas factor Z and the derivatives K_T und K_p can be derived analytically.

The isentropic exponent κ is defined as ratio of pressure fluctuations and corresponding fluctuations of the density at isentropic conditions and may be calculated, for example, by measurements of the velocity of sound [16, 17]

$$\kappa = -\frac{v}{p}\left(\frac{\partial p}{\partial v}\right)_S = \frac{\varrho \cdot w_{\text{crit}}^2}{p} \quad \text{with} \quad w_{\text{crit}} = \sqrt{\frac{1}{-\left(\frac{\partial v}{\partial p}\right)_{\text{is}}}} \tag{5.29}$$

For an ideal gas the isentropic exponent according to Eq. (5.29) can be transferred into the ratio of specific heat capacities at constant pressure and volume (see Table 5.1). There are various calculation procedures in the literature for approximate

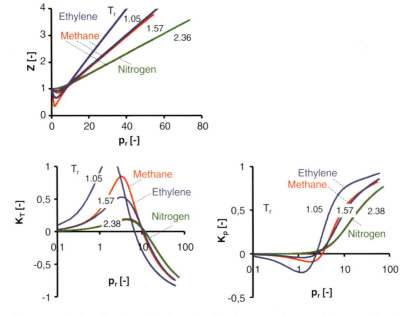

Figure 5.2 Real gas factor Z and its derivatives for ethylene, methane, and nitrogen as functions of the reduced stagnation pressure and reduced stagnation temperature.

solutions, for example, isentropic exponent for ideal gases (cf. Table 5.1) and the calculation with an equation of state.

In general, the isentropic exponents take very high values at high reduced pressures; the description of this factor on the assumption of an ideal gas behavior is then no longer permissible.

Figure 5.2 shows the real gas factor, the derivation of this factor with respect to temperature and pressure, and the isentropic coefficient for a real gas versus the reduced inlet stagnation pressure.

5.3.6
Comparison of the Nozzle Flow/Discharge Coefficient Models

Three nozzle flow models have been considered to calculate the sizing coefficient C_g: the analytical ideal gas model according to ISO 4126-1 (ISO), the analytical real gas model (REAL), and the numerical solution of the thermodynamic laws (1D-num).

In Figure 5.3 an example is given of the discharge of ethylene through a nozzle with fixed inlet conditions of 250 MPa (2500 bar) and 100 °C and 400 °C at various back pressures. Depicted is the sizing coefficient versus the reduced back pressure p_{rb}, that is, the back pressure divided by the thermodynamic critical pressure of ethylene 5.0418 MPa (50.418 bar). As expected, the sizing coefficient increases on lowering the reduced back pressure, starting from an inlet reduced back pressure of 48.585 (equal to 250 MPa). At the maximum of the sizing coefficient the critical pressure ratio in the nozzle throat is reached. According to the ISO model the maximum of the sizing

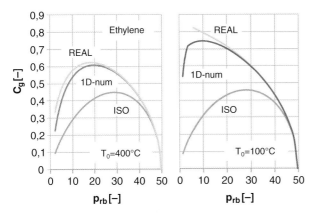

Figure 5.3 Sizing coefficient versus reduced back pressure of ethylene flowing through a nozzle with an inlet pressure of 2500 bar and inlet temperatures of 100 °C and 400 °C.

coefficient is calculated to be 0.45 at a reduced back pressure of about 28, corresponding to a fluid-dynamic critical pressure of about 14 MPa or 1400 bar. There is almost no effect of the inlet temperature. In contrast, the REAL and the most precise 1D-num models are significantly influenced by the inlet temperature. With increasing inlet temperature the sizing coefficient gets larger and the critical pressure is shifted to lower reduced back pressures. At 400 °C (Figure 5.3, left diagram), the maximum sizing coefficient calculated with the 1D-num model reads 0.61 at a reduced back pressure of 19.7 (994 bar), and at 100 °C (Figure 5.3, right diagram) it reaches a value of 0.75 at a reduced back pressure of 10.1 (507 bar). In the second case almost the thermodynamic critical temperature is established in the nozzle throat. The temperature decreases from 100 °C to less than 10 °C. Hence, the accuracy of the REAL model is reduced. Whereas at an inlet temperature of 400 °C, that is, far from the thermodynamic critical temperature even in the nozzle throat, the REAL model gives comparable results, although the pressure drop within the nozzle is very large.

The maxima of the sizing coefficients (dimensionless critical mass flow rates) at various reduced inlet pressures and two reduced inlet temperatures are given in Figure 5.4. According to EN-ISO 4126-1, with an assumption of an ideal behavior of the gas, the sizing coefficient is almost constant. Considering real gas effects with the 1D-num model, considerably different values can be found: sizing coefficients of about 63% larger values compared to EN-ISO 4126-1 at an inlet pressure of 2500 bar ($p_{r0} = 50$, $T_0 = 300\,°C$) are calculated and a 4% larger value was found at 100 bar inlet pressure. Comparable results would be found with other gases. Overall, the type of gas generally has a smaller effect on the sizing coefficient than the state conditions at the entrance.

The REAL and the 1D-num nozzle flow models were validated with measurements performed on a high-pressure nozzle with nitrogen at a test facility of BASF. The measurements were conducted with nitrogen at inlet pressures of up to 1000 bar [5]. Furthermore, the calculations agree very accurately with the values in DIN EN ISO 9300, based on many measurements on nozzles performed worldwide at pressures below 200 bar.

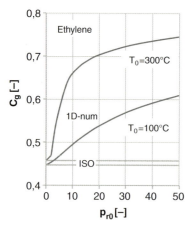

Figure 5.4 Sizing coefficient at critical flow conditions in the throat versus reduced inlet stagnation pressure calculated according to EN-ISO 4126-1 and a nozzle flow model including real gas effects for ethylene at inlet stagnation temperatures of 100 °C and 300 °C.

In general, the sizing coefficients presented herein are valid if no condensation occurs up to the nozzle throat (the narrowest flow cross section of the valve). Therefore, pressures and temperatures at the inlet and the nozzle throat should be above the saturation line of the fluid considered. Two-phase flow conditions are considered by Schmidt in [27].

5.4
Sizing of Safety Valves Applying CFD

Several authors [18–22] have successfully studied the flow through a single valve by means of CFD, but only with air at a low inlet pressure and a fixed lift. Real gas effects have not been accounted for and highly accurate data for validation of the model have not been published. Hence, Beune et al. [8, 23] investigated the flow capacity and characteristic of a commercial proportional spring-loaded high-pressure safety valve manufactured by BASF SE for liquid and gas service, Figure 5.5. The valve discharge coefficient was measured, giving values between 0.266 and 0.308 at the nominal disk lift of 1 mm. The 3D flow through the valve was calculated by means of the commercial software package ANSYS CFX [24]. To validate the results in comparison with experimental data, a high pressure test facility was constructed at BASF.

5.4.1
High Pressure Test Facility and Experimental Results

The BASF test facility in Ludwigshafen is designed to measure the valve capacity and characteristics, such as opening pressure, overpressure, blowdown, lift, and operating stability, in accordance with EN ISO 4126-1. It consists of four vessels and can be used for liquid and gas service (Figures 5.6 and 5.7). The storage vessel B1 was

5.4 Sizing of Safety Valves Applying CFD

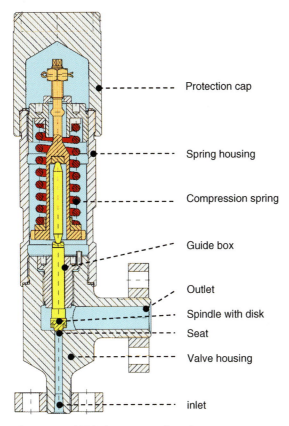

Figure 5.5 BASF high pressure safety valve.

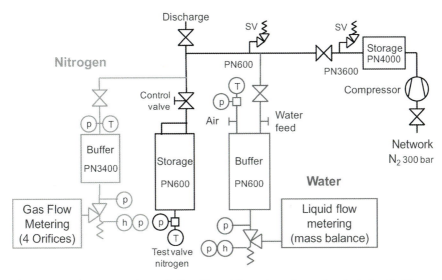

Figure 5.6 High-pressure valve test facility with partially used parts in gray for valve tests with either water or nitrogen.

Figure 5.7 Picture of the high-pressure test facility at BASF, Ludwigshafen.

connected to a 400 MPa (4000 bar) compressor and the vessel B2 working as buffer for valve tests with nitrogen. At a test run the control valve R is opened to push nitrogen from the buffer into the pressure vessel B4 with the test valve on the opposite end. During the blow off test the pressure in the buffer falls continuously while the pressure in the vessel B4 in front of the valve stays constant.

The disk lift is measured with a specially designed protection cap of the spring housing containing a magnetostrictive non-contacting linear displacement sensor (Figure 5.8). This technique enables elevated pressures inside the spring housing during the valve test to be allowed for, which leads to an additional pressure force on

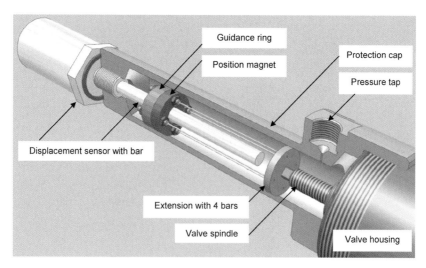

Figure 5.8 Magnetostrictive non-contacting linear displacement sensor setup to measure the disk lift.

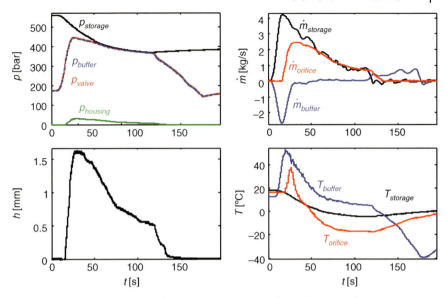

Figure 5.9 Measured quantities of a safety valve experiment for a valve test with nitrogen [8].

top of the valve disk. The sensor can cope with severe valve vibrations with an inaccuracy of less than 0.01 mm.

The flow force acting upon the valve disk can be derived from the deflection of the compression spring at steady flow conditions or with an additional unique force sensor that acts as a mechanical stop at variable maximum disk lift and still allows for accurate force measurement when the spindle moves to this stop [8].

For liquid service a mass flow rate of up to 40 kg s^{-1} with a maximum uncertainty of 2–3.5% can be measured with a mass balance behind the safety valve where the liquid is collected in a tank. Before the liquid flows into the collecting tank it passes several grids vertically located in a cylinder to decelerate the jet flow from the safety valve and only minimally disturb the reading of the mass balance. Gas mass flow rates are measured at the atmospheric pressure side behind the test valve by means of parallel located orifices in accordance to EN ISO-5167 [25]. To cover the measurement range of 0.02–10 kg s^{-1} four orifice stages are used with an inaccuracy less than 1%. During each test data are collected automatically.

After each valve test series the geometry of the valve seat and that of the valve spindle are measured on a 3D-coordinate machine with an inaccuracy of 0.9 μm. Hence, even small geometric differences caused by mechanical wear up to a few tens of a millimeter are taken into account in both experimental and numerical analysis.

Figure 5.9 shows the distribution of all measured quantities of a safety valve experiment with nitrogen. First, the buffer vessel is filled from the storage vessel (negative mass flow rate) until the set pressure of the valve is reached. Sixteen seconds later, the safety valve opens at 30 MPa (300 bar) measured just before the test valve and in the buffer vessel. The maximum disk lift of this normally opening high-pressure safety valve is reached at around 30 s and remains constant for 4 s at a lift of 1.61 mm. At this time the pressure in the valve spring housing increases up to 3.0 MPa (30 bar). The temperature in the storage vessel continually decreases until the test valve closes again. Furthermore, due to a fast pressure rise in the buffer vessel, its temperature increases from 13 to 52 °C and drops immediately after opening of the safety valve. The same temperature variations are observed for the orifice, which are on average 20 degrees lower due to the flow expansion in the safety valve. At almost steady state conditions, that is, small variations of pressure, disk lift, and mass flow rates, sample ranges of the measurements have been selected. Then, the disk is deflected to a certain lift by the compression spring and is in force equilibrium with the flow force. The disk floats on the passing flow without any restrictions. Typical averaging times are between 1 and 100 s depending on the inlet pressure of the valve.

5.4.2
Numerical Model and Discretization

In ANSYS CFX the conservation laws of mass, momentum, and energy are numerically solved in all 3 geometrical dimensions and with respect to time. For turbulence modeling, the shear stress transport (SST) model [26] has been chosen, and real gas fluid properties are calculated employing the Soave–Redlich–Kwong EoS [10]. To keep numerical stability during the iterative solution in a pressure/temperature window of 0.01 to 10 000 bar and 100 to 6000 K, process look-up tables with the thermodynamic properties have been generated. In detail, the following parameters have been defined: specific heat at constant pressure, specific gas volume, specific heat at constant volume, pressure-specific volume derivative at constant temperature, speed of sound, specific enthalpy, specific entropy, dynamic viscosity, and thermal conductivity. The compressibility factor Z and specific heat capacity at constant pressure have been fitted to data from IUPAC (2008) [12].

For the complex valve geometry, an unstructured hexahedral grid of 1.1 million hexahedral grid cells is used, Figure 5.10. The densest meshed lines are the seat and the lifting-aid. The mesh is regular in the tangential direction with the advantage of locally increased densities and high mesh quality of the planar axisymmetrical mesh. After revolving the planar mesh, the remaining half-cylinder volume around the inlet symmetry axis becomes a structured mesh extruded from a planar unstructured mesh at the top surface of the truncated spindle cone. The inlet and outlet are modeled as openings at specified values of the stagnation inlet pressure and temperature and, in addition, static pressure and temperature at the outlet, both at subsonic conditions. The walls of the high-speed flow are considered adiabatic.

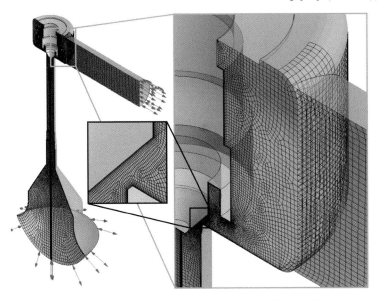

Figure 5.10 Computational domain of a steady safety valve flow simulation.

5.4.3
Numerical Results

Figure 5.11 shows the results of the numerical simulations for set pressures between 6.2 MPa (62 bar) and 25.1 MPa (251 bar) performed with nitrogen. Mass flow rates and flow forces are given as a function of the relative valve lift. The calculated mass flow rates show deviations from experimental data below 3.6%. Applying this data, the discharge coefficient for a 1D numerical real gas nozzle flow model (see Section 5.2.4) is almost independent of the operating pressure in the range from 7.3 MPa (73 bar) to 45.3 MPa (453 bar). All relevant real gas effects influencing the mass flow rate are accounted for in the model. In contrast, the disk forces show a deviation between 8 and 11% for the three low-pressure measurement points and around 14% for the two high-pressure points.

Figure 5.12 shows the Mach number, pressure, and temperature distribution of the left part of the symmetry plane of the same safety valve. In the left plot Mach numbers of up to 3.5 are present in the supersonic flow area with repetitive plumes, where the last plume is detached from the other ones. This shape is similar to overexpanded nozzle flows, where various reflected waves form a diamond pattern throughout the free jet flow. The limited space in the housing prevents the flow from further expansion, so that a second large expansion area with a strong shock at the outlet is necessary to increase the entropy to balance with the thermodynamic state with lower potential energy at the outlet. For visibility reasons the pressure contours in the middle of Figure 5.12 are plotted on a logarithmic scale with an increased minimum value of 1 MPa (10 bar) instead of the computed minimum absolute pressure of 7.4 kPa (0.074 bar) at the secondary large supersonic area in the outlet. Due to the elevated pressure in the valve housing, the

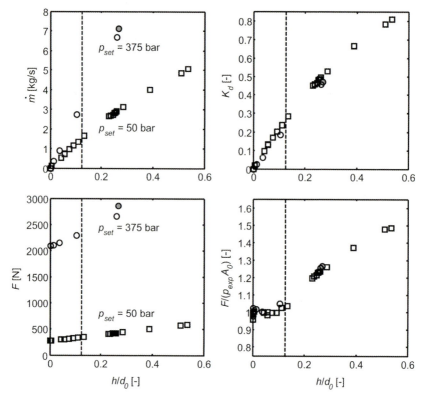

Figure 5.11 Comparison of experimental and computed mass flow rates and disk forces with nitrogen of safety valve test series at set pressures of 62 and 251 bar and temperatures between 274 and 306 K. The dashed line corresponds to the nominal disk lift of 1 mm.

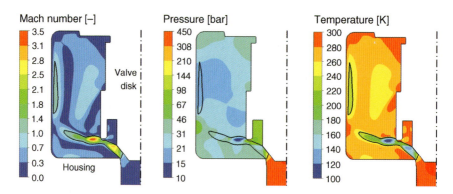

Figure 5.12 Contour plots of Mach number, pressure and temperature of left part of symmetry plane of simulation for nitrogen, $p_{inlet} = 452.8$ bar and $T_{inlet} = 306$ K at $h/d_0 = 0.22$. The black line in all three figures represents the sonic flow line at which $Ma = 1$. The pressure contour plot is logarithmic.

minimum pressure in the supersonic flow area equals 0.8 MPa (8 bar). The flow expands from the inlet pressure of 45.28 MPa (428.5 bar) to a pressure of 7.5 MPa (75 bar) in the cavity of the lifting-aid and to 15–21 MPa on the tip. In the valve tests also the pressure of the spring housing was measured. At the measurements considered here this pressure was constant, so that dynamic effects and pressure losses that would occur in the small gap connecting the valve housing with the spring housing do not have to be taken into account. As a result only the connecting face between the gap with a tolerance of 0.1 mm and the valve housing is defined as a separate wall boundary condition. At the highest operating pressure of 45.28 MPa (452.8 bar), in this valve test this measured pressure is equal to 3.4 MPa (34 bar). The numerically obtained pressure is 2.7 MPa (27 bar).

To gain more insight into development of the flow force when the valve starts to open, the dynamic opening of a safety valve is simulated by the inclusion of fluid–structure interaction (FSI) in the numerical method. In this way, the flow force when the valve is closed can be directly compared with the small force changes when the valve is open. The acceleration of the spindle with disk during valve opening and closing is given by Newton's law:

$$m_{\text{Spindle}} \cdot h'' = F_{\text{Flow}} - k_{\text{Spindle}}(h + h_0) - m_{\text{Spindle}} \cdot g \tag{5.30}$$

where m_{Spindle} is the equivalent mass of the moving parts of the valve and g the gravitational constant. A series of simulations with 17 multiple moving meshes were necessary to preserve the mesh quality parameters orthogonality, expansion, and aspect ratio during mesh deformations and to account for the complex valve geometry. Starting from a disk lift of 0.01 mm, each numerical grid was predefined with a disk lift larger by a factor of 1.5 compared to the previous one. Figure 5.13

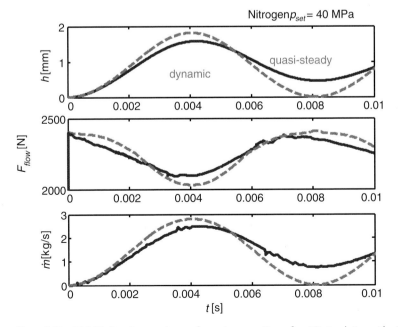

Figure 5.13 Disk lift, flow force and mass flow rate versus time of an FSI simulation with nitrogen gas; solid: dynamic; dashed: quasi-steady.

shows the results of the transient simulations of the axisymmetrically modeled flow of nitrogen through the safety valve. The simulation starts on the predefined mesh with a disk lift of 0.01 mm, an initial spring force based on the set pressure of 40 MPa (400 bar), and a spring stiffness of 253 280 N m^{-1}, a constant 10% overpressure at the inlet of 44 MPa (440 bar), an outlet pressure of 0.1 MPa (1 bar), an equivalent mass of the moving components of 0.7662 kg, and a simulation time step of 2×10^{-6} s. The dashed lines represent the results of quasi-steady simulations, which are converted to the time domain by solving Newton's law with values of the disk force interpolated between the steady-state solutions starting at 0 s, an initial velocity of 0 m s^{-1} and an initial spring displacement that is exactly the same as that used in the transient (solid lines) simulation with multiple meshes. It can be seen that the force continually decreases up to 15% at 2 mm disk lift. This behavior is in agreement with experiments. Since the disk lift has been measured with a frequency of 100 Hz, not enough measurement data are available for comparison of this spring–mass system with experimental data. Moreover, it can be seen that the static simulation does not have any damping. A damping term in the equation of motion Eq. 5.30 would have the form $\beta \cdot m_{Spindle} \cdot h'$ with $\beta > 0$ on the left-hand side. From a physical point of view, the damping accounts for flow history effects, friction, and the inertia of the valve spindle and disk. This is equal to zero in the case of a quasi-steady simulation. The simulations show that the difference between the flow forces in the two approaches correlates with the velocity of the disk. Indeed, the negative disk velocity between 0.004 s and 0.008 s, which is higher in the dynamic case than in the quasi-steady case, coincides with a flow force which is higher in the dynamic than in the quasi-steady simulation.

5.5
Summary

State of Technology for sizing safety valves is an analytical method based on a nozzle flow/discharge coefficient model assuming an ideal gas behavior (ISO model). See for example, EN-ISO 4126-1 [1]. This method has been extended to real gases (REAL model). In addition to the ideal gas model, an EoS to calculate the real gas factor and its derivatives with respect to pressure and temperature as well as the isentropic coefficient for real gases are necessary. The REAL model gives far better results as long as the (linear) averaged parameters do not behave in a strongly non-linear way. It is not recommended to apply the REAL model at conditions within the safety valve close to the thermodynamic critical point of a gas. Here, the numerical solution of the first and second law of thermodynamic is indispensable (1D-num model). The 1D-num model gives the most precise results but is the most complex model.

With ethylene as an example, a 63% higher sizing coefficient and, hence, larger mass flow rates have been calculated with the 1D-num model compared to the ISO model. The results are highly dependent on the inlet conditions, whereas the ISO model predicts almost constant values. Consequently, a constant discharge coefficient for a safety valve even at real gas conditions can only be expected if the real gas behavior is considered within the nozzle flow model. The current standard ISO 4126-1 cannot be recommended if the flow in the safety valve is strongly affected by real gas effects. In this case the REAL or 1D-num model should be applied.

The REAL and the 1D-num nozzle flow models were validated with measurements performed on a high-pressure nozzle with nitrogen at inlet pressures of up to 1000 bar. In addition, the calculations agree very accurately with the values in DIN EN ISO 9300, based on many measurements on nozzles performed worldwide at pressures below 200 bar. Moreover, Beune has intensively studied the flow behavior through high-pressure safety valves by numerical simulations with ANSYS CFX. He underlined the precision of the REAL and the 1D-num gas nozzle flow models. Additionally, he verified experimentally for a certain valve type that the discharge coefficient of the safety valve in combination with the REAL or the 1D-num nozzle flow model can be regarded as constant, as is presently stated in EN-ISO 4126-1.

In the upcoming revision of the current technical standard EN-ISO 4126-1 it is recommended to limit the application range of prescribed nozzle flow model, to specify the calculation of the property data (real gas factor and isentropic coefficient), and to extend the sizing method for applications outside the limits of the current ideal gas model.

In practice, an ideal behavior of gases is often assumed at moderate pressures when sizing a safety valve for gas service. Real gas behavior is only assumed at a very high pressure, for example, at a pressure of more than 100 bar. In general, the real gas behavior is rather determined by the proximity of the thermodynamic critical point. With the reduced thermodynamic pressure and the reduced thermodynamic temperature, the deviation from ideal behavior can be described much better than with the absolute values of pressure and temperature. If the reduced pressure and the reduced temperatures at the entrance of the nozzle exceed $\frac{p}{p_c} > 0.5$ or $\frac{T}{T_c} > 0.9$, the deviations from the ideal behavior are usually no longer tolerable.

A 3D numerical simulation (CFD) of the flow through a safety valve even at very high pressure is possible, but much more effort is required than applying the nozzle flow/discharge coefficient models. Hence, sizing has to be based on the simpler models. The advantage of CFD modeling is the prediction of stable valve operation and the adjustment of a particular valve to particular sizing conditions at best. Improvements are still necessary when calculating the force balance within the valve. Nevertheless, even at the current stage of development, CFD modeling enables minimization of the high cost of high-pressure experiments and allows a glimpse at conditions where measurements not possible or not desired.

Overall, CFD modeling has the potential to replace current experimentally driven development strategies for new safety valves quite soon. In the far future, it may replace the nozzle flow/discharge coefficient models and make experiments dispensable.

List of symbols

a	Variable of the equation of state of real gases	$[m^5 kg^{-1} s^{-2}]$
A_{throat}	Narrowest flow cross section in the nozzle	$[m^2]$

A_{seat}	Valve seat cross section	[m²]
b	Variable of the equation of state of real gases, co-volumes	[m³ kg⁻¹]
c_p	Specific heat capacity at a constant pressure	[J kg⁻¹ K⁻¹]
c_v	Specific heat capacity at a constant volume	[J kg⁻¹ K⁻¹]
C_g	Sizing coefficient for gases including real gas effects	[-]
C_{gISO}	Sizing coefficient for gases according to EN-ISO 4126-1	[-]
h	Specific enthalpy	[J kg⁻¹]
h_t	Specific stagnation enthalpy	[J kg⁻¹]
H	Enthalpy	[J]
$K_{d,g}$	Discharge coefficient of a safety valve for gases/vapors	[-]
K_P	Pressure gradient of the real gas factor	[-]
$\langle K_P \rangle$	Mean pressure gradient of the real gas factor (average between inlet and nozzle throat)	[-]
K_T	Temperature gradient of the compressibility number	[-]
$\langle K_T \rangle$	Mean temperature gradient of the real gas factor (average between inlet and nozzle throat)	[-]
M	Molar mass	[kg kmol⁻¹]
n	Number of measurement data	[-]
p	Pressure	[bar]
p_b	Counter-pressure	[bar]
p_r	Reduced pressure	[-]
$Q_{m,SV}$	Dischargeable mass flow rate through a safety valve	[J kg⁻¹ K⁻¹]
$Q_{m,OUT}$	Mass flow to be discharged from a pressurized system	[J kg⁻¹ K⁻¹]
$Q_{m,nozzle}$	Dischargeable mass flow through an ideal frictionless nozzle	[J kg⁻¹ K⁻¹]
R	Specific gas constant	[J kg⁻¹ K⁻¹]
S	Parameter in the equation of state	[J kg⁻¹ K⁻¹]
s	Specific entropy	[J kg⁻¹ K⁻¹]
T	Temperature	[K]
T_r	Reduced temperature	[-]
v	Specific volume	[m³ kg⁻¹]
V	Volume	[m³]
w	Flow velocity	[m s⁻¹]
z	Height	[m]
Z	Real gas factor	[-]
$\langle Z \rangle$	Mean value of the real gas factor (average between inlet and nozzle throat)	[-]
$a(T)$	Variable for computing the equation of state of real gases	[-]
χ	Parameter of the equation of state of real gases	[-]
δ	Parameter of the equation of state of real gases	[-]
ε	Parameter of the equation of state of real gases	[-]
γ	Parameter of the equation of state of real gases	[-]
η	Pressure ratio	[-]
φ	Velocity coefficient	[-]
κ	Isentropic exponent	[-]

$\langle \kappa \rangle$	Mean adiabatic exponent (average between inlet and nozzle throat) [-]
λ	Parameter of the equation of state of real gases [-]
μ	Dynamic viscosity [Pa s]
Π	Exponent for real gases
$\langle \Pi \rangle$	Mean value of exponent for real gases (average between inlet and nozzle throat)
ϱ	Density [kg m^{-3}]
ω	Acentric factor [-]

Subscripts

None	Static condition
0	Inlet stagnation condition
c	critical (thermodynamic)
crit	critical (fluid dynamic), state in narrowest cross section of the flow channel
CFX	calculated with the program ANSYS-CFX
is	isentropic (loss-free and adiabatic)
g	gas
num	numerical
exp	experimental
r	reduced magnitude
t	total
throat	nozzle throat
SV	safety valve

References

1 EN ISO-4126 (2010) Safety devices for protection against excessive pressure], Part 1: Safety valves, Beuth Verlag GmbH, Berlin.

2 Johnson, R.C. (1964) Calculations of real gas effects in flow through critical flow nozzles. *Journal of Basic Engineering*, **86** (21), 519–526.

3 Baurfeind, K. and Friedel, L. (2003) Berechnung der dissipationsbehafteten kritischen Düsenströmung realer Gase. *Forschung im Ingenieurwesen*, **67** (6), 227–235, DOI: 10.1007/s10010-002-0096-x.

4 Thomson, L. and Buxton, O.E. (1979) Maximum isentropic flow of dry saturated steam through pressure relief valves. *Journal of Pressure Vessel Technology-Transactions of the ASME*, **101**, 101–113.

5 Schmidt, J., Peschel, W., and Beune, A. (2009) Experimental and theoretical studies on high pressure safety valves: sizing and design supported by numerical calculations(CFD). *Chemical Engineering & Technology*, **32**, 252–262.

6 Rist, D. (1996) *Dynamik realer Gase*, Springer-Verlag, Berlin Heidelberg.

7 Sigloch, H. (2007) *Technische Fluidmechanik*, Springer Verlag, 6. Auflage.

8 Beune, A. (2009) Analysis of high pressure safety valves. PhD thesis, TU Einhoven, The Netherlands.

9 Redlich, O. and Kwong, J.N.S. (1949) On the thermodynamics of solutions. V. An equation of state. Fugacities of gaseous solutions. *Chemical Reviews*, **44**, 233–244.

10 Soave, G. (1972) Equilibrium constants from a modified Redlich-Kwong equation of state. *Chemical Engineering Science*, **27**, 1197–1203.

11 Peng, D.Y. and Robinson, D.B. (1976) A new two-constant equation of state. *Industrial & Engineering Chemistry Fundamentals*, **15**, 59–64.

12 IUPAC (2008) International Union of Pure and Applied Chemistry: IUPACChemical Data Series. http://www.iupac.org/.

13 Dohrn, Ralf (1994) *Berechnung von Phasengleichgewichten [Calculation of Phase Equilibria]*, Vieweg Verlag, Braunschweig.

14 Gmehling, J. and Kolbe, B. (1992) *Thermodynamik*, VCH Verlagsgesellschaft, 2. Auflage, Weinheim.

15 Poling, B.E., Prausnitz, J.M., and O'Connell, J.P. (2000) *The Properties of Gases and Liquids*, 5th edn, McGraw-Hill Inc., US.

16 Baehr, H.-D. (1967) Der Isentropenexponent der Gase H_2, N_2, O_2, CH_4, CO_2, NH_3 und Luft für Drücke bis 300 bar [The isentropic exponent for the gases H_2, N_2, O_2, CH_4, CO_2, NH_3 and air for pressures up to 300 bar], BWK Brennst.-Wärme-Kraft vol. 19 No. 2, pages 65–68.

17 DIN EN ISO 9300 (2003) *Durchflussmessung von Gasen mit Venturidüsen bei Kritischer Strömung*, Beuth Verlag, Berlin.

18 Zahariev, T. (2001) Berechnung der Durchströmung und der Kennwerte von Sicherheitsventilen, PhD thesis. Magdeburg, Germany: Otto-von Quericke-University Magdeburg.

19 Bredau, J. (2000) Numerische Strömungsberechnung und experimentelle Strömungsvisualisierung in der Pneumatik, PhD thesis. Dresden, Germany: TU Dresden.

20 Bürk, E. (2006) Numerische Strömungsberechnung ergänzt durch experimentelle Methoden zur Untersuchung pneumatischer Strömungen, PhD thesis. Dresden, Germany: TU Dresden.

21 Föllmer, B. and Schnettler, A. (2003) Challenges in designing API safety relief valves. *Valve World*, **10**, 39–43.

22 Moncalvo, D., Friedel, L., Jörgensen, B., and Höhne, T. (2009) Sizing of safety valves using ANSYS CFX-Flo R_. *Chemical Engineering & Technology*, **32**, 247–251.

23 Beune, A., Kuerten, H., and Schmidt, J. (2011) Numerical calculation and experimental validation of safety valve flows at pressures up to 600 bar. *AICHE Journal*, **57** (12), 3285–3298.

24 ANSYS (2006) Manual ANSYS CFX, Release 11.0. Canonsburg.

25 EN ISO 5167 (2003) Measurement of fluid flow by means of pressure differential.

26 Menter, FR. (1994) Two-equation eddy-viscosity turbulence models for engineering applications. *AIAA Journal*, **32**, 1598–1605.

27 VDI Wärmeatlas (2012), Chapter L2.4 "Berechnung kritischer Massenströme". VDI-Verlag Düsseldorf.

6
Water Hammer Induced by Fast-Acting Valves – Experimental Studies, 1D Modeling, and Demands for Possible Future CFX Calculations

Andreas Dudlik and Robert Fröhlich

6.1
Introduction

Fast-acting valves are commonly used to interrupt the liquid flow in pipelines in case of an emergency. Their sudden activation leads to high accelerations within the fluid, which result in pressure waves in the case of a compression and cavitation bubbles in case of a decompression. In both cases, mechanical loads on the pipe walls, equipment such as valves and pumps, and support structures are generated ('pressure surges'). These effects are called water hammer or cavitation hammer, and they are relevant to the safety of pipeline installations, since the induced forces can put at risk the operation of the whole pipe system and can damage system components, causing, for example, pipe deformation, pipe leakage, erosion of pumps and fittings, noise, and damage to pipe supports [1].

Cavitation significantly changes the water hammer waveform. 1D-Water hammer equations developed for pure liquid flow are not valid for two-phase transient flow. Transient cavitating flow usually occurs as a result of low pressures during transients. There are two basic types of transient cavitating flow in piping systems [2, 3]:

- two-component two-phase transient flow (gaseous cavitation; free gas in liquid flow), and
- one-component two-phase transient flow (vaporous cavitation; column separation).

Transient gaseous cavitation occurs in fluid flows with free gas either distributed throughout the liquid (small void fraction) or trapped at certain positions along the pipe and at boundaries (large void fraction) [2]. Gas may be entrained in a liquid due to gas release during low-pressure transients, cavitation, or column separation. Transient gaseous cavitation is associated with dispersive and shock waves. The pressure-dependent wave speed in a gas-liquid mixture is significantly reduced. Gas release takes several seconds, whereas vapor release takes only a few milliseconds. The effect of gas release during transients is important in long pipelines in which the wave reflection time is in the order of several seconds.

Transient vaporous cavitation (including column separation) occurs in pipelines when the liquid pressure drops to the vapor pressure of the liquid. A negligible amount of free and released gas in the liquid is assumed during vaporous cavitation [4, 5]. The water hammer wave propagates at a constant speed as long as the pressure is above the vapor pressure. Cavitation may occur as a localized vapor cavity (large void fraction, column separation) and/or as distributed vaporous cavitation (small void fraction) [2, 6, 7]. A localized (discrete) vapor cavity may form at a boundary (pipeline component), at a high point along the pipe, or at an intermediate section of the pipe (intermediate cavity) if two low-pressure waves meet [7]. Distributed vaporous cavitation occurs when a rarefaction wave drops the liquid pressure in an extended section of the pipe to the vapor pressure of the liquid. Pressure waves do not propagate through an established mixture of liquid and vapor bubbles. The inability of pressure waves to propagate through a vapor bubble zone is a major feature distinguishing the flow with vaporous cavitation from the flow with gaseous cavitation. Both the collapse of a discrete vapor cavity and the movement of the shock wave front (interface separating the liquid and liquid–vapor mixture) into a vaporous cavitation zone lead to condensation.

Practical implications (accidents, operational difficulties) of transient cavitating pipe flow have led to intensive laboratory and field research starting at the end of the nineteenth century [8].

Column separation experiments have been performed worldwide in at least 34 experimental apparatuses (pipe diameter 0.01–0.1 m; length 10–230 m) and 7 industrial installations (up to a pipe diameter of 2.0 m; length 100–10 000 m) [1, 2, 9]. Bergant and Simpson [7] compared results from numerical column separation models and measurements for rapid valve closures in a sloping-pipe laboratory apparatus.

The discrepancies between the measured and computed results found by temporal and global comparisons were attributed to approximate modeling of transient cavitating flow phenomena along the pipeline (distributed vaporous cavitation zones, actual number and position of intermediate cavities), resulting in slightly different timing of cavity collapse and superposition of the waves. In addition, discrepancies also originated from discretization in the numerical model (the unsteady friction term being approximated as a steady-state friction term [10]) and uncertainties in measurement.

Dudlik *et al.* [6] developed a method for the prevention of cavitation hammer by an innovative pipework modification. For this modification, additional valves (e.g., check valves) are installed at a specific distance downstream of the original valve. The additional valve prevents the cavitation bubble (formed due to a fast closing of a shut-off valve) from fast collapsing.

The occurrence of column separation can have a significant impact on fluid system behavior [3]. High pressures may occur when vapor cavities collapse, and the engineering implications are therefore significant. Transients of this type can lead to severe accidents [11–13]. A classic example from the literature is the accident at oigawa hydropower station in Japan in 1950 [11]. A fast-valve closure due to the draining of an oil control system during maintenance induced catastrophic water hammer pressure that split the penstock open. The resultant release of water generated a low-pressure wave, resulting in substantial column separation that caused crushing of a significant portion of the upstream pipeline.

When, due to the response of the safety valve on a pressure vessel, a gas blowdown into the relief pipe has occurred, the pressure loss in the partly filled pipeline is expected to be higher than that in the empty pipe. This leads to the protection of the possibly too small mass discharge from the vessel and to loads on the pipe supports that are considerably too high. An experimental investigation was performed to examine transient pipe flow [14].

In case of blowdown of gases or vapor, accumulation of liquid in discharge pipes results in liquid plugs and acceleration of these in the pipe system. Forces which exceed the design criteria of the pipe supports will be induced in the turning points of the pipe and in cross-section strictures. Due to vertical pipe sections as well as increase in the operational pressure, the formation of liquid plugs will be favored.

The full understanding of water hammer phenomena requires the modeling of the fluid dynamics processes coupled with the mechanical behavior of the solid structures, which is called Fluid–Structure Interaction (FSI).

In this chapter, we describe experiments conducted with cavitation-induced pressure pulses at different temperatures, primarily to identify the influence of temperature and flow parameters (such as steady-state velocity and pressure) in order to identify a 'basic requirement' for possible CFX modeling. For this, comparison with 1D-modeling is discussed in the context of a possibly better prediction using 3D-modeling strategies.

Future modeling and experimental investigations will place emphasis on so-called spontaneous condensation hammer (condensation-induced water hammer, CIWH).

6.2
Multi-Phase Flow Test Facility

The test rig design includes two 240 m pipes at high and low levels (difference in height: 10 m) and inner diameters of 54 mm and 108 mm, respectively (see Figure 6.1) [15, 16]. There are several flanges along the pipe to allow the realization of other pipework geometries and to insert pieces of glass pipe to make the flow visible. Pieces that have contact with media are made of stainless steel, so that corrosive media can be used.

Two details (circled) are given Figure 6.1. The left one shows the fast-acting valve with a fast conductivity sensor, and the right one shows a fast pressure transducer (in reality built-in from below) in the horizontal pipe sections.

Measured data are transmitted from the plant via optical fiber and saved in a transient recording station with professional software.

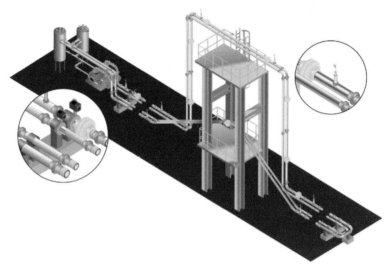

Figure 6.1 Perspective view of the pilot plant pipework (PPP).

The interconnection of the pressure vessel (maximum pressure: 40 bar), storage tank (volume: 3.5 m³), heat exchanger, compressor, and pumps enables experiments to be carried out with different temperatures or pressures up to 140 bar [17]. In the last 6 years, several types of experiments have been carried out where multiphase flow occurs within each scenario.

Experiments (approx. 2000 in total) were performed using a wide range of parameters. In scenario A, the parameters were varied as follows [6, 15, 18]:

- Steady-state (initial) liquid velocity (v_0): 1, 2, 3,4, and 5 ms^{-1}
- Valve types: globe valve, ball valve, centric and eccentric butterfly valve
- 2 axial positions of the closure valve
- Valve closing time (t_S): 200, 350 and 1000 ms
- Total axial length of the pipe system (L): 190, 240 m
- Pipe bridge included/bypassed by straight pipe
- Viscosity (η): 1, 10, 50, 80 cSt

The liquid pressure and the wave velocity are monitored with pressure transducers (P01–P23). Steam/air and liquid distribution in the cross-sectional area of the pipe is measured by wire-mesh sensors [17]. For low steady-state liquid velocities ($v_0 < 2.5$ ms^{-1}), the flow profile downstream of the valve is monitored with a high-speed camera [15, 19].

Liquid flow velocity is measured with an ultrasonic application [16]. The time resolution of all measurement systems described above varies between 1 and 10 kHz. For validation of FSI codes the force on pipe supports (fixed points 1,2, and 3) the displacement and frequency response are monitored fast as well. In the current test series the impact of increased temperature on water hammer and cavitational hammer is being investigated.

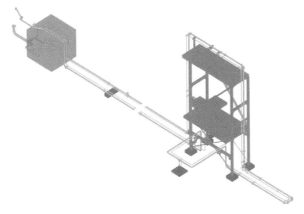

Figure 6.2 New construction of pipe bridges 1 and 2, extensions for FSI experiments.

6.3
Extension of Pilot Plant Pipework PPP for Software Validation

The existing plant was constructed for a pressure of up to 40 bar and temperatures up to 100 °C. Thermohydraulic effects should be included in numerical column separation models at above approximately 60 °C [20]. Therefore the PPP had to be reconstructed to fit the new test conditions with temperatures varying from 20 to 180 °C. The reconstruction includes the replacement of measuring technique, seals, and pumps. The pressure vessel was dated, and a 150 kW boiler is now included for heating the water.

The middle section of the PPP with the pipe bridge was reconstructed and equipped with expansion compensation for DN 100 pipe (3 m × 2 m × 2 m; see Figure 6.2). The expansion compensation is integrated in order to simulate the stream process which is relevant in the power plant technology.

FSI effects are expected in the newly constructed pipe bridge 2 sections (see Figure 6.2), where pipe support is quite elastic, as in typical pipe support conditions in power plants.

Pipe supports are used which are common, for example, for nuclear power plants (NPPs). The whole piping system is insulated with mineral wool and covered with sheet metal.

6.4
Experimental Set-Up

The experimental set-up, including the main measuring positions, is given in Figure 6.3. The total test pipe length (from FP1 to FP3) was reduced from 215 to 137 m in order to simulate typical power plant piping [14].

Hot water is pumped into the circuit from the pressurized vessel B1 into the test pipe section of 110 mm inner diameter. When the closure valve at position P01 is closed rapidly (while the pump keeps running), pressure waves are induced in the

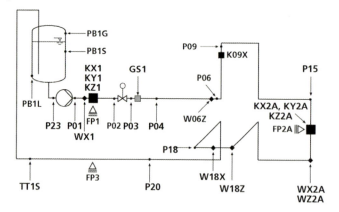

Figure 6.3 Experimental set-up for experiments according to scenario 1.

whole pipe system and measured by fast pressure transducers (P01–P23). Forces on pipe supports are measured (FP1–3) as well as displacements (W1–18). Phase and temperature distribution are measured with a newly developed wire mesh sensor and local void probes (GS). A description of the wire mesh sensor used and the experimental results obtained is given in the next section. Further information about the measuring instruments used is available in Ref. [5].

6.5
Experimental Results

The results of experiments conducted at UMSICHTs PPP including thermohydraulics and structure behavior are presented and discussed in the next section. To get more information about the cavitational hammer in various test parameters, the results of void fraction measurements from the same experiments are also presented and discussed.

6.5.1
Experimental Results – Thermohydraulics

The pressure/time history for position P03 (see Figure 6.3) at different temperatures for ambient vessel pressure is shown in Figures 6.4–6.6. At $t = 0$, the valve is closed quickly (within about 60 ms). Due to inertia, the water downstream of the valve continues to flow with nearly steady state velocity. Hence, system pressure decreases rapidly to saturation pressure, and vapor bubbles are formed, near the valve, that could fill the whole cross section of the pipe. Due to the higher pressure at the downstream vessel and friction, the liquid flow is decelerated and then reversed in the direction of the closed valve.

These generated vapor bubbles re-condense, and, when the water reaches the closed valve, a cavitational hammer pressure peak is observed (about 45 bar).

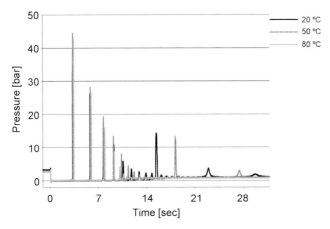

Figure 6.4 The pressure time history at P03 as a function of temperature, $v_0 = 4\,\mathrm{ms}^{-1}$.

Meanwhile, a second cavitational bubble has been created on the top of the pipe bridge, and this collapses after about 15–18 s (see Figure 6.4).

Hence, the water plug is oscillating between the valve and the top of the bridge, with saturation pressure at both ends unless there is no re-condensation. Finally, after 16 s, further cavitational hammers are observed because of the collapse of the vapor bubble on the bridge. The pressure peaks are damped due to friction, fluid structure interaction, and de-aeration of dissolved air in the water.

For the first cavitational hammer peak at P03, the increase of system temperature from 20 to 80 °C leads to a shift of re-condensation to increasing time with increasing temperature, while the height of the peaks keeps nearly constant. The time shift could be due to the following effects:

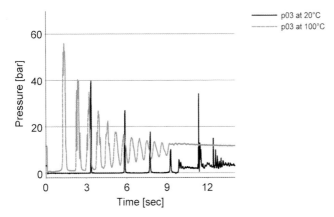

Figure 6.5 Pressure time history of P03 at different temperatures and different initial pressures at 20 °C; 1 bar (Test 00137) and 100 °C; 10 bar (Test 00277).

6 Water Hammer Induced by Fast-Acting Valves

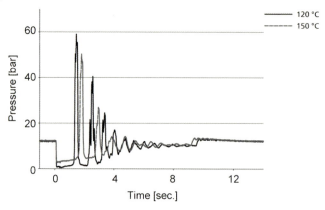

Figure 6.6 Pressure time history at P03; initial velocity: 4 ms^{-1}; vessel pressure: 10 bar.

The higher temperature leads to early and rapid de-aeration of non-condensables. Therefore the gas/vapor bubble becomes bigger. Hence, the acceleration time for the return of the liquid to the valve is higher (higher pressure peak). On the other hand, the increase of free gas in the vapor bubble should damp the cavitational hammer. Looking more closely at Figure 6.4, one can see that the higher pressure peaks increase slightly with increasing temperature.

This may indicate that the damping effect of free gas depends on the two-phase flow regime of liquid and gas (see Refs. [1, 3, 5]): finely dispersed air bubbles in a continuous liquid phase damps water hammer much more than single big bubbles, for example, in slug flow.

On the top of the bridge, the formation of vapor bubbles is suppressed with increasing temperature due to increasing saturation pressure (though gas bubbles are more easily created). At the system temperature of 80 °C, there is nearly no more condensation hammer measurable.

When the temperature is increased further, system pressure was also raised in order to achieve an initial single phase flow in the whole pipe system. For a vessel pressure of 10 bar, the pressure time history at P03 is shown in Figure 6.5 compared to conditions of 1 bar and 20 °C. Due to the higher system pressure, the first cavitational pressure peaks increase from 40 bar to about 58 bar.

The higher peak also occurs much faster because of the higher downstream vessel pressure at the end of the pipe (40 → 50 bar). Due to the faster re-condensation of the vapor bubbles, friction loss decreases; thus the first cavitational hammer increases for about 8 bar. Furthermore, after re-opening of the valve at $t = 11$ s, there is no collapse seen as at 1 bar conditions because the second bubble on the bridge was no longer formed. Also, saturation pressure is no longer reached as often at higher temperature. Therefore, the amount and frequency of pressure peaks increases.

On increasing the temperature from 120 °C (test 00307) to 150 °C (test 00329) while keeping the vessel pressure constant (10 bar), saturation pressure increases while the cavitational hammer decreases due to the lower speed of sound in liquid.

The slowly increasing level of the saturation pressure during the first two seconds (low pressure period) is an effect of the de-aeration of the dissolved air. Furthermore, due to increasing air content and friction, the number of peaks is lowered and they are damped.

6.6
2 Case Studies of Possible Future Application of CFX

Example 6.1: Improvement of Water Hammer Calculations Using Dynamic Valve Charateristics

The experimental set-up is given in the Sections 6.2–6.5 above. Figure 6.7 shows results of measuring results in comparison to 1D modeling results. Calculation was performed using MOC (method of characteristics) software.

Due to quick valve shut-down, the pressure rises from steady state pressure to 35 bar. Friction effects lead to an additional peak up to 42 bar (abs). Using manufacturer's valve characteristics, the calculated pressure peak reaches 29 bar when the valve is completely closed.

In reality, the 'dynamic' characteristic of the valve is quite different. From Figure 6.7 it can be seen that the pressure maximum is reached though the eccentric butterfly valve is still partly open!

Using a 'dynamic' characteristic obtained by transient measurements, the prediction of the first water hammer peak is greatly improved (35 bar). Friction and branched reflection effects were not modeled. Therefore, the measured peak is not fully calculated.

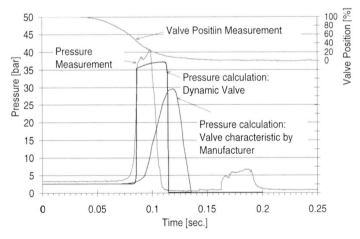

Figure 6.7 Measured and calculated results (1D) for a quick valve shut down; liquid velocity: 3 ms^{-1}.

Manufacturers' valve characteristics are determined from 100 to about 15–20% opening.

Otherwise, for water hammer calculations, it would be very helpful to have information about the full range of valve positions. This could be a very good application for CFX, a commercial (ANSYS) computational fluid dynamics program – a combination of 1D and 3D-modeling:

- 3D modeling of valve and valve characteristics
- 1D modeling using CFX results.

Example 6.2: : Improvement of Water Hammer Calculations in Hydropower Plants

This is an actual example of a case of damage caused by water hammer. On August 17th 2009, a severe accident in the hydropower station of Sayano Sushenskaya (6.4 GW output) in eastern Siberia, 76 dead, occurred (Figures 6.8 and 6.9). The flow of water through the Kaplan turbine underwent rapid deceleration, probably caused by a large piece of wood in the turbine blade. This resulted in a water hammer in the penstock and a stall downstream of the turbine, followed by a cavitation hammer.

Figure 6.8 Hydropower station of Sajano Shushenskaja.

Figure 6.9 The same hydropower station after the accident.

This caused one of the turbines to be abruptly pulled out of its foundation, whereupon it broke through the roof of the hall. The water entry into the turbine and generator room then caused a transformer explosion.

6.6.1
1D Modeling of Kaplan Turbine Failure

Due to the lack of knowledge of Kaplan turbine characteristics in two-phase flow, water hamer calculations were performed by using a 1D software code. The turbine was modeled by a valve, using a pressure loss curve given by the manufacturer for single-phase flow and extrapolating the curve for 2-phase flow at saturation pressure.

Figure 6.10 shows the used simulation schematic. The reservoir feeds into the system at a level of 203.0 m. The turbine inlet is at about 173.4 m, and the outlet at about 168.6 m. An approx. 2 m long pipe with diameter equal to that of the turbine blade (9.223 m) simulates the situation at the turbine outlet (calculation of loads). The draft tube has its lowest point at K003 at 155 m. The outlet is at 172.0 m with a water level of 176.3 m.

6.6.2
Simulation Results – Closing Time 10 s, Linear

Figure 6.11 shows the development of pressure over time with linear closing of the control surface within 10 s. After approx. 6 s, vapor pressure is reached, and first pressure peaks of about 35 bar occur after 16 s. After 50 s the system levels off at a one-phase level.

However, if one refines the increment of the calculation to 0.001 s, one gets considerable higher pressure peaks (as seen in Figure 6.12).

The breakup of the first cavitational hammer at 16 s shows that, with an increment of 0.001 s, a first short peak in the inlet of the draft tube is calculated. The width of this first high peak is 0.04 s, which, with a speed of sound of 1000 m s^{-1}, equates to

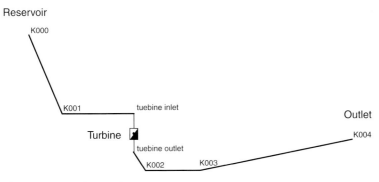

Figure 6.10 Simulation schematic.

6 Water Hammer Induced by Fast-Acting Valves

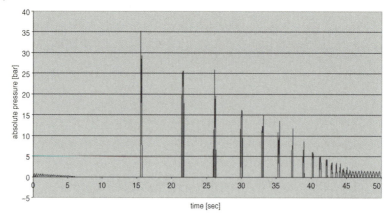

Figure 6.11 Development of pressure over time with linear closing of control surface within 10 s, increment 0.01 s.

the reflection time of the used short 2 m pipe (see Figures 6.13 and 6.14). The overall width of the peak of 0.185 s, on the other hand, equates to the reflection time of the whole draft tube (axial length: 92 m).

6.7
Possible Chances and Difficulties in the Use of CFX for Water Hammer Calculations

Table 6.1 shows typical parameters and uncertainties that very seriously influence the quality of prediction of pressure surges mentioned above.

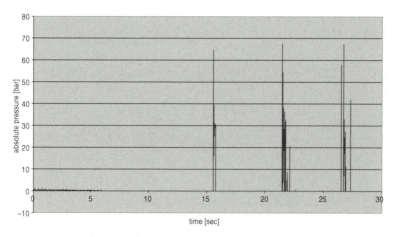

Figure 6.12 Development of pressure over time with linear closing of control surface within 10 s, increment 0.001 s.

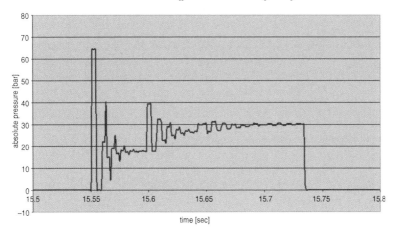

Figure 6.13 Development of pressure over time with linear closing of control surface within 10 s, breakup of 1st cavitational hammer.

6.7.1
Benchmark Test for Influence of Numerical Diffusion in Water Hammer Calculations

The benchmark test represented in Figure 6.15 is built up as follows:

- no friction
- rigid pipe
- valve closing within one calculation step
- initial temperature: 20 °C
- initial system pressure: 80 bar (always single-phase conditions)
- initial velocity: 5 m s^{-1}

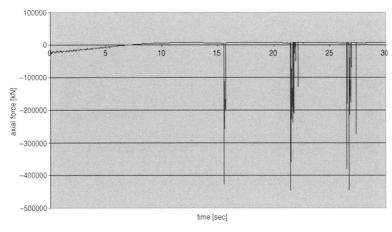

Figure 6.14 Development of force over time with linear closing of control surface within 10 s.

Table 6.1 Boundaries, characteristics of material and medium, and possible application for CFX modeling.

	Description	Characteristics of structure/ medium/codes/ boundaries	User knowledge	Impact on water hammer calc.	Conditions for possible CFX application
A	Density	m	+ +	+ +	
	Compressibility	m	+ +	+ +	
	Viscosity	m		0	
	Amount of dissolved gas	m	Good	+	+ +
	Amount of free gas	m	–	High	0 (?)
	Two-phase flow pattern	m	- -	Very high	– (?)
B	Valve char.	b	Bad	High/very high	+ +
	Pump char.	b	Good	High	+ +
	Valve closing/ opening char.	b	Good	Very high	
	Pump dynamics (start up, failure)	b	Good	Very high	
	Valve operation (regulation, closing, flow control etc.)	b	Very good	Very high	
	Valve leakage	b	- -	+/+ +	
C	Pipe material	s	Very good	Very high	
	Pipe diameter	s	Very good	Medium	
	Pipe thickness	s	Very good	Medium	
	Pipe support	s	*	High	- -
	Pipe stiffness	s	Very good	High	+
	Pipe geometry (low/high points)	s	Very good	Very high	+ +
	Fittings	s	Very good	+	+ +
	Pipe complexity (branched/single pipe etc.)	s	Very good	High	- -
D	Numerical solving/ diffusion	c	+ +	+/+ +	+ +(?)
	Spatial discretization/grid	c	+ +	+ +	+ +
	CPU time	c	0	–	– (?)

Legend: - -: very bad; –: bad; 0: medium; +: good; + +: very good; * Knowledge is good with respect to (nuclear) power plant pipe support, bad with respect to standard supports with clearance.; (?): to be discussed.

Using these conditions, the Joukowsky theory, here implicated in an MOC software, gives the theoretical solution of the problem (see Figure 6.16). All computer codes show more or less big variations of theoretical solutions. This is caused by numerical diffusion of the software solver.

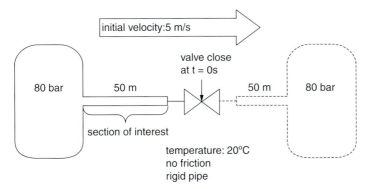

Figure 6.15 Benchmark test for investigation of numerical diffusion.

Using a 'dynamic' characteristic obtained by our own transient measurements, the prediction of the water hammer peak is greatly improved (35 bar). Friction and branched reflection effects were not modeled.

6.8
CFD – The Future of Safety Technology?

Water and cavitation hammer experiments caused by fast-closing valves were performed at Fraunhofer UMSICHT test rig in Oberhausen, Germany. The (often) large number of parameters and boundaries for 1D and 3D water hammer calculation were considered. Two case studies of the possible use of CFX in water hammer modeling were presented and discussed, including possible difficulties such as numerical diffusion.

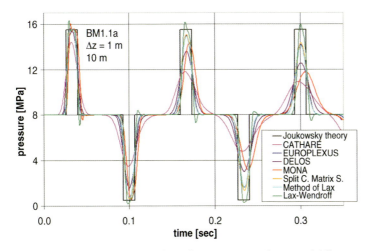

Figure 6.16 Results of benchmark test for investigation of numerical diffusion.

To our concern, the use of CFX in water hammer modeling, while improving the calculation of valve, pump and turbine characteristics under transient flow conditions, especially when reaching two-phase flow regimes, can often predict much higher water hammer peaks.

References

1 Dudlik, A. (2000) Vergleichende Untersuchungen zur Beschreibung von transienten Strömungsvorgängen in Rohrleitungen, UMSICHT-Schriftenreihe Band 20.

2 Dudlik, A., Prasser, H.-M., and Schlüter, S. (1999) Visualization of cavitating liquid flow behind fast acting valve, ECCE 2 - Second European Congress of Chemical Engineering - Montpellier 05.-07.10.1999, paper CDROM 11270003.pdf.

3 Wylie, E.B., Streeter, V.L., and Suo, L. (1993) *Fluid Transients in Systems*, Prentice Hall, Englewood Cliffs, USA.

4 Hansson, I., Kerdinskii, V., and Morch, K.A. (1982) On the dynamics of cavity clusters. *Journal of Physics D: Applied Physics*, **15** (9), 1725–1734.

5 Wylie, E.B. (1984) Simulation of vaporous and gaseous cavitation. *Journal of Fluids Engineering, ASME*, **106** (3), 307–311.

6 Dudlik, A., Schlüter, S., and Weinspach, P.-M.(May 18–20 1999) Water hammer & cavitation in long-distance energy pipeworks - measurement, simulation and prevention. 7th International Symposium on District Heating and Cooling, Lund Sweden.

7 Bergant, A. and Simpson, A.R. (1999a) Pipeline column separation flow regimes. *Journal of Hydraulic Engineering, ASCE*, **125** (8), 835–848.

8 Joukowsky, N. (1900) Über den hydraulischen Stoss in Wasserleitungen. Memoirs de l'Academie Imperiale de St.-Petersbourg. Classe Physico-Mathematique, St. Petersburg, Russia, 9 (5) (in German).

9 Bergant, A. (1992) Kavitacijski tok med prehodnimi režimi v cevnih sistemih (Transient cavitating flow in pipelines). PhD Thesis, University of Ljubljana, Ljubljana, Slovenia (in Slovene).

10 Bergant, A. and Simpson, A.R. (1994) Estimating unsteady friction in transient cavitating pipe flow, in *Water Pipeline Systems* (ed D.S. Miller), Mechanical Engineering Publications, London, England, pp. 3–16.

11 Bonin, C.C. (1960) Water-hammer damage to oigawa power station. *Journal of Engineering for Power, ASME*, **82**, 111–119.

12 Parmakian, J. (1985) Water column separation in power and pumping plants. *Hydro Review, Summer*, 85–89.

13 Almeida, A.B. (1992) Accidents and incidents: an harmful/powerful way to develop expertise on pressure transients, in *Hydraulic Transients with Water Column Separation* (eds E. Cabrera and M.A. Fanelli), Fluid Mechanics Group, Universidad Politecnica de, Valencia, Valencia, Spain, pp. 379–400.

14 Dudlik F A, Schlüter, S., Hoyer, N., and Prasser, H.-M. (12–14 April 2000) Pressure surges - experimental investigations and calculations with software codes using different physical models and assumptions. 8th International Conference on Pressure Surges, The Hague, The Netherlands, pp. 279–328.

15 Dudlik, A., Schlüter, S., and Prasser, H.-M. (1997) *Transiente Strömungen in Rohrleitungen*, in: VDI-Bericht Nr. 1359, VDI-Verlag.

16 Dudlik, A. and Schlüter, S. (12–14 April 2000) Dynamic valve characteristics-measurement and improved calculation of pressure surges. 8th International Conference on Pressure Surges; The Hague, The Netherlands, pp. 105–114.

17 Prasser, H.-M., Böttger, A., and Zschau, J. (1997) A New Wire-Mesh Tomograph for Gas-Liquid Flows, held on: Frontiers in

Industrial Process Tomography II, Delft, The Netherlands, April 9-12, proceedings, pp. 109–112.

18 Dudlik, A. and Schlüter, S. (1998) Druckstöße und Kavitationsschläge infolge schnell schließender Armaturen, Industriearmaturen Nr. 3, September, 195–200.

19 Homepage: http://www.rohrleitungstechnologie.de/.

20 Hatwin, P., Henwood, G.A., and Huber, R. (1970) On the collapse of water vapor cavities in a bubble analogue apparatus. *Chemical Engineering Science*, **25**, 1197–1209.

7
CFD-Modeling for Optimizing the Function of Low-Pressure Valves

Frank Helmsen and Tobias Kirchner

International standards for low-pressure venting (pressures up to 1.034 barg), such as ISO 28300 (API 6th Edition), provide guidance on how to determine flow performance curves for conservation vents [1]. The flow measurement concept ensures that conservative data are collected during the process of directly venting to atmosphere. However, some of these flow testing facilities do not consider design specifics of certain piping structures, so that impacts resulting from different flow conditions, as, for example, from dip tubes connected below the valve, are reflected in the valve's working behavior.

An inlet pressure drop greater than the set pressure minus the reseating pressure (blow down) will result in unstable behavior of the valve pallet, which potentially results in destruction. CFD modeling combined with expert knowledge of fluid dynamics enables the effective provision of a safe estimation of the result of such effects and reduces the number of experiments needed for flow optimization and safe application of the valve in a process unit.

Since October 2008, Braunschweiger Flammenfilter GmbH (PROTEGO®) has been working with the CFD-FLO Module from ANSYS Version 11. In order to evaluate the application of this CFD tool, we calculated the mass flow of a vent (Figure 7.1a) for which we had conducted flow testing in our TUV-certified flow test facility. We experimented with different prism layers (Figure 7.1b) as well as different mesh sizes and compared these results with each other (Table 7.1). As boundary conditions we defined the inlet with pressure impact, and the opening was defined at atmospheric conditions.

The modeling results showed hardly any difference for the calculated mass flow. All results were in between 0.407 and 0.428 kg s^{-1}. The measured mass flow was 0.411 kg s^{-1} for an inlet pressure of 66 mbar. As well as the flow performance, the flow force has an important impact on the vent pallet for weight- and spring-loaded vents. The results of the flow force were very close to each other when prism layers and a mesh with a large amount of elements were used. For a mesh with average-sized or small amounts of elements the calculated force decreased considerably. From this we draw the conclusion that the flow force calculation reacts more sensitively than the mass flow calculation toward small mesh sizes without prism

Process and Plant Safety: Applying Computational Fluid Dynamics, First Edition. Edited by J. Schmidt.
© 2012 Wiley-VCH Verlag GmbH & Co. KGaA. Published 2012 by Wiley-VCH Verlag GmbH & Co. KGaA.

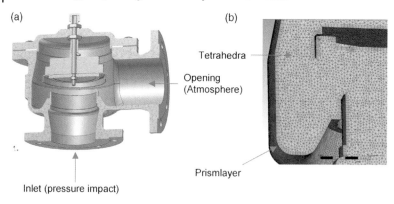

(a) (b)

Tetrahedra

Opening (Atmosphere)

Prismlayer

Inlet (pressure impact)

Figure 7.1 (a) Design of vent, (b) Mesh and prism layer.

layers. Hence, prism layers, even though they only consist of rows, should always be generated.

Based on these findings, the existing vent was flow optimized through CFD modeling (Figure 7.2a and b). Through the optimization process we increased the existing flow from 0.41 kg s^{-1} to a calculated flow of 0.71 kg s^{-1} kg s^{-1}. In the end the actual measured flow was found to be 0.75 kg s^{-1}.

One example in which the modeled results differed by more than 10% from the calculated results was the optimization of a vents pallet regarding its blow down, Figure 7.3a. The goal was to allow the vent to reach full lift and a high flow performance after just 10% pressure increase above set pressure and still have a small blow down. To achieve this, the flow force on the vent pallet was only allowed to be a little greater than the mass x gravity of the vent pallet at each lift position. The computed force path (blue curve, Figure 7.4) for all lift positions resulted in 64 N to 67 N, which showed sufficient safety regarding mass x gravity of the vent pallet of 55 N. Flow testing of a prototype device on our TÜV-certified flow test facility resulted in a vent performance which indicates an insufficient flow force. For this reason we performed a force measurement of the flow on the vent pallet for different lift positions. The result (red curve, Figure 7.4) confirms this assumption. Up to a lift of

Table 7.1 Comparison of different mesh and prism layer.

Total Elements	Tetrahedra Elements	Prism Layer	Prism Layer Elements	Pyramid Elements	Mass Flow	Force on Vent Pallet
10.842.809	9.678.539	5	1.164.270	63	0.422 kg s^{-1}	71.9 N
4.455.200	4.107.745	3	347.455	79	0.414 kg s^{-1}	73.5 N
713.915	656.166	1	57.749	25	0.407 kg s^{-1}	71.9 N
10.388.545	10.388.545	—	—	—	0.414 kg s^{-1}	71.7 N
4.453.449	4.453.449	—	—	—	0.428 kg s^{-1}	69.2 N
742.528	742.528	—	—	—	0.409 kg s^{-1}	64.5 N

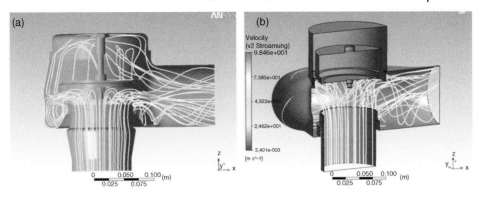

Figure 7.2 (a) Starting geometry, (b) Optimized geometry.

30 mm the computed values and the measurement match. After that the measured force decreases in comparison to the computed value. This result was surprising to us as the modeling was conducted with the same boundary conditions (only the lift was varied) and mesh settings. Also, refining of the mesh or changes of the boundary condition did not result in a significant improvement in the computed results. To reproduce the force path we needed to include the upstream and downstream geometry of the valve, Figure 7.3b (yellow curve, Figure 7.4). Including the entire tank system (Figure 7.3c) into the modeling only led to minor changes to the computation (green curve, Figure 7.4). In general we experienced that for α-values close to 1 (ideal nozzle) small disturbances become more important. Hence, inlet and outlet geometries increase in importance and cannot be ignored.

In another project we investigated how a different piping structure upstream or downstream of the vent influences the vent's flow and opening characteristic (Figure 7.5). This is important, as standards such as ISO 16852 ask for testing of combined pressure vacuum valves with flame arresters as an entire system [2]. We used a 10% full lift type technology vent with a set pressure of 2.36 mbarg. This vent

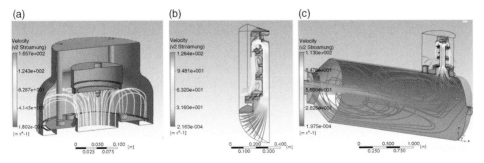

Figure 7.3 (a) End-of-line vent, (b) End-of-line vent with inlet and outlet geometry, (c) End-of-line vent including entire tank.

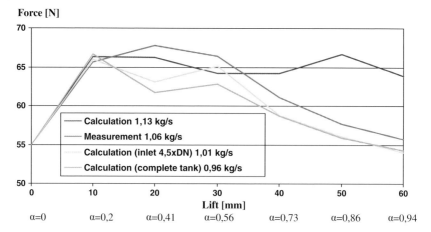

Figure 7.4 Lift as a function of force and alpha value.

reaches its fully open position at 2.6 mbarg ($=1.1 \times 2.36$). At a set pressure of 2.36 mbarg the force of the pallet equals 0.344 N. The first two half-section calculations of the single vent with and without upstream T-pipe delivers the expected results. The mass flow and the flow force resulting on the pallet were smaller for the upstream T-pipe but large enough to comply with the 10% technology. After rotating the T-pipe by 90° (calculation needs to be conducted with full model), an additional reduction of the mass flow and flow force was projected. This reduction in flow force reacting on the vent pallet results in a computational result which shows that the 10% overpressure to reach full lift is not achieved. Hence, the mass flow is reduced even more. As this result was not expected and we wanted to exclude the computational difference between the half-section model and the full model we repeated the first two calculations with the full model. However, this then led to a tendency even to increase the effect of the rotated T-pipe part. A summary of the results is shown in Table 7.2.

To verify the computational results, a prototype model of the vent and T-pipe was manufactured (Figure 7.6). We then tested the 3 different set-ups on the TUV-certified flow test facility. The measured mass flow at 66 mbarg was up to 10% below the calculated values and differed for the differently oriented T-pipe by 0.107 and 0.104 kg s^{-1}.

It was also our goal to show an insufficient flow force onto the vent pallet for a T-pipe rotated by 90° through a volume flow diagram (Figure 7.7).

In a double logarithmic diagram, a vent which reaches full lift will show a straight line (green and black line). A vent in which the pallet does not reach full lift will show a bend (red line). From this we were able to prove through plotting the results in a pressure–flow diagram that the flow force was not sufficient to make the pallet reach full lift.

This experimental set-up demonstrates that the working characteristic of weight- and spring-loaded valves depends on the installed position, and this needs to be considered for safe operation of a process plant. These results explain why standards

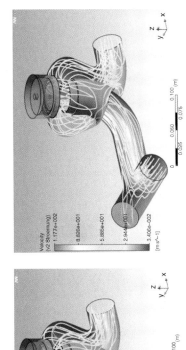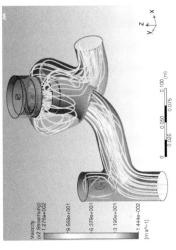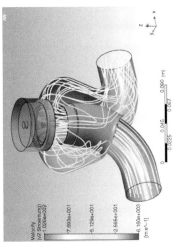

Figure 7.5 Different orientation of geometries.

7 CFD-Modeling for Optimizing the Function of Low-Pressure Valves

Table 7.2 Percent values are related to the force of the vent pallet of 0.344 N.

Orientation of T-Pipe	Without	Without	0°	0°	90°
Type of calculation model	Half	Full	Half	Full	Full
Force on vent pallet at 2.6 mbarg	0.423 N (+ 23%)	0.434 N (+ 26%)	0.354 N (+ 3%)	0.402 N (+ 17%)	0.325 N (− 5%)
Flow performance 66 mbarg	0.124 kg s^{-1}	0.118 kg s^{-1}	0.119 kg s^{-1}	0.119 kg s^{-1}	0.109 kg s^{-1}

modelling: 0,124 kg/s   modelling: 0,118 kg/s   modelling: 0,109 kg/s
measurement: 0,111 kg/s measurement: 0,107 kg/s measurement: 0,104 kg/s

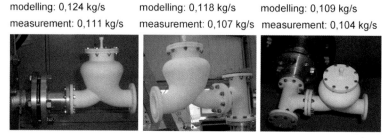

Figure 7.6 Prototype model of vent and T-pipe section. modeling: 0.124 kg s^{-1}; modeling: 0.118 kg s^{-1}; modeling: 0.109 kg s^{-1}; measurement: 0.111 kg s^{-1}; measurement: 0.107 kg s^{-1}; measurement: 0.104 kg s^{-1}.

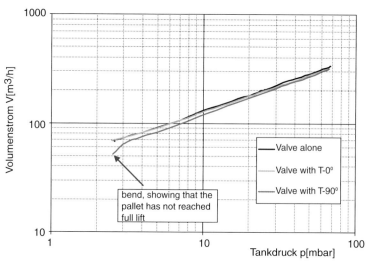

Figure 7.7 Volume flow tank pressure diagram.

such as ISO 16852 require flow testing of low-pressure valves which are combined with flame arresters [2]. Adding the pressure drop of the single components does not necessarily provide accurate results. CFD modeling is an effective tool to determine the effects which different pressure-drop-inducing parts can have on the working characteristics of vents.

References

1 ISO 28300:2008 Petroleum, petrochemical and natural gas industries – Venting of atmospheric and low-pressure storage tanks.

2 ISO 16852:2008 Flame arresters – Performance requirements, test methods and limits for use.

Part Three
Fire and Explosions – are CFD Simulations Really Profitable?

8
Consequences of Pool Fires to LNG Ship Cargo tanks

Benjamin Scholz and Gerd-Michael Wuersig

8.1
Introduction

Natural gas is carried by special seagoing ships if economic transportation over long distances by pipeline is not feasible. Because of the reduction in volume, the gas is transported in a liquid state – liquefied natural gas (LNG) takes up about one six hundredth of the volume of the gas. Large-scale LNG carriers ($130\,000\,m^3$ up to $270\,000\,m^3$) are designed for the intercontinental transport of LNG. A fleet of approximately 350 ships is in service and continues to experience an upward trend [7]. Natural gas is widely regarded as a bridge to a low carbon future. Because of the increased transport volumes of natural gas, the Sandia National Laboratories have evaluated the potential dangers to the surroundings of spilled LNG in 2004. This study [9] relies on standards applied for risk analysis of land terminals and does not consider the ship's structure.

In this context concerns regarding the possibility of cascading failures have been raised for Moss Type LNG carriers if polystyrene is used for insulation [4]. To assess the safety and possible consequences for a ship, the Society of International Gas Tankers and Terminal Operators (SIGTTO) decided in 2006 to form a working group to evaluate the response of LNG carrier cargo tanks exposed to a large enveloping pool fire. The possible fire could result from spillage of LNG and subsequent ignition. With the background of the 11th September 2001, fire hazards associated with marine accidents or intentional acts have also been considered [4].

On behalf of Germanischer Lloyd the authors contributed to the work of SIGTTO with the analysis of the ship response to large fire caused by LNG spill. The result of this work is represented in the following.

In the maritime industry two different containment systems for the carriage of LNG have been established – membrane tanks and self-supporting tanks. The NO 96 and Mark III membrane systems of Gaztransport & Technigaz (GTT) are the most used membrane tank types. This containment system consists of a cryogenic liner which is directly supported by the ships hull. To prevent any harmful effects

Process and Plant Safety: Applying Computational Fluid Dynamics, First Edition. Edited by J. Schmidt.
© 2012 Wiley-VCH Verlag GmbH & Co. KGaA. Published 2012 by Wiley-VCH Verlag GmbH & Co. KGaA.

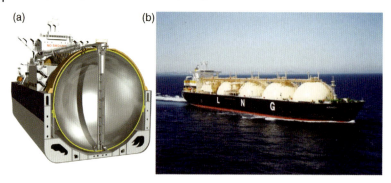

Figure 8.1 Moss tank system (a) and Moss type LNG carrier (b).

of the cryogenic fluid on the ship's structure an insulation layer including a secondary barrier is placed between the liner and the ship's structure. The advantage of the membrane tanks is that they use the ship's hull efficiently, and they have less void space between cargo and the double-hull structure than other containment systems.

The most used independent tank design is the Moss type. The Moss containment system is characterized by a spherical tank which is supported by a single cylinder, the skirt. Figure 8.1 shows the arrangement of this spherical tank within a ship structure. The weather cover welded to the main deck protects the insulation, located directly on the tank steel, from ambient influences. Self-supporting tanks are more robust and have greater resistance against sloshing loads than the membrane systems.

LNG is transported at a low pressure and the corresponding saturation temperature of approximately $-163\,°C$. Despite the quality of the insulation and the type of the containment system there is always a heat flux into the cargo tank, which leads to evaporation known as natural boil-off. For a closed tank with cryogenic liquid gases inside, the internal pressure increases due to this boil-off. An engulfing fire of the ship is the worst case scenario from the point of view of boil-off. To avoid a rupture of a cargo tank, pressure relief valves have to be installed.

Gas Carriers are built according to the requirements of the International Code for the Construction and Equipment of Ships Carrying Liquefied Gases in Bulk (IGC Code). This code provides international standards for the transport of liquefied gases by sea. The purpose of the code is to minimize the risk to the ship, its crew, and the environment caused by the hazardous properties of the products involved [3].

The requirements for sizing the pressure relief valves are included within this code, which assumes a fire engulfing the tank with a constant average heat flux of $108\,kW\,m^{-2}$. It should be noted that, in a special study conducted by the National Academy of Science in 1973 [8] on behalf of the US Coast Guard, this value has been considered to be an average value even for fires with much higher local peak heat fluxes. This heat flux of $108\,kW\,m^{-2}$ is also used in applicable API and NFPA standards [1, 3, 5, 13].

8.2
Evaluation of Heat Transfer

The air gap between the weather cover and the insulation system of a Moss-type LNG carrier is of particular importance when considering the flow of heat from outside the ship into the insulation system (Figure 8.1). Because of this gap, all three modes of heat flow – conduction, radiation, and convection – are involved in the process of heat transfer from an enveloping fire into the cargo.

Experiments on LNG pool fires [12] have measured local peak heat fluxes of 300 kW m^{-2} at the outer surface. From the same experiments it was noted that inside the fireball the combustion is incomplete due a shortage of air, which results in lower temperatures and therefore lower heat fluxes. Therefore the CFD calculations presented here consider a heat flux range of 88 kW m^{-2} up to 300 kW m^{-2} as the emissive power of an LNG pool fire [10]. The fire itself has not been modeled; instead, simplified scenarios were used to model the heat flux into the ship's structure above the waterline.

8.2.1
Simplified Steady-State Model (One-Dimensional)

Figure 8.2 shows a simplified model for the calculation of the heat transfer into the cargo. This model consists of flat plates instead of spherical walls, which is a good

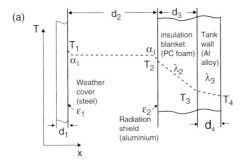

	Density	Thermal conductivity	Heat capacity	Emission coefficient	Melting temperature
Si-unit	kg m^{-3}	W m^{-1} K^{-1}	J kg^{-1} K^{-1}	-	K
Aluminium	2700	70	904	0.07	873
Insulation	26.5	0.038	1045	-	473 – 573
Air at 0°C	1.27	0.02436	1006.1	-	-
Steel	7850	44.5	475 (at 20°C)	0.7	1023

Figure 8.2 Model (a) and material properties of the spherical tank system (b) [6].

approximation because of the large radius of the sphere. The weather cover (steel), the air gap between the weather cover and the insulation (air), the insulation covered by an aluminum vapor barrier, and the inner tank wall are illustrated in Figure 8.2. Depending on the builder, various insulation materials are used, but polystyrene insulation is selected in this case because of its low melting temperature [6].

Depending on the temperature of the outer weather cover, the temperature distribution for a profile of the simplified Moss containment system can be calculated. The material properties used for the following heat transfer calculations are given in Figure 8.2.

In determining the temperature of the weather cover, T_1, it is noted that the heat of the engulfing fire is mostly transferred via radiation and absorbed by the weather cover. Using the equation for radiation, the flame temperature, T_o, can be calculated for a known heat flux in terms of the cover temperature by Eq. (8.1).

$$T_0 = \left[\frac{\dot{q}}{\sigma} + T_1^4\right]^{\frac{1}{4}} \tag{8.1}$$

8.2.2
Different Phases of Deterioration

The heat transfer into a Moss containment system can be broke down into three phases:

- Phase 1 - Heating up the weather cover (following Eq. (8.1)) from ambient temperature until the melting point of the insulation is reached at the insulation surface (This phase includes the heating up of the insulation to the melting temperature).
- Phase 2 - Melting of the insulation to the point of complete deterioration
- Phase 3 – Heat flow without insulation, which forms the worst case scenario due to the high expected heat transfer into the insulation.

Figure 8.3 shows the heat flux of an LNG fire as a function of the receiving wall temperature. During phase 1 the emitted heat flux into the weather cover decreases due to the warming up of this cover. This is represented by the red line (curve 1) for an initial heat flux of $300\,\text{kW}\,\text{m}^{-2}$ and by the blue line (curve 2) with a initial heat flux of $108\,\text{kW}\,\text{m}^{-2}$.

In addition, a hypothetical heat flux into a completely uninsulated tank is illustrated by the green line (curve 3) in Figure 8.3. The maximum possible heat flux into the tank wall, in steady state, can be found at the intersection of curve 3 with curve 1 and curve 2 for initial heat flux values of $300\,\text{kW}\,\text{m}^{-2}$ and $108\,\text{kw}\,\text{m}^{-2}$ respectively [6].

Regardless of the initial heat flux assumed, even with no insulation at all on the tank, the maximum heat flux into the cargo tank is about half of the initial emissive power of the fire, as illustrated in Figure 8.3.

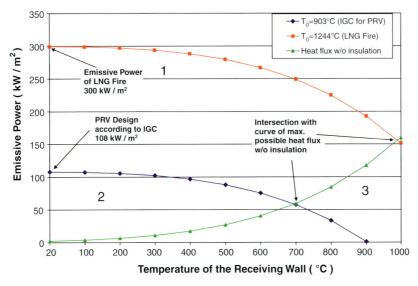

Figure 8.3 Maximum heat flux into the tank [6].

8.2.3
Possibility of Film Boiling

An LNG cargo tank insulation system is specially designed to minimize natural boil-off by keeping the heat flux levels low. Under these conditions the heat transfer between the submerged wall of the tank and the pool of boiling LNG will be via natural convection only, with evaporation taking place on the free surface. If the insulation system were removed the heat flux would increase and the phenomenon of nucleate boiling would occur. This is characterized by bubbles forming on the heated inner wall of the tank, detaching, and rising to the surface. During this phase the tank wall is at almost the same temperature as the cargo.

If the heat flux was further increased, the transition from nucleate to film boiling would occur. In this case a film of methane vapor would separate the tank wall from the cooling LNG. The heat flux into the fluid decreases significantly and the surface tank wall temperature increases. This consequence of this can be that the wall is overheated and may collapse.

Figure 8.4 shows the heat fluxes for nucleate and film boiling for methane and the temperature difference between the temperature of the wall and the saturation temperature of the liquid. It can be seen that for a heat flux of $20\,\text{kW}\,\text{m}^{-2}$ up to $300\,\text{kW}\,\text{m}^{-2}$ into the fluid the wall temperature will adjust between 7 K and 15 K above saturation temperature of the fluid. In the region of $300\,\text{kW}\,\text{m}^{-2}$ the transition to film boiling occurs and the temperature difference increases to unacceptable values. Even considering the worst case scenario where there is no insulation left on the tank surface the heat flux into the LNG will be much below $300\,\text{kW}\,\text{m}^{-2}$ (see Section 3.2). Consequently there is no chance of film boiling [6].

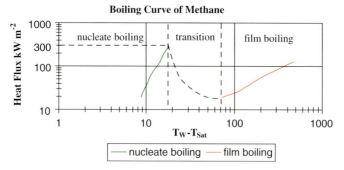

Figure 8.4 Boiling curve of methane, from Ref. [11] modified.

8.2.4
Burning Insulation

The focus of the study was to determine the heat flux into the Moss containment system. Melting of the insulation was considered in the evaluations. However, the insulation material is also combustible, and in the case of a fire the degradation of the insulation would be much greater compared to just melting of the insulation.

Nevertheless the air in the cargo hold is only able to supply oxygen for burning $21\,m^3$ of the insulation. This represents the combustion of a 5 mm thickness of the insulation, if the burning of the insulation is assumed to be uniform. Air supply from the ambient to the hold space can be excluded. The insulation system on a typical Moss-type LNG carrier is about 290 mm thick. The reduction in insulation thickness and the additional heat of combustion can therefore be neglected. From a practical point of view it seems far more likely that a local fire would burn 'holes' in the insulation. As the above explanation shows, such 'holes' will be of limited size, and therefore their effects can be neglected [6].

8.3
CFD-Calculations

CFD has been used to perform calculations of the response of a Moss containment system to fire exposure to illustrate the transient heating up of the different components. All three physical effects of the heat transfer mentioned previously in Chapter 3 have been simulated. Inside the solid components of the tank system thermal conduction has been considered [10].

For the analysis of the response of the spherical tank system under fire exposure a 2-D model has been used, which is shown in Figure 8.5.

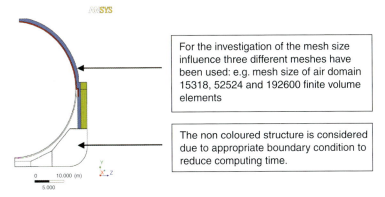

Figure 8.5 Model of the solution domain, from Ref. [10] modified.

8.3.1
Buckling Check of the Weather Cover

The importance of the weather cover in reducing the heat flux from the fire has been demonstrated in the simplified analysis discussed in Chapter 3. The American Bureau of Shipping (ABS) performed a fully coupled Finite Element thermal stress Analysis (FEA) of a typical Moss-type LNG tank weather cover. This component is subjected different levels of heat flux from an external fire; $88\,kW\,m^{-2}$, $108\,kW\,m^{-2}$, and $200\,kW\,m^{-2}$. This evaluation was carried out in order to predetermine the earliest possible point of buckling of the weather cover [2].

The ABS FEA predicted the location and time of thermal structural collapse (buckling) as determined by a large local deformation. Such deformation would occur with an abrupt drop in the Young's modulus, which for steel is at about 1023 K.

The following very conservative boundary conditions were assumed:

- a constant heat flux, independent of the weather cover temperature, was applied,
- no heat transfer was assumed from the weather cover into the air gap or insulation,
- the bottom of the cover, at the welded connection to the main deck, was the only heat sink, modeled with the temperature held at $373\,^\circ K$.

Under the above conditions, failure of the cover was predicted around the connection of the cover sheet and top platform (see Figure 8.6).

8.3.2
Checking the CFD Model

Initial calculations for the heating up of the weather cover have been carried out by GL with the same boundary conditions as used by ABS (steady state heat flux, adiabatic

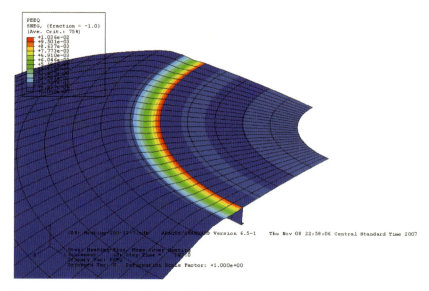

Figure 8.6 Location of thermal collapse (buckling), [2].

wall). Figure 8.7 shows the heating up of the weather cover for a heat flux of 88 kW m^{-2} for the calculations done by ABS (curve 1) and GL (curve 2). With these boundary conditions the cover takes 798 s to heat up to 1023 °K (ABS: 755 s). The calculations carried out by GL confirmed the results of ABS and vice versa.

The comparison between ABS's Finite Element Model and GL's CFD model is based on the same initial heat fluxes (Figure 8.7), which are kept constant for the complete calculation. Further calculations (described in Ref. [10]) reduced the heat flux with increasing weather cover temperature (see Eq. (8.1)). The temperature difference between fire and weather cover decreases with increasing weather cover temperature. This prolongs the heating up of the cover, shown by curve 3. The cover

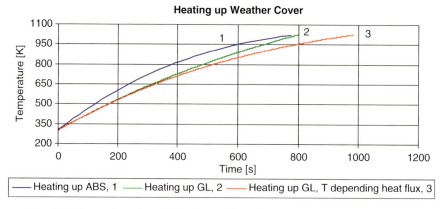

Figure 8.7 Comparison of data for heating up the weather cover (ABS and GL).

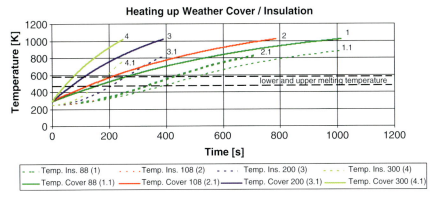

Figure 8.8 Heating up of the weather cover and insulation, from Ref. [10] modified.

takes 980 s to heat up to 1023 °K, which extends the buckling of the structure by approximately 200 s (Figure 8.7).

8.3.3
Temperature Evaluation of Weather Cover/Insulation

The transient simulations include the complete heating up of the weather cover until buckling of this component starts. The CFD analysis is related to the first two phases of the incident (see Section 8.2.2). The starting point for the simulations is the normal operation conditions regarding temperature of the ship's structure and inside the containment system (ambient temperature 20 °C, liquid temperature of −163 °C). For the simulations, a fully liquid-wetted inner tank wall and empty ballast water tanks are assumed.

In Figure 8.8 the heating up of the weather cover is shown for different initial heat fluxes: $88\,kW\,m^{-2}$, $108\,kW\,m^{-2}$, $200\,kW\,m^{-2}$, and $300\,kW\,m^{-2}$.

The solid lines represent the heating up of the weather cover and the dashed lines the heating up of the insulation. The gradient of the curves depends on the initial heat flux, which is absorbed by the weather cover. The increase of the temperature on the surface of the insulation follows the corresponding temperature increase with a time delay [10].

8.3.3.1 Temperature Distribution inside the Insulation

Due to the transient calculations, the absorption of radiation and the corresponding temperature increase on the insulation surface are calculated for a fire scenario. For this case the temperature distribution in a cross section of the insulation is shown in Figure 8.9 for different initial heat fluxes. The position $x = 0\,m$ corresponds to the position of the tank wall.

The gray graph shows the temperature distribution under normal operating conditions. The other colored graphs illustrate the temperature distribution in the cross section of the insulation at the beginning of melting on the surface.

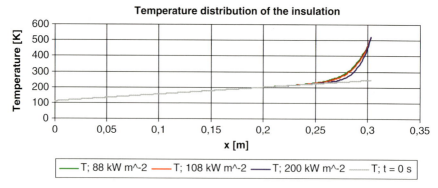

Figure 8.9 Temperature distribution inside cross section of the insulation, from Ref. [10] modified.

Regardless of the magnitude of the initial heat flux only the outer 40 mm of the insulation thickness is affected at the given time span (time until the outer surface temperature reaches the melting point).

This confirms that even a ship fully engulfed in a pool fire will not have an increased heat flow into the LNG containment system until the insulation system is almost completely destroyed [10].

8.3.3.2 Hold Space Temperature Distribution During Incident

The CFD heat transfer analysis includes the full cross section of the ship, as shown in Figure 8.5, and did not focus only on the weather cover and the insulation. Figure 8.10

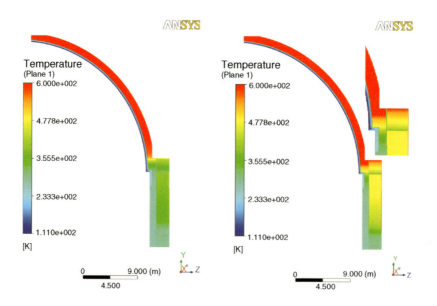

Figure 8.10 Temperature distributions in the sphere, [10].

shows the temperature distribution at the beginning of the insulation melting (left side) and at the thermally caused buckling of the weather cover (right side).

At both time steps, regardless of the initial heat flux, there is a temperature layering in the air gap between weather cover and insulation. This effect is also visible inside the ballast water tanks. The maximum temperatures are arranged in the area of the sphere. The temperature of the insulation surface follows the weather cover with a time delay, but no significant natural convection has developed within the air gap. Therefore, a temperature exchange via natural convection between the colder skirt and the sphere during the heating up is not possible. Such a natural convection would help to transport some of the heat to the lower and cooler parts of the air gap.

Furthermore, the calculation indicates that the skirt will not be heated to critical temperatures with respect to structural integrity. The skirt is protected by the ballast tanks, which have been assumed to be empty for the calculations (would be the case of a fully laden ship). Consequently, failure of the tank support structure can be excluded as a possibility, even for very high heat fluxes into the cover. Due to the rotational symmetry of the sphere the two-dimensional calculations are sufficient for the evaluations reported in Ref. [10].

8.3.4
Results of CFD Calculation in Relation to Duration of Pool Fire Burning According to the Sandia Report

The Sandia study [9] clearly demonstrates that a large breach in the ship will cause a rapid spill and set up a large diameter pool of LNG. The larger the pool the shorter will be duration of the fire as the fire runs out of fuel. On the other hand, a fire of long duration can only occur as the result of a small breach in the ship, where the outflow of LNG is replenishing the fire. These fires will of course have a much smaller pool diameter.

The Sandia Report provides tables with probable pool diameters as a function of the breach size assumed in the ship. For breach sizes smaller than $3\,m^2$ the pool diameter is calculated to be less than 300 m, which is about the length overall of a full size Moss-type LNG carrier. So the fully engulfing fire we have considered can only come from a breach size greater than $3\,m^2$. The average burning duration for various breach sizes is indicated in Tables 10 and 14 of the Sandia report, for accidents and intentional acts respectively.

Figure 8.11 combines the results of study [10] to the findings of the Sandia report. It relates the temperature of buckling of the weather cover to the tank breach size and consequently the duration of the fire. The duration of the fire is indicated by the full size of the bars in Figure 8.11. The height of the bars indicates the duration of the fire. The Sandia report concluded that collision will lead to holes up to $2\,m^2$ in diameter, whereas a $5\,m^2$ hole is the most likely hole size for very severe intentional acts.[1]

The colors in Figure 8.11 are used to specify the time required to heat up the weather cover to the temperature for different heat fluxes. The green color represents

---

1) Hole sizes represent the diameter of a circular hole of area equal to that of the actual hole.

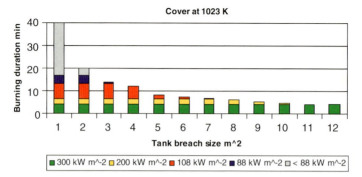

Figure 8.11 Heating up of the weather cover and tank breach size, from Refs. [9] and [10] modified.

the initial heat flux of 300 kW m^{-2}. The other colors indicate the additional time for the lower heat fluxes to reach the buckling temperatures.

Figure 8.11 demonstrates the following:

- With an initial heat flux of 300 kW m^{-2} a time of 4.16 min is needed to heat up the weather cover to the buckling temperature. Holes smaller than 10 m^2 would burn for that time.
- A time (green plus yellow) of 6.51 min to reach the buckling temperature is required for a fire with an initial heat flux of 200 kW m^{-2}. This burning time is available with LNG spills caused by a breach size smaller than 6 m^2.
- For hole sizes above 3 m^2 the weather cover cannot reach the buckling temperature due to a fire with an initial heat flux of 108 kW m^{-2} (green, yellow plus red).
- For a heat flux of 88 kW m^{-2} (green, yellow, red, plus blue) caused by breach sizes less than 3 m^2 the weather cover can reach the melting temperature.

Similarly, Figure 8.12 shows the minimum time, related to the tank breach size, until the deterioration of the insulation starts. The assumed fire load of 108 kW m^{-2} corresponds to the basic assumptions of the IGC-Code and API-520.

The green bar indicates the time required to reach the minimum assumed melting temperature of 473 K. The yellow bar indicates the heating up from 473 K to 573 K,

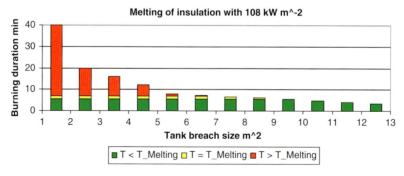

Figure 8.12 Melting of insulation caused by a heat flux of 108 kW m-2, from Refs. [9] and [10] modified.

which is the upper value of the melting range. The red bar displays the period which remains for deterioration of the insulation until the fire is expired. Only breach sizes in the ship with a red bar will create a fire which will last long enough to affect the insulation system. This means that fires from hole sizes above 5 m² will expire before the insulation system is affected.

The study reported in Ref. [10] has concluded that at the beginning of insulation melting, only the outer surface has reached the melting temperature. In addition to the heat required to melt the insulation, further heat is required to raise the temperature of the insulation system.

The total time to increase the temperature of the insulation to the melting temperature can be calculated based on the complete insulation thickness. For a typical polystyrene system, with a thickness of 0.29 m, the heating up from an average temperature of 273 K to the melting temperature requires two to three times the energy which is necessary for melting. This additional time is not included in Figure 8.12.

Taking full account of the heat that will be needed, the calculations carried out show that for an initial heat flux of 108 kW m^{-2} a time of at least 28 min is required to heat up the polystyrene to the melting temperature and an additional 14 min are necessary to melt all the insulation (melting rate 2.1 cm min^{-1}). Figure 8.12 shows that only fires with an average heat flux of 108 kW m^{-2} from holes below 5 m² burn long enough to start the melting of the insulation. No fire lasts long enough to destroy the insulation completely. In the case of an initial heat flux of 300 kW m^{-2} and with a melting rate of 3 cm min^{-1} it will take 18 min to heat up the insulation to 473 K and an additional 9 min for melting. Figure 8.13 shows that only holes below 1 m² create a fire lasting long enough.

The heat flux into the spherical LNG cargo tank system will increase only if the insulation is nearly completely deteriorated. If the incident gets to that point, then the LNG will start boiling, and consequently the pressure inside the tank will increase and cause the pressure relief valves on the tank to open.

The study [10] has determined that with an initial heat flux of 108 kW m^{-2}, complete deterioration will not occur before the insulation is affected for at least 47.5 min by this heat flux (Figure 8.12). Even in the worst case scenario considered, with an initial heat flux of 300 kW m^{-2}, the insulation system can be expect to last at least 29.5 min. Figure 8.13 shows that only holes below 1 m² create a fire lasting long

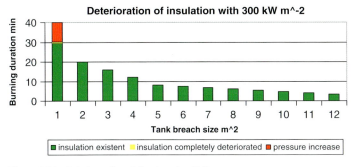

Figure 8.13 Relation of pool fire duration [14].

enough to destroy the insulation. Such damage will not be able to create an engulfing fire. Therefore only parts of the insulation will be damaged and not the entire insulation as assumed in Ref. [4].

8.3.5
CFD – the Future in Safety Technology?

CFD modeling is becoming an essential part of the evaluation of complex hazard scenarios. The presented work is an example of the current contribution of CFD to the evaluaton of such scenarios. The full solution of the above-described problem would cover the simulation of the spill, the liquefied gas evaporation, the gas dispersion, the fire after ignition, and the effects on the ship in one simulation. Such a complex calculation is not possible in an acceptable computing time with the current state-of-the-art technology. To evaluate such a complex scenario the simulation has to be broken down according to the different effects in order to get a result.

Whether the described complex simulation will ever be possible with increased computing power and further development of CFD software is hard to predict. Without a doubt such capabilities would benefit the analysis of safety-relevant hazard scenarios by reducing the requirement for the extremely expensive tests which are currently necessary.

8.4
Conclusions

1) An engulfing fire is not able to destroy the complete polystyrene insulation of a Moss Tank.
2) With a 300 kW m^{-2} initial heat flux:

   a) Only spills from holes of size below 2 m^2 will burn long enough to destroy the insulation thickness. But the tank will not be engulfed by the fire in these cases.
   b) The minimum time period to destroy the insulation will be 29 min after the fire is started (not 10 min as assumed in Ref. [4]).

3) With the more realistic sizing heat flux of the PRVs (108 kW m^{-2}):

   a) For large engulfing fires with breach sizes of 5 m^2 or greater the fire duration is too short to heat up the weather cover to the buckling temperature (1023 K).
   b) Only for leak sizes smaller than 3 m^2 is the heat flux able to heat up the weather cover above 1023 K.
   c) Even with a small hole size of 1 m^2 the fire does not burn long enough to completely destroy the insulation system.

4) There is no increase in heat flow into the cargo tank until virtually all the insulation is destroyed.
5) Dangerous tank wall overheating of an LNG-filled Moss Tank can be excluded even if the insulation is destroyed completely.

References

1. API RP 520: Sizing, Selection and Installation of Pressure-Relieving Devices in Refineries: Part I – Sizing and Selection, Edition: 7th, American Petroleum Institute, 01-Jan-2000.
2. (November 2007) American Bureau of Shipping: Coupled Thermal-Stress Analysis on Cover Dome in Moss-Type LNG Carrier. Technical Report, TR-2007-020.
3. Germanischer Lloyd: Rules for Classification and Construction, Part I – Seagoing Ships, Chapter 6 – Liquefied Gas Carrier, Edition 2008.
4. Havens, J. and Venart, J. (2008) Fire performance of LNG carriers insulated with polystyrene foam. *Journal of Hazardous Materials*, available online at www.sciencedirect.com.
5. Heller, F.J. (1983) Safety Relief Valve Sizing: API Versus CGA Requirements Plus A New Concept For Tank Cars, Article.
6. Kabelac, S., Würsig, G., and Freund, M. (2007) *Thermodynamic Boundary Condition*, Helmut Schmidt Universität, Hamburg.
7. (February/March 2008) LNG world shipping Journal, Statistics.
8. National Academy of Science (1973) *Pressure-Relieving Systems for Marine Cargo Bulk liquid Containers, Committee on Hazardous Materials*, Division of Chemistry and Chemical Technology National Research, Council, Washington, DC.
9. Sandia National Laboratories: Guidance on Risk Analysis and Safety Implications of a Large Liquefied Natural Gas (LNG) Spill over Water, Printed December 2004.
10. Scholz, B. (2008) *CFD Modellierung des Wärmeeintrags in LNG Schiffstanks, Diplomarbeit*, Universität Rostock, Lehrstuhl Strömungsmechanik.
11. Sciance, C.T., Colver, C.P., and Sliepcevich, C.M. (1967) Pool Boiling of Methane between Atmospheric Pressure and the Critical Pressure. *Advances in Cryogenic Engineering*, **12**.
12. SIGTTO; Report On The Effects Of Fire On LNG Carrier Containment Systems, First Edition 2009.
13. Würsig, G. (1998) Comparison of IGC-Code, CGA and API regulations for relief valve sizing for tanks with liquefied gases under fire conditions, Technical Report.
14. Würsig, G., Gaughan, J., Scholz, B., Sannes, L., Kabelac, S., and Leder, A. (2009) *Effects of Enveloping Pool Fires on LNG tank Containment Systems*, Gastech, Abu Dhabi.

9
CFD Simulation of Large Hydrocarbon and Peroxide Pool Fires

Axel Schönbucher, Stefan Schälike, Iris Vela, and Klaus-Dieter Wehrstedt

9.1
Introduction

Accidental fires in process industries often occur as pool fires, which are hazardous to people and adjacent objects because of thermal radiation, largely sooting plumes and formation of other combustion products [1–3]. In addition to experimental pool fire tests, numerical investigation of these fires using Computational Fluid Dynamics (CFD) codes is becoming more important. In regard to safety, thermal radiation is one of the main interests in pool fire research. The Surface Emissive Power (SEP) is a key parameter to characterize thermal radiation emitted by a fire. The derived quantity SEP is usually defined as the heat flux due to thermal radiation in relation to flame surface A_F. As well as the SEP, the temperatures T and irradiances E of pool fires are of particular interest. CFD simulation of large pool fires is helpful for a better understanding of fire dynamics and to reduce the number of large-scale experiments. To predict the thermal radiation from the jet engine fuel JP-4 ($d = 2, 8, 16$ and 25 m) and DTBP (di-*tert*-butyl peroxide, $d = 1.12$ and 3.4 m) pool fires CFD methods are used, and the CFD results are compared with experiments.

9.2
Governing Equations

The rate of change of extensive properties, such as mass, momentum, energy or species mass, of a given quantity of matter can be described by conservation laws [4–6]. Assuming the control volume to become infinitesimally small leads to a differential form of the species mass equation:

$$\frac{\partial \varrho_i}{\partial t} = -\mathrm{div}(\varrho_i \vec{u}) - \mathrm{div}\,\vec{j}_i + M_i r_i. \tag{9.1}$$

In a combustion process with diluted gases, a consistent diffusion coefficient for all species is suitable, while thermo-diffusion (Soret effect) and pressure diffusion can be neglected, and mass flux density is given by Fick's 1st law [7]. Setting the relationships $\sum_i \varrho_i = \varrho$, $\sum_i j_i = 0$ and $\sum_i M_i r_i = 0$ into Eq. (9.1) leads to the continuity equation:

$$\frac{\partial \varrho}{\partial t} = -\mathrm{div}(\varrho \vec{u}). \tag{9.2}$$

The momentum conservation equation is given by:

$$\frac{\partial (\varrho \vec{u})}{\partial t} = -\mathrm{div}(\varrho \vec{u}\vec{u}) - \sum \varrho_i F_i - \mathrm{div}\bar{\bar{p}}. \tag{9.3}$$

The right hand side of the equation expresses the forces acting on the fluid, including surface forces (pressure, normal, and shear stress etc.) and body forces (gravity, centrifugal, Coriolis forces, and electromagnetic forces). The pressure tensor $\bar{\bar{p}} = -p\vec{n} + \bar{\bar{\tau}}$ is based on an empirical correlation and is separated into a pressure and hydrostatic part.

The energy conservation equation is given by:

$$\frac{\partial}{\partial t}(\varrho e) = -\mathrm{div}\left((\varrho \vec{u} e + \vec{u} p) + \vec{q}_{\mathrm{con}}\right) - \bar{\bar{p}}\,\mathrm{grad}\,\vec{u} - \sum_i ((j_i + \varrho_i \vec{u}) F_i) + \mathrm{div}\,q_{rad} \tag{9.4}$$

The conductive heat flux $\vec{q}_{\mathrm{con}}$ is determined by Fourier's law, while pressure terms and the Dufour effect can be neglected in pool fire simulation [7].

9.3
Turbulence Modeling

Pool fires are turbulent non-premixed fires with a small momentum above the pool surface. Turbulent flows are highly unsteady, three-dimensional, contain much vorticity, and increase the mixing of momentum. There are six possibilities to predict turbulent flows, which all have their specific advantages and disadvantages. Large Eddy Simulation (LES) has been the most promising approach to model turbulence in pool fires [8, 9]. The CPU (Central Processing Unit) time of LES lays between the CPU time of a Direct Numerical Simulation (DNS) and a simulation based on Reynolds-Averaged Navier–Stokes (RANS) turbulence models. In a first conceptual step an arbitrary scalar quantity ϕ is filtered and separated into two terms [10]:

$$\phi = \hat{\phi} + \tilde{\phi} \tag{9.5}$$

While several filter types, such as gaussian filter, box filter, or sharp spectral filter, are available, the general filtering operation normalized over the whole domain can be defined as:

$$\hat{\phi} = \int G(x - x^*, \Delta) \phi(x^*) \mathrm{d}^3 x^*. \tag{9.6}$$

With the approach Eq. (9.6) it is possible to calculate the filtered structures $\hat{\phi}$ of a turbulent velocity field directly, while the CPU time-intensive closure of the small structures (fine scale $\tilde{\phi}$) is accomplished by the sub-grid model. The fine scale $\tilde{\phi}$ can be modeled by the Smagorinsky–Lilly model with dynamic calculation of the Smagorinsky coefficient [11]. The Smargorinsky–Lilly model is a zero-equation model based on the mixing length hypothesis.

9.4
Combustion Modeling

For description of turbulence interaction with combustion, statistical methods such as the use of probability density functions (pdf) is appropriate. For detailed chemistry and description of all source term variables, an assumed pfd approach or a pdf transport approach can be used. All pdf variables are a function of the random vector Φ, which leads to a joint pdf with state space Ψ [12, 13]:

$$P(\Psi)d\Psi \equiv \text{Prob}\,[\Psi \le \Phi < d\Psi]. \tag{9.7}$$

A function for use as pdf must have the following conditions:
Exlusion from negative values

$$P(\Psi) \ge 0, \tag{9.8}$$

scaling

$$\int_{-\infty}^{\infty} P(\Psi)d\Psi = 1 \tag{9.9}$$

and conditions for $\Psi \to \pm\infty$

$$\lim_{\Psi \to -\infty} P(\Psi) = \lim_{\Psi \to \infty} P(\Psi) = 0. \tag{9.10}$$

Using an assumed pdf approach, the structure of pdf is forced and quality of the solution depends on the accuracy of approximation with real distribution [14]. Advantages are the closure of the chemical source term and the relatively low computional costs. In practice β-pdf based on the Γ function are often used together with the mixture fraction ξ approach because of its flexibility to generate absolute maxima:

$$P(\hat{\xi}) \equiv \frac{\Gamma(\beta_1+\beta_2)}{\Gamma(\beta_1)\Gamma(\beta_2)} \hat{\xi}^{\beta_1-1}(1-\hat{\xi})^{\beta_2-1}, \tag{9.11}$$

with

$$\beta_1 = \bar{\xi}\left[\frac{\bar{\xi}(1-\bar{\xi})}{\xi'^2} - 1\right], \tag{9.12}$$

$$\beta_2 = (1-\bar{\xi})\left[\frac{\bar{\xi}(1-\bar{\xi})}{\xi'^2} - 1\right]. \tag{9.13}$$

Mixture fraction ξ and mixture fraction variance ξ' are solved for equilibrium chemistry as separate differential equations [7, 15]. With increasing mixing velocity, one or more reactions will differ from equilibrium. An advancement of the equilibrium chemistry models can be achieved by calculating the first non-equilibrium reaction and connecting the drift from equilibrium with a parameter of a laminar counterflow flame. The important parameter is the flame stretch, b, which is connected with scalar dissipation $\chi = 2D(\text{grad}\,\xi)^2$ [16]:

$$b = 2\pi D \left[\frac{\text{grad}\,\xi \cdot \text{grad}\,\xi}{(\xi^+ - \xi^-)^2}\right] \cdot \exp 2\left\{\text{erf}^{-1}\left[\frac{\xi - \frac{1}{2}(\xi^+ - \xi^-)}{\frac{1}{2}(\xi^+ - \xi^-)}\right]\right\}^2. \tag{9.14}$$

This equation describes the fact that for each flame stretch scalar dissipation χ can be big or small. Consequently, the scalar dissipation is an adequate parameter to describe deviation from equilibrium. Scalar properties in the flame are then functions of mixture fraction of stretched laminar counterflow flames. Thus, the turbulent flame can be described as an ensemble of many small laminar flames (flamelets), which have a distribution of scalar dissipation velocity, so that the velocity field within the flame is changed by eddy movement [17]:

$$-\varrho D\,\text{grad}^2\xi \frac{\partial^2 w_i}{\partial \xi^2} = M_i r_i. \tag{9.15}$$

Using Favre separation, the averaged reaction rate is given by:

$$\overline{M_i r_i} = \frac{1}{2}\bar{\varrho} \int_0^1 \int_0^\infty \chi \frac{d^2 w_i}{d\xi^2} \tilde{P}(\chi, \xi)\,d\chi\,d\xi. \tag{9.16}$$

Knowing the probability density function $P(\chi, \xi)$, Eq. (9.16) can be solved. Generally, the parameters χ and ξ are treated as statistically independent, with a product approach $P(\chi, \xi) = P(\chi) \cdot P(\xi)$. As described above for $P(\xi)$, usually a β-pdf is used, while for $P(\chi)$ a logarithmic normal distribution can be applied [18].

9.5
Radiation Modeling

Neglecting scattering, and making gray gas assumption, the radiative transport equation (RTE) can be defined as [19, 20]:

$$\frac{dL}{ds} = -a_{\text{eff}}(s)L(s) + a_{\text{eff}}(s)L_b(s), \tag{9.17}$$

where the effective absorption coefficient a_{eff} can be modeled in different ways.

One way is to use the radiation models OSRAMO II and stochastic OSRAMO III, both based on organized structures in pool flames [21, 22]. To determine average effective absorption coefficient a_{eff} of JP-4 pool fires, a four-step discontinuity function (Figure 9.1) is used, which includes the experimentally determined

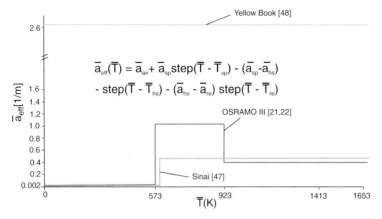

Figure 9.1 Four-step discontinuity function of effective absorption coefficient a_{eff}.

organized structures of the fire absorption coefficients: $a_{\text{air}}(T) = 0.02\ m^{-1}$, $a_{\text{sp}}(T) = 1.035\ m^{-1}$, $a_{\text{hs}}(T) = 0.404\ m^{-1}$, $a_{\text{re}}(T) = 0.380\ m^{-1}$.

Another way to model the effective absorption coefficient is defined by Eq. (9.18) [23, 24]:

$$-a_{\text{eff}} = \frac{1}{s}\ln(1-\varepsilon_m). \tag{9.18}$$

The total emissivity ε_t is the sum of the gas and soot emissivities:

$$\varepsilon_m = \varepsilon_g + \varepsilon_s, \tag{9.19}$$

where the total emissivity ε_m is calculated as a weighted sum of gray gas model (WSGGM) [25]:

$$\varepsilon_m = \sum_{i=1}^{N} k_{m,i}(T)[1-\exp(-a_i s)]. \tag{9.20}$$

In Eq. (9.20) $k_{m,i}$ and a_i are the temperature-dependent emission weighting factor and the absorption coefficient for the ith gray gas component, respectively, and s is the path length. Provided that the sum of coefficients $k_{m,i}$ is equal to unity, coefficients $k_{m,i}$ and a_i are obtained from a fit to total emissivity. The soot absorption coefficient a_s can be estimated by [26]:

$$a_s = \frac{3.72 f_s C_0 T}{C_2}, \tag{9.21}$$

with the constants

$$C_0 = \frac{36\pi n k_w}{(n^2 - k_w^2 + 2)^2 + 4n^2 k_w^2},\ C_2 = 1.4388 \cdot 10^{-2}\ m\ K \tag{9.22}$$

and the optical constants $n = 2.52$ and $k_w = 1.46$ based on the measurements of Dalzell and Sarofim [27]. The soot volume fraction f_s is determined by different soot models, for example, the Magnussen model [28], the Tesner model [29], the Lindstedt model [30], or the Moss-Brookes model [31].

Of special engineering interest are the incident radiation G, radiative heat flux q_{rad}, and radiative source term div q_{rad}, which are derived from the following integrals and are discretized, for example, with Discrete Ordinates model (DO) [32, 33]:

$$G = \int_{4\pi\Omega_0} L(\vec{s})d\Omega \approx \sum_m w_m L_m, \quad q_{rad} = \int_{4\pi\Omega_o} L(\vec{s})\vec{s}d\Omega \approx w_m L_m \vec{s},$$

$$\text{and} \quad \text{div } q_{rad} = a_{eff}(4\pi L_b - G). \tag{9.23}$$

The Discrete Transfer model (DT, [34]) and the Monte Carlo model (MC, [35]) are also used in the present chapter.

9.6
CFD Simulation

In CFD simulation, a domain is represented by a 3-D hexahedral block structured mesh [36–39] (Figure 9.2). The fuel is assumed to be already evaporated, and the fuel vapor entering the domain from the inlet has a constant temperature of $T = T_b$ and an experimentally determined constant mass flux. The inlet is surrounded by a low rim and an adiabatic ground area. The remaining boundary conditions are set as 'opening pressure' (CFX) and 'pressure outlet' (FLUENT) at a relatively large distance from the pool to achieve open boundary conditions. With increasing axial und vertical distance from the pool, cell size increases.

The time steps vary, depending on sufficient convergence from $t = 10^{-6}$s to $t = 10^{-4}$s, depending on the Courant-Friederich-Levy (CFL) criterion. In the pool fire simulations presented here, the following sub-models are used:

- Assumed pdf approach with laminar flamelets containing up to 112 species and 800 elementary reactions and eddy dissipation model.
- Lindstedt, Tesner, or Moos-Brookes soot model.
- Monte Carlo, Discrete Transfer, and Discrete Ordinates models for radiation.

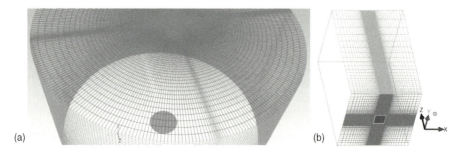

Figure 9.2 Block structured grid for simulation of (a) JP-4 (b) DTBP pool fires.

Table 9.1 Starting and boundary conditions of CFD pool fire simulation.

Starting Conditions		Boundary Conditions	
Quantity	Value		
Mass fraction N_2	0.743	Upper end face	$p = p_a$
Mass fraction O_2	0.231	+ Lateral area	298 K
Mass fraction Ar	0.012	'Pressure outlet'	Open boundary conditions
Mass fraction CO_2	0.001		
Mass fraction H_2O	0.013	Lower end face	Adiabatic
p_a	1013.25 hPa	+ Pool rim	Heat flux to pool rim
$p - p_a$	0	'Wall'	$\dot{q} = 0$
Flow velocity	$u_x = u_y = u_z = 0$		
Temperature	298 K	Pool	Experimentally determined
Gravitational acceleration	9.81 m s^{-2}	'Mass flow inlet'	Mass burning rate
Mixing fraction f	$f = 0$		$p = p_a$, $T = T_b$

The coupling between thermal radiation and soot reactions is described either by a modified effective absorption coefficient or a weigted sum of gray gases approach. For JP-4 pool fires, a four-step discontinuity function is used which includes the experimentally determined organized structures of the fire: effective reaction zone, hot spots, and soot particles. Again in DTBP pool fires, a weighted sum of gray gases model is applied for coupling thermal radiation and soot reactions.

The governing equations were solved by an iterative solution method with either coupled or segregated solvers, for example, the pressure correction methods SIMPLE (Semi-Implicit Methods for Pressure-Linked Equations). The starting and boundary conditions are listed in Table 9.1.

The main purpose is to determine the temperature T, Surface Emissive Power SEP and irradiance E. The simulations are started with a two-equation model based on the eddy viscosity hypothesis, such as $k-\varepsilon$ with a buoyancy correction term to reach a certain flame height which refers to the developing stage of the fire. Assuming the flame to be developed, further simulation is continued by using Scale-Adaptive Simulation (SAS) and Large Eddy Simulation (LES). CFD simulation is carried out with commercial software ANSYS CFX 11 [40] and ANSYS FLUENT 12 [41].

9.7
Results and Discussion

9.7.1
Flame Temperature

A quantitative description of JP-4 pool fire dynamics ($d = 16$ m and 25 m) is shown in Figure 9.3 by the simulated temperature fields. In these fields the flame pulsation is noticeable. The flame pulsation is connected with a formation and rising of vortices

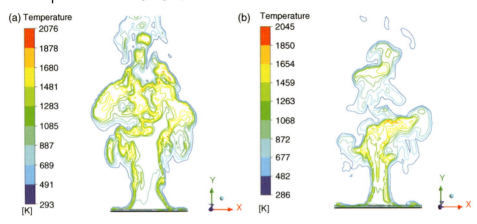

Figure 9.3 CFD-predicted instantaneous temperature fields of JP-4 pool fire (a) $d = 16$ m; (b) $d = 25$ m.

which greatly affects the location of the maximum flame temperature. Inside the vortices the temperatures are significantly higher (2100 K $\leq T \leq$ 1800 K) than in the outside area (1200 K $\leq T \leq$ 1800 K). Hot flame gases have a significantly lower density and therefore a rise in hot volumes is observable with increasing time. As a consequence, the flame surface is cooled down by air entrainment, so that the temperature in the upper part of the flame is lower. The comparison of small ($d = 2$ m) and large scale ($d \geq 16$ m) JP-4 pool fires shows that the maximum temperatures in the flames are more and more shifted in the direction of the pool surface because of the frequent occurrence of hot spots near the pool surface. Analyzing the instantaneous results, it can be seen that in case of a JP-4 pool fire with $d = 2$ m the maximum temperatures are visible in the upper part of the flame (due to the more frequent occurrence of hot spots). In the case of a JP-4 fire with larger pool diameter ($d = 16$ m and 25 m), maximum temperatures are located near the pool and the inside of the lower vortices.

Pulsation of flames occurs in all simulations, although with different frequencies. This has an influence on the SEP of the flame and also on the irradiance to surrounding areas and people. In Figure 9.3, the constriction of the flame between the clear combustion zone and the plume zone is visible. Up to the borderline which separates the combustion zone from the upper plume zone higher temperatures are found. Above this the flame temperature significantly decreases.

JP-4 pool fires show a decreasing maximum time-averaged flame temperature $\bar{T}$ with increasing pool diameter. Temperatures $\bar{T}$ of 1280 K ($d = 2$ m), 1250 K ($d = 8$ m), 1230 K ($d = 16$ m) and 1200 K ($d = 25$ m) are obtained by CFD.

The CFD-predicted instantaneous temperature fields of DTBP pool fires are shown in Figure 9.4.

For DTBP pool fires, CFD maximum axial time-averaged flame temperatures [$\bar{T}$ of 1410 K ($d = 1.12$ m) and 1520 K ($d = 3.4$ m)] (Figure 9.5) agree well with the experimental results [1500 K ($d = 1.12$ m) and 1580 K ($d = 3.4$ m)].

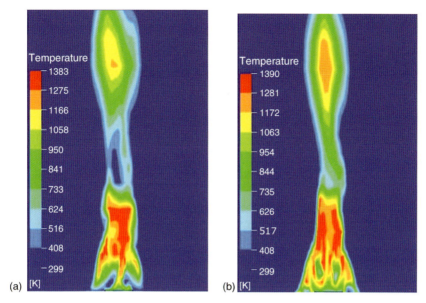

Figure 9.4 CFD predicted instantaneous temperature fields of DTBP pool fires (a) $d=1.12$ m; (b) $d=3.4$ m.

9.7.2
Surface Emissive Power (SEP)

The SEP is a derived quantity, and its value depends on flame surface and flame shape. Especially the value of the flame height, but also atmospheric transmission and general experimental errors are import. The SEP of a fire can be obtained by CFD simulation in three ways. In the first way the SEP is predicted by the radiative heat flux q_{out} leaving each grid cell placed on the flame surface A_F. The value of q_{out} can by

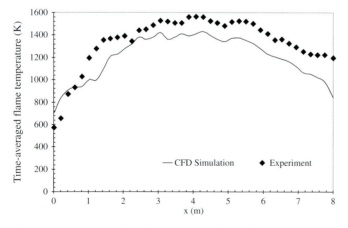

Figure 9.5 Time-averaged axial temperature profiles of DTBP ($d=3.4$ m).

obtained by calculating the component of radiative flux q_{rad} that is normal ($\vec{n} \perp A_F$) to the cell surfaces which define the flame surface A_F:

$$SEP = q_{out} = (1-\varepsilon_m)q_{in} + \varepsilon_m \sigma T^4, \quad (9.24)$$

with q_{in} defined as

$$q_{in} = \int_{\vec{s}\cdot\vec{n}<0} L\vec{s}\cdot\vec{n}d\Omega. \quad (9.25)$$

To get the *SEP* at the flame surface it is necessary to determine a surface A_F which represents a realistic shape of the flame. One possibility is an isosurface of temperature $(T > T_a)$.

The procedure can be described as follows:

- An instantaneous flame surface $A_{F,CFD}$ is defined as an isosurface of temperature (interior wall).
- The CFD calculated heat flux $q_{out}(t)$ is averaged over the isosurface $A_{F,CFD}$ for each time interval Δt (a usual value is $\Delta t = 0.1$ s) to evaluate area-averaged heat flux $<q_{out}(t)>$.
- The heat flux $<q_{out}(t)>$ is averaged over the steady burning time $(t \approx 10$ s), which results in a time-averaged heat flux $<\bar{q}_{out}(t)> \equiv \overline{SEP}_{CFD}$. It is assumed that a steady-state burning time of 10 s shows real burning behavior.

Figure 9.6 shows the instantaneous SEP, depending in general on pool diameter and fuel, calculated by using an isosurface of temperature $T = 400$ K for a JP-4 pool fire $(d = 16$ m$)$.

In the second way, SEP is predicted based on a q_{out} distribution over a cross section through the flame to obtain CFD-predicted thermograms (Figure 9.7 b1, b2):

- A cross section A_{xy} through the flame is defined at a distance $\Delta y = d/4$ from the pool rim.
- The CFD calculated heat flux $q_{out}(t)$ is averaged over cross section A_{xy} for each time interval Δt (a usual value is $\Delta t = 0.1$ s) to predict heat flux $<q_{out}(t)>$ distribution.
- The histograms are averaged over the steady burning time to get $\overline{SEP}$ (Figure 9.7 b3).

The instantaneous histograms show the spatial inhomogeneities and temporal deviations of a pool fire.

Thirdly, SEP_{CFD} of a fire is predicted by irradiance $\bar{E}_{CFD}(\Delta y/d)$ (see Section 9.3) with virtual wide-angle radiometers positioned near the pool rim. The irradiance $\bar{E}$ is connected with the $\overline{SEP}_{CFD}$ by the following equation:

$$\bar{E}(\Delta y/d) = \tau_{at}\alpha_E\varphi_{E,F}(\Delta y/d)\overline{SEP} \quad (9.26)$$

Setting atmospheric transmissivity τ_{at} and absorption of the receiver α_E to one, the irradiance is only a function of the view factor $\varphi_{E,F}$. For calculating the irradiance, additionally the knowledge of the flame length is necessary, which can be determined from the highest axial extension of the isosurface with a defined temperature.

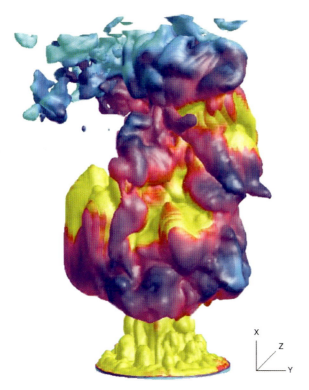

Figure 9.6 CFD-predicted SEP (x,y,t) on an isosurface of temperature $T = 400$ K for a JP-4 pool fire ($d = 16$ m).

The CFD predicted values of time averaged $\overline{SEP}$ show agreement with experimental results, especially in the case of JP-4 (Figure 9.8 and Table 9.2):

9.7.3
Irradiance

The irradiance E is determined in CFD simulation by virtual radiometers, which are defined at several points in the computational domain at different relative distances Δy from the pool rim. The radiometers have no thermal interaction and a sensitivity of 100%, so that the heat loss to the receiver element has not to be considered. The opening angle is 180°. In the resulting field of view the virtual radiometers receive the irradiance.

The receiver area element obtains all heat fluxes from computational cells in the defined field of view. In the atmospheric transmissivity, the absorption of carbon dioxide and water vapor is included. The incident radiation G from flame to computational cell with a virtual wide-angle radiometer is calculated as:

$$E_{CFD} \equiv \int_{2\pi\Omega_0} L(\vec{s}) d\Omega, \quad \text{at } \Delta y/d > 0 \tag{9.27}$$

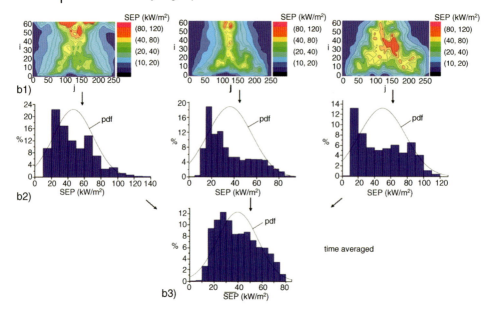

Figure 9.7 (b1) Instantaneous SEP on cross section A_{xy}, (b2) instantaneous histogram of SEP distribution, (b3) time-averaged $\overline{SEP}$ distribution.

The CFD-predicted time averaged irradiances $\bar{E}_{CFD}$ from JP-4 pool fires (Figure 9.9) for pool diameters $d = 8\,m$, $d = 16\,m$ and $d = 25\,m$ are in good agreement with experimental results.

The CFD-predicted time-averaged irradiances $\bar{E}_{CFD}$ from DTBP pool fires where $d = 1.12\,m$ and $d = 3.4\,m$ (Figure 9.10) also agree well with experimental results.

9.7.4
Critical Thermal Distances

CFD simulations are also a helpful tool to determine the critical thermal distances Δy_{cr} in relation, for example, to the limit of harmful effects ($1.6\,kW\,m^{-2}$) [46]. The critical thermal distance is the intersection of the horizontal line of $1.6\,kW\,m^{-2}$ with the irradiance $\bar{E}(\Delta y)$-curve. The irradiance $\bar{E}(\Delta y)$ can be obtained using Eq. (9.26) and the following equations for calculating the view factor φ:

$$\varphi_{E,F,h} = \frac{1}{\pi}\left[\arctan\sqrt{\frac{k+1}{k-1}} - \left(\frac{k^2-1+c^2}{\sqrt{CK}}\right)\arctan\sqrt{\frac{(k-1)C}{(k+1)K}}\right] \quad (9.28)$$

$$\varphi_{E,F,v} = \frac{1}{\pi}\left[\frac{1}{k}\arctan\frac{c}{\sqrt{k^2-1}} + \frac{c(C-2K)}{k\sqrt{CK}}\arctan\sqrt{\frac{(k-1)C}{(k+1)K}} - \frac{c}{k}\arctan\sqrt{\frac{k-1}{k+1}}\right] \quad (9.29)$$

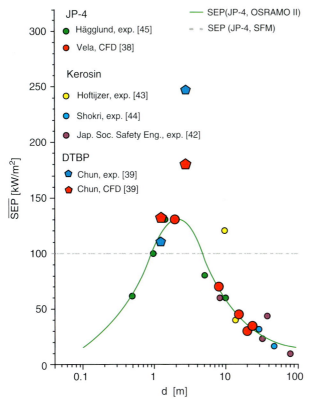

Figure 9.8 Measured and CFD-predicted SEP of hydrocarbons and DTBP pool fires as a function of pool diameter d.

$$\varphi_{E,F,\max} = \sqrt{\varphi_{E,F,h}^2 + \varphi_{E,F,v}^2} \qquad (9.30)$$

with

$$c = \frac{\bar{H}}{d/2}; \quad k = 1 + \frac{\Delta y}{d/2} \qquad (9.31)$$

Table 9.2 Measured and CFD-predicted $\overline{SEP}$ of different fuels with different d.

Fuel	d [m]	$\overline{SEP}_{exp}$ [kW m^{-2}]	$\overline{SEP}_{CFD}$ [kW m^{-2}]
JP-4	2	130 [42]	105
JP-4	8	70 [43]	70
JP-4	16	45 [44]	45
JP-4	20	31 [45]	32
JP-4	25	35 [45]	35
DTBP	1.12	130	110
DTBP	3.15	250	180

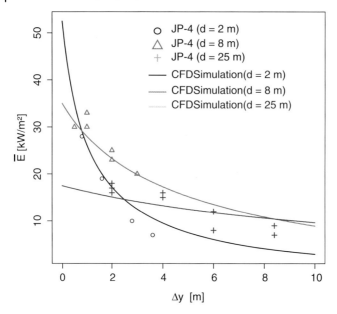

Figure 9.9 Measured and CFD-predicted time-averaged irradiances $\bar{E}_{CFD}$ from JP-4 pool fires as a function of relative distance Δy from the pool rim.

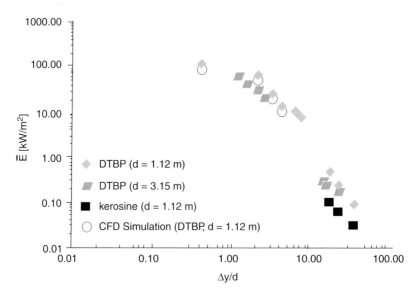

Figure 9.10 Measured and CFD-predicted time-averaged irradiances $\bar{E}_{CFD}$ from DTBP pool fires as a function of relative distance $\Delta y/d$ from the pool rim.

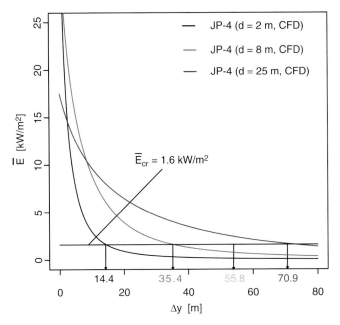

Figure 9.11 Critical thermal distances Δy_{cr} of JP-4 pool fires.

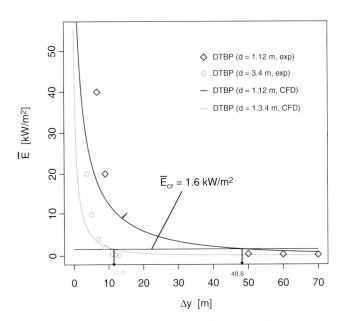

Figure 9.12 Critical thermal distances Δy_{cr} of DTBP pool fires.

$$C = (b-1)^2 + a^2 \quad K = (b-1)^2 + a^2 \tag{9.32}$$

For JP-4 pool fires, critical thermal distances of $\Delta y_{cr} = 14.4$ m ($d = 2$ m), 35.4 m ($d = 8$ m), 65.8 m ($d = 16$ m) and 70.9 m ($d = 25$ m) are predicted by CFD simulation (Figure 9.11).

For DTBP pool fires, critical thermal distances (Figure 9.12) of $\Delta y_{cr} = 11.4$ m ($d = 1.12$ m) and 48.8 m ($d = 3.4$ m) are obtained. The critical thermal distances Δy_{cr} of DTBP pool fires are by a factor of 3 to 4 higher then critical thermal distances of JP-4 pool fires.

9.8
Conclusions

The following conclusions are summarized from the results:

By CFD simulation it is possible to predict the SEP_{CFD} for hydrocarbon and organic peroxide pool fires by the following three ways:

- calculation of an isosurface of temperature defined as the flame surface,
- calculation of an SEP distribution on a cross section, and
- calculation of the irradiance $E(\Delta y/d, t)$ by virtual radiometers defined near the pool rim.

CFD simulation predicts the time-dependent irradiances $E(\Delta y/d, t)$ and time averaged irradiance $\overline{E}(\Delta y/d)$ either with virtual radiometers at different horizontal distances $\Delta y/d$ from the pool rim or through the connection with $\overline{SEP}_{CFD}$ by a view factor, atmospheric transmissivity, and absorption of the receiver.

The use of the four-step discontinuity function for the effective absorption coefficient a_{eff} of JP-4 pool fires plays an important role in predicting $\overline{SEP}_{CFD}$ with CFD simulation.

For a successful CFD simulation the use of non-equilibrium chemistry models like laminar flamelet models is necessary.

9.9
CFD – The Future of Safety Technology?

In combination with experimental results, CFD simulation is going to become an important tool for determining source terms of large open fires. For CFD simulation, the requirements of experimental data for validation of sub-models will increase in the coming years. CFD simulations will have an excellent future in safety technology.

Nomenclature

Symbol	Units	Quantity
a	[m^{-1}]	Absorption coefficient
A_F	[m^2]	Flame surface

b	[m]	Flame stretch
d	[m]	Pool diameter
D	[m^2 s^{-1}]	Diffusion coefficient
e	[kg m^{2}s^{-2}]	Total energy
E	[W m^{-2}]	Irradiance
f	[1–2]	Volume fraction
F	[kg m s^{-2}]	Body forces
G	[W m^{-2}]	Incident radiation
H	[m]	Flame length
j	(kg m^{-2}s^{-1})	Mass flux density
L	[W m^{-2}s^{-1}r^{-1})]	Radiance
M	[kg mol^{-1}]	Molecular mass
$\vec{n}$	[–]	Unit vector normal to surface
p	[pa]	Pressure
q	[W m^{-2}]	Heat flux density
r	[mol m^{-3}s^{-1})]	Reaction rate
$\vec{s}$	[–]	Unit vector in direction of s
s	[m]	Path length
SEP	[W m^{-2}]	Surface Emissive Power
t	[s]	Time
T	[K]	Temperature
u	[m s^{-1}]	Flow velocity
w	[–]	Mass fraction

Greek

ε	[–]	Emissivity
Γ	[–]	Gamma function
λ	[m]	Wavelength
φ	[–]	View factor
χ	[s^{-1}]	Scalar dissipation
ξ	[–]	Mixture fraction
ϱ	[kg m^{-3}]	Density
τ	[–]	Transmissivity
Ω	[sr]	Solid angle

Indices

at	Atmospheric
b	Boiling
con	Conduction
eff	Effective
E	Receiver
g	Gas
in	Incoming
m	mixeture
out	Outgoing
rad	Radiative
s	Soot

References

1 Balluff, C., Brötz, W., Schönbucher, A., Göck, D., and Schieß, N. (1985) *Chemie Ingenieur Technik*, **57**, 823.

2 Persson, H. and Lönnermark, A. (1951) *Tank Fire Review of Fire Incident*, Brandforsk Project 513-021 Swedish

National Testing and Research Institute, Boras, Sweden.
3 Gawlowski, M., Hailwood, M., Schalau, B., and Schönbucher, A. (2009) *Chemical Engineering & Technology*, **32**, 207.
4 Bird, R.B. and Granham, M.D. (1998) *The Handbook of Fluid Dynamics*, CRC Press, New York.
5 Prandtl, L. (2002) *Führer Durch die Strömungslehre*, Vieweg-Verlag, **11**.
6 Oertel, H., Böhle, M., and Dohrmann, U. (2009) *Strömungsmechanik*, Vieweg-Teubner, **5**.
7 Warnatz, J., Maas, U., and Dibbel, R.W. (2006) *Combustion*, 4th edn, Springer-Verlag, Berlin Heidelberg.
8 Ferziger, J.H. and Peric, M. (1997) *Computational Methods for Fluid Dynamics*, Springer-Verlag, Berlin Heidelberg, 2. ed.
9 Sagaut, P. (2004) *Large Eddy Simulation for Incompressible Flows*, Springer-Verlag, Berlin Heidelberg.
10 Pope, S.B. (2000) *Turbulent Flow*, Cambridge University Press.
11 Lilley, D.K. (1993) NASA Contract Report 93–75. NASA Langley Research Center.
12 Mathieu, J. and Scott, J. (2004) *An Introduction to Turbulent Flows*, Springer-Verlag, Berlin Heidelberg.
13 Gerthsen, C. (1993) *Physik*, Springer-Verlag, Berlin Heidelberg.
14 Gerlinger, P. (2005) *Numerische Verbrennungssimulation*, Springer-Verlag, Berlin Heidelberg, 2. Aufl.
15 Gutheil, E. and Bockhorn, H. (1987) *Physicochemical Hydrodynamics*, **9**, 525.
16 Dahn, W.J.A. and Bish, E.S. (1993) *Turbulence and Molecular Process in Combustion* (ed T. Takeno), Elsevier, New York.
17 Peters, N. (1987) 21th Symp. (Int.) on Combustion, 1231.
18 Liew, S.K., Bray, K.N.C., and Moss, J.B. (1984) *Combustion and Flame*, **56**, 199.
19 Siegel, R., Howell, J.R., and Lorengel, J. (1993) *Wärmeübertragung durch Strahlung*, Teil 3 (ed. U. Grigull), Springer-Verlag, Berlin-Heidelberg.
20 Modest, M. (1993) *Radiative Heat Transfer*, McGraw-Hill Series in Mechanical Engineering, New York.
21 Fiala, R., Göck, D., Zang, X., and Schönbucher, A. (1991) *Chemie Ingenieur Technik*, **63**, 760.
22 Fiala, R., Göck, D., Zhang, X., and Schönbucher, A. (1992) *TÜ*, **33** (Teil 1), 137.
23 Krishnamoorthy, G. (2010) *Journal of Hazardous Materials*, **182**, 570.
24 Snegirev, A.Y. (2004) *Combustion and Flame*, **136**, 51.
25 Lallemant, N., Sayre, A., and Weber, R. (1996) *Progress in Energy and Combustion Science*, **22**, 543.
26 Hottel, H.C. (1954) *Heat Transmission* (ed. W.H. McAdams), McGraw-Hill, New York, 3. ed.
27 Dalzell, W.H. and Sarofim, A.F. (1969) *Journal of Heat Transfer*, **91**, 100.
28 Magnussen, B.F. (1998) *Modelling NOx and Soot Formation by the Eddy Dissipation Concept*, First Topic Oriented Technical Meeting, Internation Flame Research, Foundation, Amsterdam.
29 Tesner, P.A., Snegirova, T.D., and Knorre, V.G. (1971) *Combustion and Flame*, **17**, 253.
30 Leung, K.M., Lindstedt, R.P., and Jones, W.P. (1991) *Combustion and Flame*, **87**, 289.
31 Brookes, S.J. and Moos, J.B. (1999) *Combustion and Flame*, **116**, 486.
32 Raithby, G.D. and Chui, E.H. (1990) *Journal of Heat Transfer*, **112**, 415.
33 Mathur, S.R. and Murthy, Y.Y. (1999) *Journal of Thermophysics and Heat Transfer*, **13** (4), 467.
34 Lockwood, F.C. and Shah, N.G. (1981) 18th Symp. (Int.) on Comb., 1405.
35 Howell, J.R. (1968) *Advances in Heat Transfer*, vol. **5** (eds J.P. Jartnett and T.F. Irvine), Academic Press., New York.
36 Vela, I. (2009) CFD prediction of thermal radiation of large, sooty, hydrocarbon pool fires. PhD. Thesis, Universität Duisburg-Essen.
37 Chun, H. (2007) Experimentelle Untersuchungen und CFD-Simulationen von DTBP-Poolfeuern. PhD. Thesis, BAM Dissertationsreihe Band 23.

38 Vela, I., Chun, H., Mishra, K.B., Gawlowski, M., Sudhoff, P., Rudolph, M., Wehrstedt, K.-D., and Schönbucher, A. (2009) *Forschung im Ingenieurwesen*, **73**, 87.

39 Chun, H., Wehrstedt, K.-D., Vela, I., and Schönbucher, A. (2009) *Journal of Hazardous Materials*, **167**, 105.

40 (2008) ANSYS CFX 11.0, User Guide, EAE Technology.

41 (2002) ANSYS FLUENT 12, User Guide. AEA Technology.

42 (1982) Japan Institute for Safety Engineering, Report on Burning of Petroleum Fires.

43 Hoftijzer, R.W.G. (1980) *TNO-report*. 9092, **6**, 1.

44 (1999) Society of Fire Protection Engineers, Engineering Guide for Assessing Flame Radiation to External Targets from Pool Fires. Bethesda, Maryland.

45 Hägglund, B. and Persson, L.E. (1976) FOA Rapport C30126-D6 (A3).

46 (2007) Installation and equipment for liquefied natural gas - Design of onshore installations; German version EN 1473.

47 Sinai, Y.L. (2000) *Fire Safety Journal*, **35**, 51.

48 (1997) Methods for the calculation of physical effects ('Yellow Book'), CPR 14E (Part 2, Chap. 6: Heat flux from fires, 6.1 - 6.130), 3. Aufl.

10
Modeling Fire Scenarios and Smoke Migration in Structures
Ulrich Krause, Frederik Rabe, and C hristian Knaust

10.1
Introduction

Mathematical modeling of physical phenomena is an attempt to create tools which can help

- to predict relevant parameters of an event at its final stage,
- to predict the course of a process with time, and
- to quantify the influence of system parameters, boundary conditions, material properties and so forth on the process.

Applied to fire modeling the aim is either to prevent the occurrence of fires or to prevent or at least mitigate their potentially harmful consequences or to be able to reconstruct the chain of events of a fire. The first two aims, of course, are far more frequently made use of than the last one.

Mathematical modeling is based on the following principles:

- to identify the physical (and/or chemical, biological, etc.) phenomena which determine the process under investigation,
- to find suitable mathematical descriptions of these phenomena (a 'mathematical model').

In general, a mathematical model for a natural phenomenon consists of a set of mathematical expressions which allow a closed solution to be obtained.

To ensure a sufficient accuracy of the predictions undertaken with the model and to reach an acceptable level of uncertainty, the model has to undergo a process of verification and validation.

In the construction sector, fire protection concepts based on fire scenario modeling have gained wider and wider application during the recent decade [1].

For a fire, the underlying physical phenomena are the transfer of mass, momentum, and energy in a thermodynamic system, including the occurrence of chemical reactions [2]. Hence, a mathematical model has to be based on equations which describe these effects. For a general case these are the balance equations for mass,

Process and Plant Safety: Applying Computational Fluid Dynamics, First Edition. Edited by J. Schmidt.
© 2012 Wiley-VCH Verlag GmbH & Co. KGaA. Published 2012 by Wiley-VCH Verlag GmbH & Co. KGaA.

momentum, and energy transfer of continuum mechanics. These equations are given in Annex 1.

The set of general balance equations is, however, not mathematically closed. This means that a solution of the system of equations needs further equations: the so-called closure laws. These closure laws describe, for example, source terms of the balance equations, material properties, the chemical reaction model, and so on. Closure laws may also exist in the form of sub-models, for example, of chemical reactions, turbulence, and so on.

The full set of balance equations for mass, momentum, and energy transfer in a thermodynamic system can only be solved numerically; that is to say, the 'true' solution is approximated by a numeric scheme. Nowadays, this is done using suitable computer codes, for example, based on the principles of computational fluid dynamics (CFD) [3].

Therefore, in fire modeling we have also to deal with numerics, namely with numerical solutions for partial differential equations. The different CFD codes for fire modeling contain different numerical schemes of which the user has at least to know the advantages, disadvantages, and limitations.

However, in earlier years a detailed numerical analysis of fire phenomena was impossible. Therefore, one attempted to simplify the system of equations to a level at which an analytic solution of the remaining equations was accessible.

This restricted the phenomena to be investigated to a limited number of special cases, for which the simplified model was valid. Nevertheless, this method allowed us to study the principal features of fires, for example, temperature and velocity effects in a fire plume (see Section 10.5). It is still of value to apply this method, because it can give a quick and useful insight into the main effects of a fire.

It is extremely important that users of fire models are aware about the validity of the models, their range of application, and their limits and restrictions. Computer codes containing fire models must not be used as black boxes. Therefore requirements exist not only for the fire models or the computer codes but also for the users. Valuable guidance to this problem was given by Janssens [4] and in the ASTM standards 1355a [5] and 1895 [6].

At current state of development, the main areas of application for fire modeling are

- prediction of building fire scenarios,
- prediction of thermal loads on structures,
- prediction of smoke ventilation,
- prediction of the effect of automatic extinction systems, and
- egress modeling.

During recent years, in many countries building regulations have undergone a profound change from the so-called prescriptive codes to performance-based design. This means that building authorities are encouraged to accept fire protection concepts which may deviate from existing regulations when the same level of safety is achieved by other means. Accepted approvals for fire safety concepts are the methods of fire safety engineering.

These are computational methods which are based on physical principles and which demonstrate that the targets of fire safety in a specific project to which they are applied are reached.

10.2
Hierarchy of Fire Models

Fire models may be ranked according to their level of sophistication. Sophistication means

- the degree to which the models approximate reality,
- the complexity,
- the range of validity,
- the level of generalizability.

The model hierarchy is depicted in Figure 10.1.

Simple empirical models merely give a functional fit to measured data: application to cases which are not within the range of experimental parameters is not possible.

Analytical models are derived from the general equations for mass, momentum, and energy transfer. The balance equations (in the form of differential equations) are reduced to forms which are accessible to direct integration. Hence, the analytical models are based on physical principles; however, the complexity of these models is low.

Zone models separate the fire room (or rooms) into at least two zones: an upper zone where the layer of hot fire gases (or 'smoke') is located and a lower smoke-free

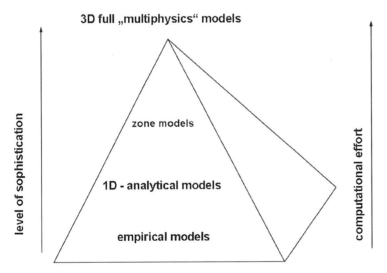

Figure 10.1 Hierarchy of fire models according to their level of sophistication and computational effort.

zone. The zones are at different temperatures, but the temperature within one zone is assumed to be uniform. The balance equations are used in the form of ordinary differential equations (ODEs).

In most zone models an energy balance and a mass balance for the flux of gas are detemined. A momentum equation is not considered. Instead, the flow velocity of the gas above the flame zone is computed by an empirical approach, a so-called plume model. This is explained in more detail in Section 10.4.

Full 'multiphysics' models solve the complete set of balance equations for mass, energy, and momentum transfer. The balance equations are in the form of partial differential equations.

'Multiphysics' is a term possibly causing some confusion, as there is, of course, only one kind of physics. Multiphysics means that different physical phenomena like heat transfer, mass transfer, momentum transfer, chemical reactions, fluid–structure interaction and so on can be analyzed simultaneously during the same run of the computer code used. Other names in use for the same class of models are CFD (computational fluid dynamics) models or field models. In the following text we will use the term CFD models.

CFD models compute the evolution with time of local distributions of the physical quantities under investigation. Hence, much more detailed information can be obtained from CFD models than from analytical models or zone models. Therefore, in the hierarchy of models according to Figure 10.1, CFD models rank at the top.

10.3
Balance Equations for Mass, Momentum, and Heat Transfer (CFD Models)

In this section the basics of CFD models are explained. The balance of a physical quantity Φ at an infinitesimal volume element dV is considered. As an example, this is shown for the fluid mass m in Figure 10.2.

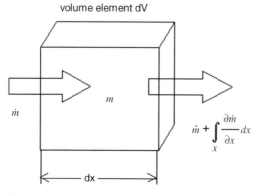

Figure 10.2 Balance of the fluid mass m across the length dx in an infinitesimal volume element.

The mass flux enters the volume element from the left and leaves it to the right. Passing the volume element along the distance dx the quantity $\dot{m}$ experiences the variation $\int \frac{\partial \dot{m}}{\partial x} dx$.

For the variation of the mass of fluid within the volume element, one obtains Eq. (10.1).

$$\frac{dm}{dt} = \dot{m} - \left(\dot{m} + \int_x \frac{\partial \dot{m}}{\partial x} dx \right) = - \int_x \frac{\partial \dot{m}}{\partial x} dx \qquad (10.1)$$

Eq. (10.1) expresses the fact that the variation of m with time is balanced by the fluxes of m across the boundaries of the volume element.

Applying a Gaussian transformation to Eq. (10.1) one obtains the form

$$\int_V \left(\frac{\partial \varrho}{\partial t} + \frac{\partial \varrho u_i}{\partial x_i} \right) dV = 0 \qquad (10.2)$$

This is the continuity equation in integral form. The integral form of the balance equations is used in CFD codes using finite volume solution schemes (e.g., FLUENT, ANSYS CFX, OpenFoam).

Differentiating Eq. (10.2) one obtains

$$\frac{\partial \varrho}{\partial t} + \frac{\partial \varrho u_i}{\partial x_i} = 0 \qquad (10.3)$$

Equation (10.3) is the continuity equation in differential form. This form is used by CFD codes based on finite difference solution schemes (FDS).

An analoguos treatment may be applied to the other quantities: momentum and energy.

In differential form the general balance equation underlying a CFD model is:

$$\underbrace{\frac{\partial}{\partial t}(\varrho \phi)}_{\text{accumulation}} + \underbrace{\frac{\partial}{\partial x_i}(\varrho u_i \phi)}_{\text{convection}} = \underbrace{\frac{\partial}{\partial x_i}\left(\Gamma_\phi \frac{\partial \phi}{\partial x_i} \right)}_{\text{diffusion}} + \underbrace{S_\phi}_{\text{source term}} \qquad (10.4)$$

Equation (10.4) is a general formulation for the variation of a quantity ϕ with space and time in a flow field. ϱ is the fluid density, u_i is the velocity vector in tensor notation, Γ_ϕ is the transport coefficient of molecular transport and S_ϕ is the volumetric rate of the source term with respect to ϕ.

Φ may either be 1 – then we get the mass balance, or the components of the velocity vector u_i – then we get the momentum balance, or the enthalpy h – then we get the energy balance.

Note that we need a momentum balance for each space direction i.

The complete set of equations is given in annex 1.

10.4
Zone Models

Zone models are a simplified form of fire models separating a room fire into mostly two zones. These are the hot gas and smoke layer zone above the fire and the cold smoke-free zone underneath the first zone.

The thermodynamic properties are assumed to be constant across each zone but are in general different from one zone to the other. The heat and mass transfer from one zone to the adjacent one is based on the following balance equations

$$\frac{dH}{dt} = \dot{Q} + p \cdot \dot{V} + \dot{H}_{in} - \dot{H}_{out} \tag{10.5}$$

$$\frac{dm_k}{dt} = \sum_k \dot{m}_k \tag{10.6}$$

From Eq. (10.5) it is obvious that the enthalpy content within one zone is balanced by the heat produced, the energy of volumetric change, and the incoming and outgoing enthalpy fluxes.

$\sum_k \dot{m}_k$ is the sum of all incoming and outgoing mass fluxes for zone k.

In contrast to CFD models, the balance equations in zone models are ordinary differential equations (ODEs), which can be solved with much less computational effort compared to PDEs.

A momentum balance is not solved in a zone model. Instead, the fire zone is modeled using a plume model as will be explained in Section 10.5. A plume describes the flux of smoke and fire gases above the flame zone under consideration of the entrained air.

There are numerous zone model computer codes on the market. Widely used are CFAST [7] developed at the US NIST and MRFC from the Technical University of Vienna, Austria [8]. The application of a zone model to the trailer house discussed in the present paper has been studied by Knaust [9] and is depicted in Figure 10.3.

10.5
Plume Models

Plume models are empirical or semi-empirical models which allow calculation (without using computers) of the mass flux and the temperature of the buoyant flow of gases above a fire (plume).

The basic idea is that of an ideal plume with the following presuppositions

- the fire is a point source,
- the buoyancy forces are a result of the difference in density between the hot gases above the flame and the cold air in the environment,
- the alteration in density within the plume along the height is neglected (Boussinesq approximation, not to be confused with the Boussinesq approximation according to the Reynolds stress model !).

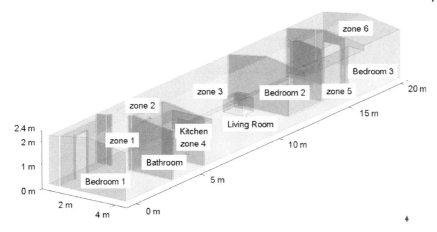

Figure 10.3 Geometry of the trailer house used in NIST's fire experiment and application of a multi-zone model to fire modeling (from [9]).

A wide variety of plume models have been developed, mostly differing in some empirical factors, depending on the experiments to which the equations were fitted. One of the most widely used plume models is that by Heskestad, quoted in [9]. The principle is depicted in Figure 10.4.

In contrast to the ideal plume, Heskestad supposed that the fire is not a point source but is extended across a certain area. Using a fixed angle (15°) of the plume, a virtual origin may be constructed. The basic equations of the Heskestad model are given in Table 10.1.

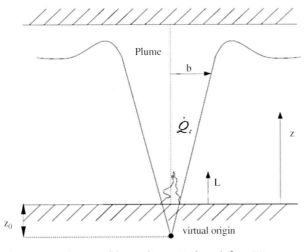

Figure 10.4 Plume model according to Heskestad (from [9]).

Table 10.1 Plume equations from Heskestad, quoted in [9].

Quantity	Equation	Range of validity
flame height	$L = 0.235\dot{Q}^{2/5} - 1.02D$	$7 < Q^{2/5}/D < 700\,\mathrm{kW}^{2/5}\,\mathrm{m}^{-1}$
virtual origin	$z_0 = 0.083\dot{Q}^{2/5} - 1.02D$	z_0 negative for large flame areas with respect to the heat released z_0 positive for large heat release with respect to the flame area
plume radius	$b = 0.12(T_0/T_\infty)^{1/2}(z-z_0)$	—
temperature difference in plume axis	$\Delta T = 9.1\left(\frac{T_\infty}{gc_p^2\varrho_\infty^2}\right)^{1/3}\dot{Q}_c^{2/3}(z-z_0)^{-5/3}$	—
flow velocity in plume axis	$u = 3.4\left(\frac{g}{c_p T_\infty \varrho_\infty}\right)^{1/3}\dot{Q}_c^{1/3}(z-z_0)^{-1/3}$	—
mass flux	$\dot{m}_p = 0.071 Q_c^{1/3}(z-z_0)^{5/3} + 1.92\cdot 10^{-3}\dot{Q}_c$	$z > L$
mass flux	$\dot{m}_p = 0.0056\dot{Q}_c\frac{z}{L}$	$z < L$

10.6
Computational Examples

The computational examples reported here are linked with an experiment in a so-called trailer house undertaken by the Building and Fire Research Laboratory (BFRL) of the US National Institute for Standards and Technology (NIST) [10]. The trailer house is a pre-fabricated single-story family home consisting of living room, three bedrooms, kitchen and a bathroom, as shown in Figure 10.3. The original idea of BFRL's experiments was to test the response of smoke detectors on a developing fire set up on an armchair in the living room.

10.6.1
Isothermal Turbulent Flow through a Room with Three Openings

The first computational example deals with an isothermal flow in the living room of the trailer house. Apart from the windows (which were considered to be closed in this example), the room has three openings. Half of the left side, which is the connecting floor to bedroom 2, was considered to be the flow inlet, while the two doors at the right were considered as pressure outlets. The sketch of the boundaries as set in the CFX pre-processor is shown in Figure 10.5.

The velocity of the air inflow was set to $1\,\mathrm{m\,s}^{-1}$; the gage pressure at the outlets was set to zero. Computations were performed on an unstructured grid with 567 000 cells.

Figure 10.6 depicts a comparison of the steady-state flow pattern in the x–z plane at a height of $y = 1.2$ m in the empty room using a Large Eddy Simulation (LES) with a Smagorinsky subgrid-scale turbulence model and a Reynolds-averaged Navier–Stokes (RANS) k-ε turbulence model. Note that in LES computations a transient computation has to be performed to obtain a steady-state solution.

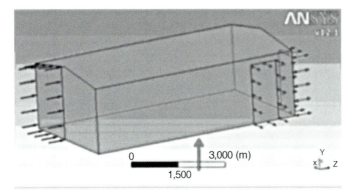

Figure 10.5 Flow boundaries at the living room for the example of isothermal turbulent flow.

Though in general the computational results show only minor differences, the LES/Smagorinsky turbulence model reveals more details in the structure of vortices appearing in the recirculation zone. The reason is that the k-ε model presumes a homogeneous and isotropic turbulence field, and this has a 'smearing' effect on the structure of vortices.

The same example was computed using the Fire Dynamics Simulator (FDS), Figure 10.7. Due to the use of a finite difference solver, FDS can only work on structured grids with uniform edge lengths of individual volume elements. This limits its applicability, especially to curved structures.

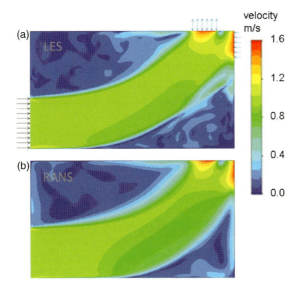

Figure 10.6 Isothermal flow in a room with three openings computed with ANSYS CFX. Flow pattern in the x-z-plane at a height of 1.2 m using a LES/Smagorinsky turbulence model (a) and a RANS/k-ε model (b).

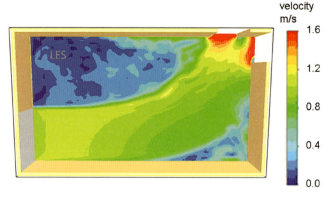

Figure 10.7 Isothermal flow in a room with three openings computed with FDS. Flow pattern in the x-y-plane at a height of 1.2 m using a LES/Smagorinsky turbulence model.

FDS uses an LES/Smagorinsky turbulence model with a fixed Smagorinsky factor C_s of 0.2, while this factor is user-defined in CFX.

The flow pattern computed with FDS shows only minor differences to the CFX solutions, especially the one using LES. However, during postprocessing it was observed that a true steady-state solution was not obtained with FDS.

Instead, the solution oscillated from time step to time step without a physical reason. The reason of these oscillations could not be clarified.

As a conclusion it can be stated that this simple case was computed with only minor differences by the different model options used. Flow velocities differed for not more than 7% of the average value from the three computations. Location of the main stream lines, vortices, and recirculation zones were in good agreement. A further comparison performed in [12] showed negligible differences between structured and unstructured grid computations for CFX. Further efforts have to be undertaken to disclose the origin of oscillations caused by the FDS solver.

10.6.2
Buoyant Non-Reacting Flow over a Heated Surface

In the second example a buoyant flow over a hot surface was investigated. The room where the event took place was the same as that in the previous example with an additional armchair located inside, as shown in Figure 10.8. The upper surface of the seat was set to a temperature of 600 °C, thus producing a buoyant flow inside the room. The red arrows point to the view planes for the postprocessing: 1 – front view, 2 – side view. In this example, all three openings of the room pressure boundaries were set with a gage pressure of zero.

Figure 10.9 shows the distribution of the average flow velocity in the front view plane after 80 seconds from the beginning of the heat release as computed by CFX (left picture) and FDS (right picture). The main flow patterns were computed similarly by both codes with minor differences. The most obvious one is the flow

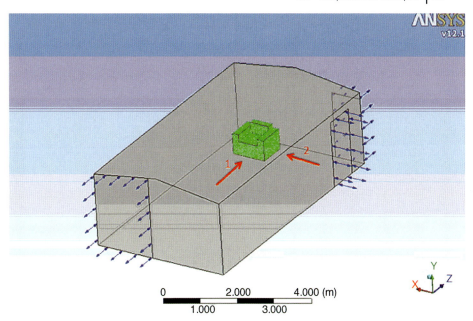

Figure 10.8 Living room with flow inlet, pressure outlets, and armchair.

detachment at the ceiling computed by FDS due to the staggered model of the ceiling. This leads to the computation of vortices which are not computed by CFX. Considering that the turbulent fluctuation velocity among other parameters influences the reaction rate computed by the mixture fraction model and the turbulent transport coefficients, it may be concluded that nonphysical computation of eddies increases the error of the simulation.

Recently, an additional function, 'smooth edges', has been introduced into FDS to match the problem of inclined geometric elements. Figure 10.10 shows the result of applying this function to the same example as that shown in Figure 10.9. An evaluation of the velocity field did not reveal any differences from the example without smoothed edges.

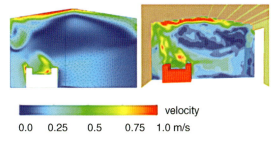

Figure 10.9 Front view of the buoyant flow above the armchair inside the living room. Left: ANSYS CFX, unstructured grid; right: FDS, structured grid, cell width 5 cm.

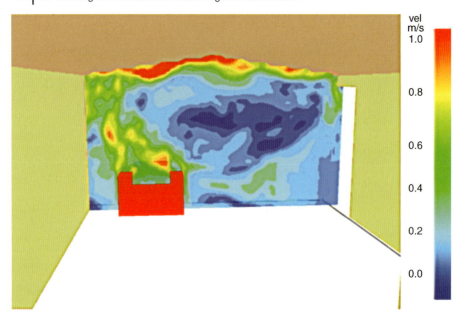

Figure 10.10 Front view of the buoyant flow above the armchair inside the living room using FDS, structured grid, cell width 5 cm, with function 'smooth edges'.

10.6.3
Simulation of an Incipient Fire in a Trailer House

The geometry of the entire trailer house as depicted in Figure 10.3 was used to study the propagation of an incipient fire and compare the results with data measured by NIST.

The origin of the fire was the armchair placed in the living room at the location indicated in Figure 10.11. The purpose of the experiment pursued by NIST was to test the response time of smoke detectors. Thermocouples were arranged inside the house along so-called trees labeled from A to F with each tree carrying 7 thermocouples at different heights. The armchair was extinguished 180 s after ignition, but the propagation of the heat wave and smoke was observed beyond this time. Figure 10.11 shows the arrangement as used in NIST's experiments together with the positions of slices used for evaluation of results.

The heat release from the armchair was modeled as a given function of mass loss per time multiplied with the calorific value of polyurethane according to Figure 10.12. Figure 10.13 shows the temperature distribution along slice 1 as computed with CFX, while Figure 10.14 exhibits the corresponding results for FDS.

Both codes computed the upward buoyant flow induced by the heat released due to the combustion of polyurethane (the fire plume) and the subsequent formation of the ceiling jet in the living room. Moreover, the heat transfer to the more distant rooms of the house, for example, bedroom 3 according to Figure 10.3, was also reflected by both codes.

10.6 Computational Examples

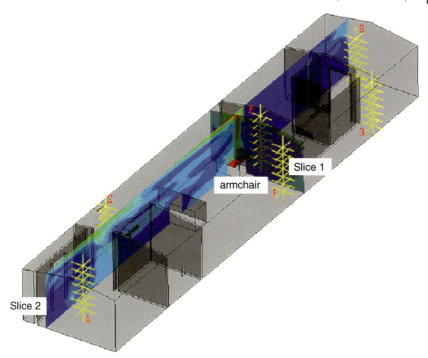

Figure 10.11 Experimental set-up of the trailer house, thermocouple trees labeled from A to F, and slices used for evaluation of results.

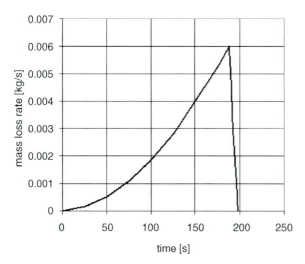

Figure 10.12 Mass loss rate vs time of polyurethane foam as part of the armchair involved in the incipient fire.

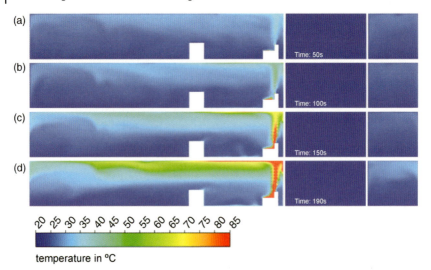

Figure 10.13 Temperature distribution along slice 2 at different times computed with CFX, from top to bottom: $t = 50\,\text{s}$, $100\,\text{s}$, $150\,\text{s}$, $190\,\text{s}$.

The ceiling jet was, however, more pronounced in the computations with FDS. A possible reason may be the above-mentioned nonphysical increase in turbulence.

Figures 10.15 and 10.16 depict the comparison of measured and computed temperatures at three different positions along tree E, which was the one located closest to the armchair. For CFX computations, agreement with the measured

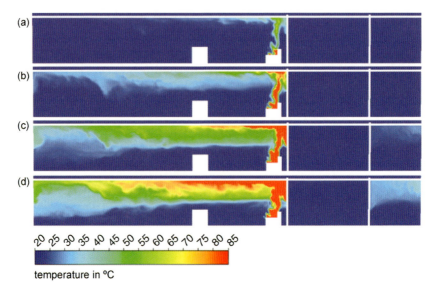

Figure 10.14 Temperature distribution along slice 2 at different times computed with FDS, from top to bottom: $t = 50\,\text{s}$, $100\,\text{s}$, $150\,\text{s}$, $190\,\text{s}$.

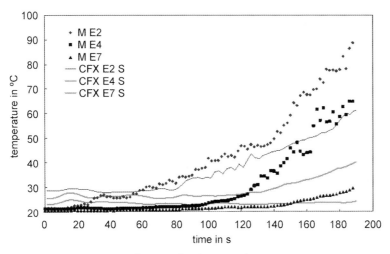

Figure 10.15 Comparison of measured and computed temperature time-curves at position E2 (CFX).

temperature is satisfactory for the first 120 s. After this, CFX under-predicts the temperatures at this position, with deviations of up to 20 K.

In the computations with FDS, strong oscillations were observed for the temperature location closest to the ceiling. The nature of these oscillations could not be clarified, but there is some evidence that they are caused by the numerical scheme for gradients in the near-wall region, since they are less pronounced at the thermocouple positions become more distant from the ceiling. A physical reason for these oscillations can be excluded, since there are no flow alterations with the frequency of the oscillations computed.

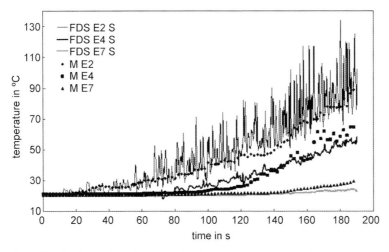

Figure 10.16 Comparison of measured and computed temperature time-curves at position E2 (FDS).

10.6.4
Simulation of Smoke Migration

The prediction of the flow of smoke through a building in the case of a fire is an essential part of the fire protection concept. The aim is to preserve a smoke-free layer along the escape routes for as long as possible to allow safe evacuation of people.

However, it is difficult to characterize smoke, because smoke is not a scalar quantity which can be directly derived from the solution of a balance equation. Instead, smoke is a mixture of numerous gaseous, liquid, and solid fire products. In the present study, however, smoke was treated as a quasi-component in which all fire products were inluded.

Figure 10.17 exhibits the distribution of the mass fraction of smoke across slice 1 in the CFX and FDS computation. The section colored red indicates the region where the smoke mass fraction is 9% or higher. The heights of the smoke-free layer computed by the two codes differ only slightly despite the remarkable differences in the computed temperatures.

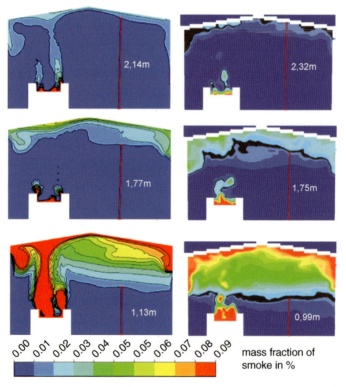

Figure 10.17 Distribution of the mass fraction of smoke across slice 1 computed with CFX (left column) and FDS (right column). From top to bottom $t = 50$ s, 100 s and 190 s from ignition. Red lines indicate the height of the smoke-free layer.

10.7
Conclusions

In analogy to physical experiments, 'numerical experiments', for example, fire simulations, also have uncertainties. The challenge is to identify the reasons for these uncertainties and to quantify them. The numerous input data and model features influence the computational result with different degrees of sensitivity.

Janssens [4] described three classes of uncertainties which have to be assessed for fire simulations:

- uncertainties in the input data,
- uncertainties in the model simplifications and assumptions,
- uncertainties of the validation experiments.

A comprehensive survey of methods for the analysis of uncertainties in engineering calculations has been given by Kmetsch [11]. However, the equation system of the type of Eq. (10.4) (see Annex 1) is even more complex than the one discussed by Kmetsch, and a sensitivity analysis is difficult to perform.

Due to the numerous assumptions and simplifications in the computational models and due to the uncertainties in the input data, it cannot be expected that computational results will be very precise. However, the main benefits from CFD calculations are

- the influence of all relevant physical phenomena and key factors on the process under investigation can be studied in detail,
- much more information is obtained than from 'rule-of-thumb' calculations or other simple calculation methods,
- scenarios with a high level of complexity according to geometry and physics can be considered.

10.8
CFD – The Future of Safety Technology?

Fire simulations based on CFD are a very powerful tool to predict (and also to reconstruct *a posteriori*) fire scenarios, and offer the possibility to extend the range of safety-related investigations further than would be possible with experiments only.

The purpose of simulations is to supplement rather than to replace experiments and to perform parameter variations to an extent that would never be possible by experiments alone.

Fire simulations enable us

- to obtain detailed quantitative information, for example, on temperatures, concentrations of species, and flow velocities in the fire room,
- to calculate the evolution of the fire with time, and
- to investigate particular effects and phenomena separately.

However, acceptance of simulations in the community of safety engineers, but also among authorities, will largely depend on the accuracy of the computed results (although simpler methods are inevitably less accurate and are nevertheless accepted).

Therefore, a key point for research during future years will undoubtedly become the assessment of uncertainties of fire simulations and the development of methods and tools for this purpose.

Annex 1: Set of Partial Differential Equations for CFD Models (differential form)

Mass balance (continuity equation):
$$\frac{\partial \varrho}{\partial t} + \frac{\partial \varrho u_i}{\partial x_i} = 0$$

Momentum balance (for each space direction):
$$\frac{\partial \varrho u_i}{\partial t} + \frac{\partial \varrho u_i u_j}{\partial x_j} = \varrho \cdot g_i - \frac{\partial p}{\partial x_i} + \eta \left(\frac{\partial^2 u_j}{\partial x_i^2} + \frac{\partial^2 u_i}{\partial x_i \partial x_j} - \frac{2}{3} \delta_{ij} \frac{\partial^2 u_k}{\partial x_i \partial x_k} \right)$$

Energy balance:
$$\frac{\partial (\varrho h - p)}{\partial t} + \frac{\partial \varrho h u_i}{\partial x_i} = a \frac{\partial^2 \varrho h}{\partial x_i^2} + \tau_{ij} \frac{\partial u_i}{\partial x_j} + \tilde{\dot{q}}$$

List of symbols

ϱ	density (kg/m³)
τ_{ij}	shear stress tensor (Pa)
η	dynamic viscosity (Pa·s)
δ_{ij}	Kronecker delta
ϕ	scalar quantity
Γ_ϕ	transport coefficient
g	gravity acceleration (m s⁻²)
h	Specific enthalpy (J kg⁻¹)
H	enthalpy (J)
$\dot{H}$	enthalpy flux (J s⁻¹)
$\tilde{\dot{q}}$	heat source density (J m⁻³ s⁻¹)
$\dot{Q}$	heat flux (J s⁻¹)
p	pressure (Pa)
S_ϕ	source per volume and time
u	flow velocity (m s⁻¹)
x	spatial coordinate (m)
t	time (s)
V	volume (m³)
$\dot{V}$	volume flux (m³ s⁻¹)

References

1 Hosser, D. (2009) Technischer Bericht 04/01 'Leitfaden Ingenieurmethoden des Brandschutzes' (Technical report 04/01 'Guideline to methods of fire safety engineering'), Vereinigung zur Förderung des deutschen Brandschutzes e.V., in German.

2 Drysdale, D. (1998) *An introduction to Fire Dynamics*, 2nd edn, John Wiley & Sons.

3 Ferziger, J. and Peric, M. (2002) *Computational Methods for Fluid Dynamics*, Springer Verlag, Berlin.

4 Janssens, M. (2002) Evaluating computer fire models. *Fire Protection Engineering*, **13**, 19–22.

5 (2005) ASTM E 1355a Standard Guide for Evaluating the Predictive Capability of Deterministic Fire Models.

6 (2007) ASTM E 1895 Standard Guide for Determining Uses and Limitations of Deterministic Fire Models.

7 Peacock, R.D., Forney, G.P., Reneke, P.A., Portier, R.M., and Jones, W.W. (1993) *CFAST, the Consolidated Model of Fire Growth and Smoke Transport*, NIST Technical Note 1299, National Institute of Standards and Technology, Gaithersburg, Maryland, USA.

8 Schneider, U. (ed.) (2008) *Ingenieurmethoden im Brandschutz, 2. Auflage*, Werner Verlag GmbH & Co KG, Köln, 978-3-8041-5238-0; 499 S.

9 Knaust, C. (2009) Modellierung von Brandszenarien in Gebäuden (Modelling of fire scenarios in buildings) PhD thesis, Technical University Vienna, Austria (in German).

10 Bukowski, R.W., Peacock, R.D., Averill, J.D., Cleary, T.G., Bryner, N.P., Walton, W.D., Reneke, P.A., and Kuligowski, E.D. (2007) *Performance of Home Smoke Alarms - Analysis of the Response of Several Available Technologies in Residential Fire Settings*, National Institute of Standards and Technology, Gaithersburg, Maryland, USA.

11 Kmetsch, T. (2004) Unsicherheiten in Ingenieurberechnungen (Uncertainties in engineering calculations) PhD thesis, Otto-von-Guericke-University, Magdeburg, Germany (in German).

12 Rabe, F. (2010) Vergleich von Computational Fluid Dynamics-Programmen in der Anwendung auf Brandszenarien in Gebäuden (Comparison of Computational Fluid Dynamics codes in the application of fire scenarios in buildings) Master Thesis, Matrikelnummer 761098, Beuth Hochschule für Technik, Berlin.

Part Four
CFD Tomorrow – The Way to CFD as a Standard Tool in Safety Technology

11
The ERCOFTAC Knowledge Base Wiki – An Aid for Validating CFD Models

Wolfgang Rodi

11.1
Introduction

CFD is used more and more in all industries involving fluid flow and has also become a major design tool in safety technology. However, it is still an emerging technology and involves many uncertainties, so that quality and trust in CFD applications is an important issue. Uncertainty can arise from numerical inaccuracies, but these are being kept more and more under control, especially as a result of ever-increasing computer power. Of greater concern is the adequacy of the underlying mathematical models describing the physical processes, especially the effects of turbulence. Direct numerical simulations (DNS) are not possible in the foreseeable future for practical calculations. Large-eddy simulation (LES) is seen as the technology of the future, and in some areas it is already used in practical applications. However, most practical flow calculations today are still carried out with statistical (or RANS) turbulence models, and it is now generally accepted that none of these models is or will be universal, that is, there exists no model that produces reliable predictions for all flows. In view of this, reliable information is required on which models work for which flows, and, based on this knowledge, guidelines need to be developed for solving individual flow problems. This is basically what was attempted in the EU Network Project QNET-CFD, in which a Knowledge Base was generated containing various components. Knowledge was collected for two types of flows, one referred to as Application Challenges (ACs) for individual industrial sectors, representing flows closely similar to those occurring in practice, where the ability of CFD to reliably predict the main design and assessment parameters is of prime interest. These flows are generally complex systems containing certain Underlying Flow Regimes (UFRs). The second type represents more generic flows (such as boundary layers, jets, step flows, etc.) for which much more detailed experimental data are available as well as better resolved numerical calculations, so that more thorough testing and validation of CFD methods and the turbulence models used can be carried out.

In the EU project, a Knowledge Base was developed in which a larger number of ACs and UFRs were stored. The members of the project came mainly from industry, but also from universities and national research organizations and from the major

European CFD vendors. Each member had to provide one AC and one UFR and had to subject the submitted contributions to a strict quality control according to a review template. The resulting contributions forming the Knowledge Base were stored at the University of Surrey, and the Knowledge Base was maintained there for a transition period in which ERCOFTAC (the European Research Community on Flow, Turbulence and Combustion) took over the Knowledge Base and turned it into an interactive Wiki. In this process, further quality control was exercised, and certain cases were taken out from the Knowledge-Base part to be made public and were stored away for eventual improvement. The remaining ACs and UFRs were grouped into Gold and Silver domains according to their quality, as explained below. These were made available to the Fluid Mechanics community at: http://qnet-ercoftac.cfms.org.uk, and ERCOFTAC continuously enhances and expands the Knowledge Base Wiki by adding new test cases.

In this paper, a brief introduction to and description of the Knowledge Base is presented: The concept, structure, and operational features are outlined, and an overview over the content is given. For details and examples of Application Challenges (ACs) and Underlying Flow Regimes (UFRs) the reader is referred to the Wiki itself.

11.2
Structure of the Knowledge Base Wiki

The navigation tree of the Wiki is shown in Figure 11.1. Through this, the various parts of the Knowledge Base can be reached; it is provided here as reference will be made to it in the following descriptions.

As mentioned already, knowledge was collected and is provided for two types of flow test cases, namely:

11.2.1
Application Challenges (AC)

These are complex flow situations closely similar to those encountered in industrial practice for which mainly global design and assessment parameters are of interest, and measurements of the details of the flow are often not available. For this category, the Knowledge Base provides files with the actual data for the geometry, the measurements, and the CFD results. The application challenges are grouped into test cases for different application areas:

- External Aerodynamics
- Combustion
- Chemical, Process, Thermal, and Nuclear Safety
- Civil Construction and Heating, Ventilation, and Air Conditioning (HVAC)
- Environmental Flow
- Turbomachinery Internal flow

Figure 11.1 Navigation tree of Wiki.

Test cases relevant to safety technology can be found mainly in the areas of *Chemical Process, Thermal, and Nuclear Safety,* and *Civil Construction* and *HVAC*.

11.2.2
Underlying Flow Regimes (UFR)

Application challenges contain Underlying Flow Regimes, which are more generic, building-block flows for which much more detailed experimental results are available, which are better resolved in numerical calculations, and for which also wider testing using different simulation and modeling approaches (RANS, LES, DES) was performed. For UFRs the experimental and computational results are mostly given only in graphical or tabular form, but in future, when new UFRs will be added, the actual data will also be stored in the Knowledge Base.

The UFRs are grouped into the following four flow types:

- Free Flows
- Flows Around Bodies
- Semi-confined Flows
- Confined Flows

Following a quality review based on the templates given in the *Quality* section of the Wiki, the Knowledge Base was partitioned for both ACs and UFRs into Gold and Silver Domains. The Gold Domain is the repository for content that has been carefully checked and therefore satisfies high quality standards. The Gold Domain is restricted to ERCOFTAC members (for information on ERCOFTAC see: www.ercoftac.org). However, the list of all ACs and UFRs, as given in the *Index* of the Wiki (reproduced below), and all the abstracts are in the public domain. The Silver Domain is the repository for less mature content which is still under discussion and open for improvement. Parts of the Silver Domain are marked as Silver Star. This indicates content which has matured to levels of quality and significance approaching Gold Standard and is made public to serve as examples of the quality found in the Gold Domain and in some cases will act to invite the addition of new CFD results obtained with alternative models. Both Silver Star and Silver test cases are fully in the public domain.

11.3
Content of the Knowledge Base

The content of each AC and UFR test case prepared according to the templates given in the *Library* section of the Wiki contains the following items:

- **Abstract** - Summary of content including a brief introduction to the flow considered.
- **Description** - Introduction to the test case and the physical phenomena involved – for ACs also the practical importance and parameters of interest, and for UFRs a review of previous studies and choice of test case.
- **Test case studies** - Description of the test case experiments with overview of experimental approach, a description of the experimental set-up and measurement techniques, overview of the measured data, and information on the measurement errors. Description of CFD simulations, again with overview of the simulations and the solution strategy, computational domain, grid and numerical accuracy, boundary conditions, physical models, in particular turbulence models used. For ACs provision of the actual experimental data and CFD results.
- **Evaluation** - Comparison of CFD calculations with test case experiments and discussion.
- **Best Practice Advice** - Recapitulation of the key fluid physics, pointers to application uncertainties, provision of direct advice concerning computational domain, boundary conditions, discretization and grid resolution, also which physical model to use for the case considered, and finally, some recommendation for future work.
- **Quality review** - This is provided only for test cases in the Gold Domain.

A good impression of the content of the Knowledge Base can be obtained by looking at the publicly available examples of ACs and UFRs in the Wiki (http://qnet-ercoftac.cfms.org.uk). At the end of the paper, lists of Application Challenges and Underlying Flow Regimes available in the Knowledge Base are given, reproduced from the *Index* of the Wiki. The color indicates the Gold and Silver Domains, and the star the publicly accessible documents in the Silver Star Domain.

11.4
Interaction with Users

Users are encouraged to make comments and suggestions on the individual ACs and UFRs and to exchange views and experiences with other users and with the Editorial Board through the *Forum* section. This also allows users to comment on and suggest improvements to the Wiki itself and to its functioning, and provides feedback questionnaires.

Users cannot directly edit or change the Knowledge Base documents, but they are encouraged to provide new content to the Knowledge Base, either by supplying modifications/extentions to existing content or by adding new contributions, that is ACs or UFRs for new test cases. How to do this is described in detail in the *Library* section, which also provides the templates according to which the new content to be submitted has to be prepared. In the *Help* section further information is given on how to contribute material, and this section also provides further useful information such as a glossary, links to other useful fluid mechanics sites, and answers to questions that may arise.

11.5
Concluding Remarks

The ERCOFTAC Knowledge Base Wiki launched in September 2009 is briefly introduced here, and guidance as to its structure and content is given. Developed from the Knowledge Base generated in the EU-Network-Project QNET-CFD, the Wiki is a unique repository of structured and quality-checked knowledge on a wide range of flow situations, both for complex flows closely matching industrial practice (Application Challenges) and for more generic, building block flows (Underlying Flow Regimes). This knowledge made available to the Fluid Mechanics community by the Wiki can be used in various ways. First, comprehensive knowledge from both experiments and numerical simulations on a wide variety of flows is provided, which forms a useful source of information on these flows. Further, guidance is given on how best to calculate the individual flows, including details of the various aspects of a calculation such as the numerical method, the resolution requirements, the boundary conditions, and the turbulence model, this guidance being based on the evidence of the results provided and discussed. Also, the test cases provide target data for users to test their own calculation methods, so helping to improve the quality of and trust in their CFD procedures, an aspect which is particularly important in the area of Safety Technology. The Knowledge Base Wiki is not static but is continuously enhanced and expanded by adding new test cases, and this is facilitated by the framework developed for storing results, in particular the templates for a common format of presentation of the information and the data. The Wiki provides an ideal tool for storing results generated in EU-funded and other projects that would otherwise be lost and for making these results available to a wide community.

AC Index

Application Area	AC number	Application Challenges	Contributor	Organisation
External Aerodynamics				
	1-01	Aero-acoustic cavity	Fred Mendonca	Computational Dynamics Ltd
	1-02	RAE M2155 Wing	Pietro Catalano, Anthony Hutton	CIRA, Qinetiq
	1-05	Ahmed body	Jean-Paul Bonnet, Remi Manceau	Université de Poitiers
	1-08	L1T2 3 element airfoil ★	Jan Vos, Anthony Hutton	CFS Engineering SA, Qinetiq
Combustion				
	2-01	Bluff body burner for CH4-HE turbulent combustion	Elisabetta Belardini	Universita di Firenze
	2-06	The confined TECFLAM swirling natural gas burner	Stefan Hohmann	MTU Aero Engines
	2-07	Confined double annular jet	Charles Hirsch	Vrije Universiteit Brussel
Chemical & Process, Thermal Hydraulics & Nuclear Safety				
	3-01	Buoyancy-opposed wall jet	Jeremy Noyce	Magnox Electric
	3-02	Induced flow in a T-junction	Frederic Archambeau	EDF - R&D Division
	3-03	Cyclone separator	Chris Carey	Fluent Europe Ltd
	3-08	Spray evaporation in turbulent flow ★	Martin Sommerfeld	Martin-Luther-Universität Halle-Wittenberg
	3-10	Combining/dividing flow in Y junction	Lewis Davenport	Rolls-Royce Marine Power, Engineering & Technology Division
	3-11	Downward flow in a heated annulus	Mike Rabbitt	British Energy
Civil Construction & HVAC				
	4-01	Wind environment around an airport terminal building ★	Steve Gilham, Athena Scaperdas	Atkins
	4-02	Flow and Sediment Transport in a Laboratory Model of a stretch of the Elbe River	Wolfgang Rodi	Universität Karlsruhe
	4-03	Air flows in an open plan air conditioned office ★	Isabelle Lavedrine, Darren Woolf	Arup
	4-04	Tunnel fire	Nicholas Waterson	Mott MacDonald Ltd
Environmental Flows				
	5-05	Boundary layer flow and dispersion over isolated hills and valleys ★	Ian Castro	University of Southampton
Turbo-machinary Internal Flows				
	6-02	Low-speed centrifugal compressor	Nouredine Hakimi	NUMECA International
	6-05	Annular compressor cascade with tip clearance	K. Papailiou	NTUA
	6-06	Gas Turbine nozzle cascade	Elisabetta Belardini	Universita di Firenze
	6-07	Draft tube	Jan Eriksson, Rolf Karlsson	Vattenfall Utveckling AB
	6-08	High speed centrifugal compressor	Beat Ribi, Michael Casey	MAN Turbomaschinen AG Schweiz, Sulzer Innotec AG
	6-10	Axial compressor cascade	Fred Mendonca	Computational Dynamics Ltd
	6-12	Steam turbine rotor cascade	Jaromir Prihoda	Czech Academy of Sciences

UFR Index

Flow Type	UFR number	Underlying Flow Regime	Contributor	Organisation
Free Flows				
	1-01	Underexpanded jet	Christopher Lea	Health and Safety Laboratory
	1-02	Blade tip and tip clearance vortex flow ★	Michael Casey	Sulzer Innotec AG
	1-05	Jet in a Cross Flow	Peter Storey	ABB Alstom Power UK
Flows around Bodies				
	2-01	Flow behind a blunt trailing edge	Charles Hirsch	Vrije Universiteit Brussel
	2-02	Flow past cylinder	Wolfgang Rodi	Universität Karlsruhe
	2-03	Flow around oscillating airfoil ★	Joanna Szmelter	Cranfield University
	2-04	Flow around (airfoils and) blades (subsonic) ★	K. Papailiou	NTUA
	2-05	Flow around airfoils (and blades) A-airfoil (Ma=0.15, Re/m=2x10^6)	Peter Voke	University of Surrey
	2-06	Flow around (airfoils and) blades (transonic)	Jaromir Prihoda	Czech Academy of Sciences
	2-07	3D flow around blades	Dirk Wilhelm	ALSTOM Power (Switzerland) Ltd
Semi-confined Flows				
	3-01	Boundary layer interacting with wakes under adverse pressure gradient - NLR 7301 high lift configuration	Jan Vos	CFS Engineering SA
	3-03	2D Boundary layers with pressure gradients (A)	Florian Menter	AEA Technology
	3-04	Laminar-turbulent boundary layer transition	Andrzej Boguslawski	Technical University of Czestochowa
	3-05	Shock/boundary-layer interaction (on airplanes)	Anthony Hutton	Qinetiq
	3-06	Natural and mixed convection boundary layers on vertical heated walls (A)	André Latrobe	CEA / DRN / Department de Thermohydraulique
	3-07	Natural and mixed convection boundary layers on vertical heated walls (B)	Mike Rabbitt	British Energy
	3-08	3D boundary layers under various pressure gradients, including severe adverse pressure gradient causing separation	Pietro Catalano	CIRA
	3-09	Impinging jet	Jean-Paul Bonnet, Remi Manceau	Université de Poitiers
	3-10	The plane wall jet ★	Jan Eriksson, Rolf Karlsson	Vattenfall Utveckling AB
	3-11	Pipe expansion (with heat transfer)	Jeremy Noyce	Magnox Electric
	3-12	Stagnation point flow	Beat Ribi	MAN Turbomaschinen AG Schweiz
	3-13	Flow over an isolated hill (without dispersion)	Frederic Archambeau	EDF - R&D Division
	3-14	Flow over surface-mounted cube/rectangular obstacles ★	Ian Castro	University of Southampton
	3-15	2D flow over backward facing step	Arnau Duran	CIMNE
	3-18	2D Boundary layers with pressure gradients (B)	Fred Mendonca	Computational Dynamics Ltd
	3-30	2D Periodic Hill Flow ★	Christoph Rapp, Michael Breuer, Michael Manhart, Nikolaus Peller	Technische Universität München, Helmut-Schmidt Universität Hamburg
Confined Flows				
	4-02	Confined coaxial swirling jets	Stefan Hohmann	MTU Aero Engines
	4-03	Pipe flow - rotating	Paolo Orlandi, Stefano Leonardi	Universita di Roma 'La Sapienza'
	4-04	Flow in a curved rectangular duct - non rotating	Lewis Davenport	Rolls-Royce Marine Power, Engineering & Technology Division
	4-05	Curved passage flow ★	Nouredine Hakimi	NUMECA International
	4-06	Swirling diffuser flow	Chris Carey	Fluent Europe Ltd
	4-08	Orifice/deflector flow ★	Martin Sommerfeld	Martin-Luther-Universität Halle-Wittenberg
	4-09	Confined buoyant plume	Isabelle Lavedrine, Darren Woolf	Arup
	4-10	Natural convection in simple closed cavity	Nicholas Waterson	Mott MacDonald Ltd
	4-11	Simple room flow	Steve Gilham, Athena Scaperdas	Atkins
	4-13	Compression of vortex in cavity	Afif Ahmed, Emma Briec	RENAULT
	4-14	Flow in pipes with sudden contraction ★	Francesca Iudicello	ESDU

12
CFD at its Limits: Scaling Issues, Uncertain Data, and the User's Role

Matthias Münch and Rupert Klein

Simulation codes for Computational Fluid Dynamics (CFD) may be considered mature engineering tools today. They are routinely used in a supportive, sometimes central, role, for example, in the design stages for the shapes of airplanes or cars or for the layout of gas turbines and internal combustion engines. They have stood the practicioners' test of time, and have been found to be very useful in many more application areas than would be appropriate to list here. In the present volume, however, we are interested in CFD applications in safety-critical contexts, and it is thus only appropriate to raise the following three more fundamental questions, which we will discuss in some detail in the later chapters:

CFD codes are based on numerical methods for the solution of partial differential equations. These methods provide discrete computational results with flow field data associated with the nodes or cells of a computational grid and with discrete levels in time. Ideally, these numerical solutions can be proven to converge to the exact solutions of the considered equations in the limit of finer and finer computational grids and smaller time steps. In contrast, most fluid flows of practical interest are turbulent, and in CFD simulations it is generally neither possible nor of interest to resolve all details of the turbulent fluctuations. Therefore, turbulent-flow CFD simulations are notoriously under-resolved relative to the smallest turbulence scale – the Kolmogoroff dissipation scale. Aren't the tools of numerical mathematics then being used way outside their regime of applicability? We discuss this question further in Section 12.1 below.

For solutions to the fluid flow equations to be uniquely defined one must specify appropriate initial and boundary conditions. This raises an immediate issue in the context of safety-related flow simulations, because the exact circumstances of a safety-critical accident are quite generally not known in advance. Therefore, any CFD simulation carried out in the course of a safety assessment can at best be considered as one sample out of an ensemble of infinitely many possible alternative scenarios. Thus, there is a need for solid criteria for the selection of such sample simulations. This issue is closely related to the quantification of the area of uncertainty for CFD simulations, and this is the theme of Section 12.2. Finally, Section 12.3 focuses on the role of the user of a CFD simulation package and on the background knowledge which a proficient user should be in command of to be in a position to deliver reliable contributions to safety assessments.

Process and Plant Safety: Applying Computational Fluid Dynamics, First Edition. Edited by J. Schmidt.
© 2012 Wiley-VCH Verlag GmbH & Co. KGaA. Published 2012 by Wiley-VCH Verlag GmbH & Co. KGaA.

12.1
Numerics and Under-Resolved Simulations

12.1.1
Numerical Discretizations and Under-Resolution

As stated in the introduction, CFD codes are based on numerical methods for the solution of partial differential equations. To be specific, let us consider the numerical discretization of a partial derivative $\partial \phi / \partial x$ as it might arise in the transport equation for the concentration of a chemical species in an inviscid flow,

$$\frac{\partial \phi}{\partial t} + u\frac{\partial \phi}{\partial x} + v\frac{\partial \phi}{\partial y} + w\frac{\partial \phi}{\partial z} = \omega. \qquad (12.1)$$

Here t denotes time, $(x, y, z) \equiv \vec{x}$ are spacial Cartesian coordinates, $(u, v, w) \equiv \vec{v}$ the associated velocity components, and ω represents a reaction source term.

The classcial approach to deriving discrete approximations to expressions involving partial derivatives uses Taylor series expansions. Consider some space- and time-dependent function $\phi(t, \vec{x})$. The three-term Taylor expansions for the values $\phi^n_{i\pm1,j,k} = \phi(t^n, x_i \pm \Delta x, y_j, z_k)$, given the function's value and its first and second partial derivatives at (t^n, x_i, y_j, z_k), read

$$\phi^n_{i\pm1,j,k} = \phi^n_{i,j,k} \pm \Delta x \left(\frac{\partial \phi}{\partial x}\right)^n_{i,j,k} + \frac{\Delta x^2}{2!}\left(\frac{\partial^2 \phi}{\partial x^2}\right)^n_{i,j,k} + o\left((\Delta x)^2\right). \qquad (12.2)$$

Here $o((\Delta x)^2)$ denotes a remainder that vanishes faster than $(\Delta x)^2$ as $\Delta x \to 0$. By subtracting these two equations one obtains a finite-difference approximation for the partial derivative

$$\left(\frac{\partial \phi}{\partial x}\right)^n_{i,j,k} = \frac{\phi_{i+1} - \phi_{i-1}}{\Delta x} + o(\Delta x) \quad (\Delta x \to 0). \qquad (12.3)$$

If the third partial derivative, $(\partial^3 \phi / \partial x^3)^n_{i,j,k}$ exists, then the error term, $o(\Delta x)$ in Eq. (12.3), can be replaced by $o(\Delta x^2)$, that is, one then knows that the error of the approximation is not only small compared with Δx as $\Delta x \to 0$, but that it vanishes as fast as Δx^2.

These statements hold *only in the limit of vanishing step size*, that is, as $(\Delta x \to 0)$, as visualized in Figure 12.1. Clearly, if the finite difference approximation of the first derivative, that is, the first term on the right of Eq. (12.3), is applied on too coarse a grid, a false representation of the slope of the curve results (long-dashed line in the figure). The short-dashed line in the figure is tangential to the graph of ϕ in x_i, and thus has the desired slope $(\partial \phi / \partial x)(x_i)$. Only if the grid is so finely spaced that within a distance of Δx the approximated curve no longer features any sizeable variation can one count on obtaining proper approximations. One then speaks of *resolved solutions*.

Unfortunately, under-resolved representations of solutions as sketched in Figure 12.1 are the rule rather than the exception in CFD for turbulent flows. See also Section 12.2.3 for a discussion of a concrete example in the context of fire safety.

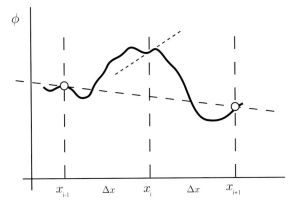

Figure 12.1 Finite differences taken on too coarse a grid: the slope of the --- corresponds with the coarse-grid finite difference approximation to the local slope in x_i (---) according to Eq (12.3).

It is granted that finite difference approximations as discussed here are not the most sophisticated numerical techniques employed in modern CFD solvers. In fact, finite-element methods or finite-volume schemes are designed to also handle solutions that may be locally not even differentiable, (see, for example, Refs [1–3]). Yet, the mathematical theory for these methods also characterizes only their behavior in the limit of infinite resolution, that is, for $\Delta x \to 0$ in the present notation. Therefore there is as little support for using these methods in under-resolved situations as there is for the finite difference techniques discussed above.

12.1.2
Turbulence Modeling

Next one may argue that any serious CFD code accounts for the under-resolution through explicit subgrid scale models which represent the net influence of the non-resolved scales on the resolved ones. Below we will distinguish between Reynolds-averaged Navier–Stokes (RANS) modeling and Large Eddy Simulation (LES) in this context, and discuss each of these approaches separately.

The distinction between RANS and LES may be explained on the basis of Figure 12.2, which shows a sketch of the typical energy spectrum of fully developed, homogeneous, and isotropic turbulence, with the kinetic energy shown as a function of the characteristic scale of turbulent eddies (red line). We observe that over an intermediate range of wavenumbers the spectrum closely follows a power law $E(k) \sim k^{-5/3}$. At the high-wavenumber end the energy decays rapidly due to viscous dissipation acting on the smallest scales of turbulence; at the low-wavenumber end the spectrum also decays because in practical situations we consider flows of finite spacial extent, and wavelengths larger than the system scale will surely not carry any energy. For this reason, the range of wavenumbers in the transition from zero energy

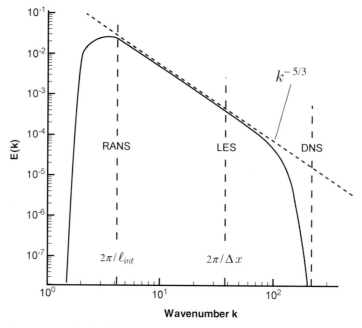

Figure 12.2 Sketch of the typical energy spectrum of fully developed homogeneous isotropic turbulence in a suitable non-dimensionalization.

to the power law cascade is associated with the 'range of non-universal scales' since these scales are influenced by the particulars of any given flow configuration and thus do not have universally valid characteristics. Now, in short, RANS models resolve a flow field only up to the integral scale of turbulence, which is roughly the largest scale within the cascade, whereas LES models resolve a much larger range of scales that corresponds to wavenumbers a long way into the similarity regime where $E(k) \sim k^{-5/3}$.

For completeness we mention direct numerical simulations (DNS), which resolve all turbulent scales but are too computationally demanding to be used in practical engineering applications. We will also not discuss models based on propagating probability density functions for the velocity fluctuations in time (see Ref. [4]), as these are again very expensive and have not made it into engineering production codes.

12.1.2.1 Reynolds-Averaged Navier–Stokes (RANS) Models

RANS models were originally introduced for the simulation of statistically stationary flows, but the technique has been extended to unsteady flows, in which case the acronym is URANS ('unsteady' RANS). The key idea behind these approaches is to consider, instead of a single flow realization, an entire ensemble of flows which have the same initial and boundary data on the large, observable, non-universal scales of the flow, but which differ in the details on small turbulent scales. Taking the statistical ensemble mean, one expects the universal small-scale fluctuations to average out,

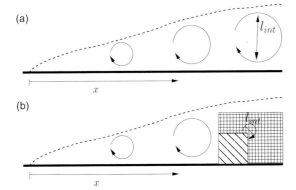

Figure 12.3 Sketch of turbulent boundary layers without (a) and with (b) an obstacle. The local integral scales of turbulence differ considerably in the two cases. For the efficient full resolution of the latter, adaptive grid refinement techniques would have to be employed.

while the non-universal characteristics from the system scale down to the integral scale, ℓ_{int}, depend on the specifics of the considered flow and need to be simulated. Thus, RANS models resolve all scales up to the integral scale of turbulence, while the effects of the entire turbulent cascade (the $k^{-5/3}$-range in Figure 12.2) are captured through some subgrid scale model. There are many alternative options in choosing a RANS-type turbulence closure scheme, and the reader is referred to pertinent textbooks (such as Ref. [4]) for details.

Of interest for the present discussion is the fact that for a proper RANS simulation to get close to numerical convergence, the integral scale of turbulence, ℓ_{int}, must be resolved. Now, the integral scale in a large-scale, nontrivial flow is not a uniquely defined quantity, but it rather depends on the local flow state. Consider, for example, the sketches of turbulent boundary layer flows in Figure 12.3. The top sketch indicates how the integral scale of turbulence increases roughly in proportion with the total boundary layer thickness. A grid resolution of about 10–15 grid points across the layer would then suffice to resolve the integral scale, so that a RANS simulation could be considered accurate.

In contrast, the second sketch shows a boundary layer within which a sharp-edged obstacle is embedded. Intense turbulence will be generated in the vicinity of the obstacle's trailing edge. The *local* integral scale of turbulence in this region will be comparable to the thickness of the boundary layer on top of the obstacle. The length of the obstacle being small compared to the length of the entire flow, this boundary layer will be much narrower than the bulk layer. Much higher resolution would be needed to correctly resolve this flow. Handling such situations efficiently will require a computational code with an adaptive computational grid that places grid points selectively in regions with rapid spatio-temporal variability.

Fully adaptive CFD simulation codes have not reached a level of development that would allow one to use them as engineering design tools on a daily basis, and even with full adaptivity, serious issues remain as regards the interplay of numerical discretization errors and the models employed to represent the subgrid scales [5].

12.1.2.2 Large Eddy Simulation (LES) Models

Large eddy simulation models follow a quite different strategy than those of the RANS type. Here the spatio-temporal grid resolution is chosen such that part of the self-similar energy cascade of a turbulent flow can be explicitly resolved. Since in that case the unresolved fluctuations cover a much smaller range than in RANS, and since they are also well into the regime of self-similarity of the fluctuation statistics, one hopes that the task of designing effective models to represent the net effects of these subgrid scales will be much less formidable than it is for RANS models (see, e.g., Ref. [6]).

While this is a very plausible point of view, it comes with the caveat that LES simulations are under-resolved in the sense of numerical approximations by design. To elucidate the point, consider the classical Smagorinsky model for the parametrization of turbulent fluxes on the scale of the computational grid cells [6]. In this model, the net effect of the unresolved scales on the resolved ones is represented by an effective turbulent friction term in the momentum equation, which for incompressible flow reads

$$\nabla \cdot \left[\nu_t \left(\nabla \vec{v} + (\nabla \vec{v})^t \right) \right]. \tag{12.4}$$

This corresponds with the standard friction term in the Navier–Stokes equations, except that the molecular viscosity, ν, is replaced with a turbulent apparent viscosity

$$\nu_t = \frac{C_s \Delta x^2}{2} \| \nabla \vec{v} + (\nabla \vec{v})^t \|, \tag{12.5}$$

where $C_s \sim 0.1 \ldots 0.2$ is the Smagorinsky constant and $\|\cdot\|$ is the Euclidian induced norm for a tensor. This closure has a direct interpretation in terms of Prandtl's 'Mischungsweg-Ansatz' (see any standard textbook on turbulence modeling).

To assess whether or not an LES simulation using the Smagorinsky model may or may not be expected to be resolved, we estimate the typical grid point-to-grid point variability of the simulated velocity field. To this end we follow Kolmogoroff's classical argument, which states that the total flow, ε, of energy through the turbulent scales has to match its dissipation at the smallest scales. Again, with the usual dimensional arguments employed in turbulence theory, and taking into account that in a LES simulation the smallest scale is that given by the grid spacing, Δx, (see, e.g., Ref. [4, 7]), one has

$$\varepsilon \sim \frac{u'^3}{\ell_{\text{int}}} =: C_s \Delta x^2 \left(\frac{u_{\Delta x}}{\Delta x} \right)^3, \tag{12.6}$$

where u' is a typical fluctuation velocity of turbulent eddies comparable in size with the integral scale, ℓ_{int}, and $u_{\Delta x}$ is a typical grid level velocity fluctuation. Thus we find that

$$\frac{u_{\Delta x}}{u'} \sim \left(\frac{1}{C_s} \frac{\Delta x}{\ell_{\text{int}}} \right)^{1/3}. \tag{12.7}$$

Except for the factor of $1/C_s$ on the right, this is the classical scaling for velocity fluctuations for eddies of scale Δx in the turbulent cascade. Thus we conclude that the

Smagorinsky model is *designed* to adhere to the theoretically predicted velocity scaling at the grid level.

Whereas this conclusion is good news for the turbulence theoretician, it is not good from the view point of the numerical analyst for the following reason. Again employing standard dimensional arguments, the typical velocity gradients induced by eddies of some general length scale ℓ may be estimated as

$$|\nabla \vec{v}|_\ell \sim \frac{u_\ell}{\ell}. \tag{12.8}$$

Replacing ℓ with Δx and considering Eq. (12.7) we have

$$|\nabla \vec{v}|_\ell \sim \frac{u'}{\ell_{\text{int}}} \left(\frac{\ell_{\text{int}}}{\ell}\right)^{2/3} \quad \text{and} \quad |\nabla \vec{v}|_{\Delta x} \sim \frac{u'}{\ell_{\text{int}}} \left(\frac{\ell_{\text{int}}}{\Delta x}\right)^{2/3}, \tag{12.9}$$

and so the characteristic gradients *increase* with decreasing scale, in contrast to the velocity fluctuations themselves, which decrease. This, together with the observation discussed above that the Smagorinsky model is designed not to distort the classical scaling within the turbulent cascade, yields the second estimate in Eq. (12.9). This, in turn, implies that the discrete gradients in a numerical approximation of the LES flow equations will be *dominated* by the grid scale fluctuations, and will thus contain a large contribution from truncation errors as discussed in Section 12.1.

We conclude that LES simulations carried out with the Smagorinsky model or with any other model that leaves the Kolmogoroff cascade essentially intact all the way down to the grid scale are persistently, and by design, always strongly under-resolved. As a consequence, there is again no sound justification in the mathematical sense for using the classical methods of numerical analysis in discretizing the LES model equations.

In closing this section, we mention recent intensive developments of *implicit* LES models. The idea behind these is to use the built-in numerical dissipation, which is part of all robust numerical methods for advection-dominated problems, implicitly as a model for turbulent dissipation, and thus to decidedly not include *any* terms in the governing equations that would explicitly model the net effects of turbulent fluctuations, (see, e.g., Refs. [8, 9] and references therein). By design, these methods rely on promoting formal truncation errors to influence the leading-order behavior of the flow, that is, they, too, build on systematic under-resolution.

Finite-element-based variational multiscale models (VMS) (see Refs [10, 11], and references therein) follow essentially the same route of modeling the unresolved scales as is done in more standard LES approaches. The key difference is in how the larger scales of the flow 'become informed' about the presence of the unresolved scales: in the VMS approach one introduces an ad hoc scale split so that there are definitely unresolved scales, as discussed above, there are intermediate scales that get resolved, and there are large scales for which the resolution should even be sufficient to approach numerical convergence in the absence of the other two scales. While this approach comes with several innovative ideas, it does not resolve the principal problem raised here of essentially under-resolved computations and their rigorous justification.

We conclude that CFD has proven to be very useful even in fully turbulent settings; yet a sound scientific basis in the sense of provably correct behavior of a CFD solver is lacking both for lack of criteria and lack of a fundamental concept that would bridge the current gap between fluid flow theory and numerical analysis. The reader may want to check the MetStröm web site [5] for recent efforts in this direction.

12.2
Uncertainties

One important difference between CFD applications in the typical case of engineering design and testing on the one hand and safety-related applications on the other lies in the uncertainties regarding the determinants of the considered flows. In typical engineering applications the idea is to use CFD to study specific details of some well-defined flow configuration so as to obtain quantitative support for important design decisions. Just as in a laboratory experiment, the flow simulation is of maximal use if all the determining influences such as initial and boundary conditions, aerothermodynamic parameters of the fluid, and location and magnitude of external forcings are exactly known.

Application scenarios for safety-related simulations are generally much different in that key determinants of the flow fields are unknown in advance. Consider, for example, the problem of proving the suitability of the fire safety measures taken in a newly erected building. Since the concrete use of the building over its lifetime is not known in advance, it is a priori unclear where a fire could become initiated, what would be the distribution of combustible materials at the time of a fire, and how much smoke would be generated, and where, in the building. These uncertainties are crucial in that the unknowns, unfortunately, dominate the behavior of the ensuing flow fields. At the same time there is hardly any possibility to substantially reduce this uncertainty in practice.

For these reasons, CFD simulations in fire safety applications have a very different role than they do in engineering design contexts: chiefly, CFD simulations allow the user to probe the vast space of possible scenarios without having to resort to unfeasible – or at least extremely costly – laboratory experiments; they cannot be used in fire safety applications to guide the optimization of a few well-defined flow scenarios.

12.2.1
Dependency of Flow Simulations on Uncertain Parameters: Basic Remarks

To the authors' knowledge there are today no proven strategies for choosing a series of computer simulations so as to fully cover the very high-dimensional space of possible scenarios. Yet, in recent years, some promising developments have led to uncertainty quantification methods which allow us to at least explore the dependency of a flow simulation on a few, say less than ten, parameters in a systematic fashion (see Ref. [12] and references therein).

In a spectral method for uncertainty quantification one considers the solution, U, of a flow problem as a function of space and time, $(t, \vec{x})$, and also of a few key determining but uncertain parameters, $\lambda = (\lambda_1, \lambda_2, \ldots, \lambda_n)$. Thus, considering, for instance, the flow of a gas with multiple chemical components, we have, in the usual mathematical notation,

$$U : \mathbb{R}^+ \times \mathbb{R}^3 \times \mathbb{R}^{n_\lambda} \mapsto \mathbb{R}^{5+n_{sp}} \\ (t, \vec{x}; \lambda) \rightarrow U(t, \vec{x}; \lambda) \,. \tag{12.10}$$

Here n_λ and n_{sp} are the numbers of uncertain parameters and the number of chemical species in the gas phase, respectively. The solution vector U would have, for example, the components density, the three components of the velocity field, temperature, and the concentrations of the chemical species.

As a concrete example of a pair of uncertain parameters, let us think of $\lambda = (\lambda_1, \lambda_2)$ as the coordinates for the location of a given fire load in the central atrium of an appartment building. As long as we consider U just as a function, as indicated in Eq. (12.10), λ is just one more independent variable for this function and nothing distinguishes it from the space and time coordinates. In our setting of flow simulations, however, λ does have a particular role: the determining partial differential equations for the flow field only involve derivatives with respect to t and $\vec{x}$ but not with respect to λ. This is why for each particular value of this parameter vector we can generate an individual flow simulation without having to know the behavior of U for any other value of λ.

In our application we would next be interested in how the smoke distribution and the thermal loads on the walls of the building would be affected by the position, λ, of the fire load. Since we cannot carry out simulations for each and every possible position, one option for scanning the space of possibilities would be a random search in the sense of a Monte Carlo simulation. Yet this may not be the most efficient choice. The dependency of the fire-induced flow field on the position of the fire load may be expected to be nonlinear, but not discontinuous or erratic. As a consequence, it should be possible to construct an appropriate representation of this dependency by a – possibly sophisticated – interpolation or expansion procedure, and this is the first ingredient of spectral methods of uncertainty quantification. One aims at constructing a representation of the λ-dependency of $U(t, \vec{x}; \lambda)$ by introducing an expansion in terms of judiciously chosen ansatz functions, $\psi_k(\lambda)$, so that

$$U(t, \vec{x}; \lambda) = \sum_{k=0}^{K} U_k(t, \vec{x}) \psi_k(\lambda). \tag{12.11}$$

As a simple example, consider a pure polynomial interpolation, so that the $\psi_k(\lambda)$ are polynomials, and the U_k the corresponding weights. If the $\psi_k(\lambda)$ are chosen to be the Lagrange polynomials for a given set of nodes $\{\lambda_k\}_{k=0}^{K}$, so that $\psi_k(\lambda_k) = 1$ and $\psi_k(\lambda_j) = 0$ for jk, then one simply has

$$U_k(t, \vec{x}) = U(t, \vec{x}; \lambda_k), \tag{12.12}$$

and the entire expansion in Eq. (12.11) can be generated straightforwardly by running a given flow solver K times with the uncertain parameter λ taking the values λ_k for

$k = 1 \ldots K$. In the language of spectral methods of uncertainty quantification, this would be called a 'non-intrusive' method in the sense that no changes in the flow code are necessary in this procedure.

If the principal dependency of the solution to a CFD problem on its parameters λ can be expressed in terms of the chosen ansatz functions, ψ_k, with at all reasonable accuracy and with not too many terms of the expansion needed, then the representation in Eq. (12.11) can be used to great advantage for safety assessment purposes.

12.2.2
Polynomial Chaos and Other Spectral Expansion Techniques

The example of a straightforward polynomial interpolation of the U–λ dependency explained in the last section does not do justice to the much more advanced state of the art. The reader is referred to Ref. [12] for a comprehensive account. In this section we will just briefly mention a few of the more recent developments which we believe may be of relevance for safety-related CFD applications.

One central motivation behind 'polynomial chaos expansions' is *uncertainty propagation*, that is, the task of translating probability distributions over the uncertain parameters into probability distributions over the solutions which they parametrize. Thus, in the example of a fire in the atrium, one may be interested in quantifying the expected value of the maximum temperature reached near some structural element of the building with respect to all possible locations of the main fire. Suppose a probability distribution for the main fire locations is given in the form of a probability density function, $p(\lambda)$. Then the expected value would be

$$\bar{T}(t,\vec{x}) = \sum_{k=0}^{K} T_k(t,\vec{x}) \int_\Omega \psi_k(\lambda) p(\lambda) d\lambda, \quad (12.13)$$

where Ω is the domain of definition of the fire load location, λ. Similarly, the mean advective flux of the concentration, c, of some chemical species, say, would read

$$\overline{c\vec{v}}(t,\vec{x}) = \sum_{k,k'=0}^{K} (c_k \vec{v}_{k'})(t,\vec{x}) \int_\Omega \psi_k \psi_{k'}(\lambda) p(\lambda) d\lambda. \quad (12.14)$$

Evaluating such expressions will be quite costly a task in general.

Yet, here comes the central idea of the 'polynomial chaos expansions'. One picks as ansatz functions $\{\psi_k(\lambda)\}_{k=0}^{K}$ a set of *orthogonal polynomials* under the p-weighted scalar product, that is, one requires

$$\langle \psi_k, \psi_{k'} \rangle \stackrel{!}{=} \int_\Omega \psi_k \psi_{k'}(\lambda) p(\lambda) d\lambda \equiv \delta_{kk'}, \quad (12.15)$$

where δ_{ij} is the Kronecker-delta symbol. In this case, the evaluation of the expected values of linear and quadratic functions becomes trivial, since we then have

$$\bar{T}(t,\vec{x}) \equiv T_0(t,\vec{x}) \quad \text{and} \quad \overline{c\vec{v}} \equiv \langle c,\vec{v} \rangle(t,\vec{x}) \equiv \sum_{k=0}^{K} c_k \vec{v}_k(t,\vec{x}). \quad (12.16)$$

Of course, it stops here, and higher degree polynomials and more general nonlinear functions still require more effort in the evaluation of statistical quantities. However, there has been a rich development of pertinent algorithms and codes in recent years that exploit basic properties of the specialized polynomial chaos expansions such as Eq (12.16) for efficiency on such more sophisticated tasks, (see Ref. [12] and references therein).

Thus far, we have only mentioned polynomial ansatz functions, $\psi_k(\lambda)$, and these will obviously come with disadvantages and problems in some cases. For instance, due to the fact that polynomial expansions will produce strong oscillations near discontinuities and rapid transitions (Gibbs phenomenon), it is somewhere between very costly and impossible to properly represent discontinuous dependencies of solutions $U(t, \vec{x}; \lambda)$ on λ by polynomials, or to enforce the positivity of, say, a chemical concentration in a polynomial expansion. Recent developments of expansions based on alternative basis functions, such as wavelets and especially the discontinuous Haar-wavelets, meanwhile provide remedies in these cases. Thus, uncertainty quantification techniques have reached quite a mature stage of development, and they are available for use in engineering applications. This is especially true thanks to the above-mentioned non-intrusive methods, which allow us to generate all the information needed in the chaos expansions by just running an unchanged standard code and making smart use of the simulation results.

The approach to uncertainty quantification outlined here will surely not cover all issues of practical interest in the context of safety-related CFD simulations. For example, the methodology is not applicable when the space of uncertain parameters has high dimension. In practice, however, only relatively few well-chosen simulations are normally executed and analyzed in engineering safety analysis anyway, and the systematics of the uncertainty quantification methodology can provide very valuable guidelines regarding which parameter variations should be tested and how to make the most of the obtained simulation results.

12.3
Theory and Practice

Fire safety engineering is an excellent example of the development and usage of computational fluid dynamics (CFD) in a safety-critical area. The spreading of smoke in buildings can induce quite dangerous situations very quickly. This motivates fire safety engineers to use, or plan on using, CFD programs as prediction tools,[1] for example, for the spreading of fire and smoke or for the thermal load on structural elements. CFD programs are meanwhile available for all interested users, the Fire Dynamics Simulator (FDS) even comes free of charge [13]. Moreover, advertising suggests easy use of such tools, simple user interfaces, and large benefits for engineers, architects, and consultants. Yet, the preconditions to be satisfied in order

---

1) For a critical discussion of 'predictability' in the context of fire simulations see Chapter 2.

to obtain reliable results from CFD models are rarely discussed. The present chapter aims at clarifying some important issues in this context.

12.3.1
Reliability of CFD Program Results

A frequently asked question is: 'How trustworthy are the results generated by a CFD program?'. Commonly, the person raising the question is worried about the correctness of the CFD program in analogy with, say, the precision of a machine tool. In fact, it seems highly appropriate to develop quality assessment procedures for safety-related CFD codes comparable to those used to certify tools used in civil engineering. Yet, even with the best available code at hand, a user lacking sufficient knowledge regarding the implemented physics, adopted model simplifications, and numerical schemes employed in the CFD program will not succeed in producing reliable results. Such knowledge is, therefore, just as important for trustworthy simulations as the availability of a well-tested CFD program.

12.3.1.1 Verification and Validation
Up to now, the validation of CFD programs for fire safety is based mainly on numerous comparisons between computational results and data from fire experiments published in the fire safety community's scientific literature, and there is a considerable number of such publications. However, on the basis of such global comparisons with fire experiments it is not possible to decide about the general applicability of a CFD code.

In fact, favorable comparison with a finite number of experimental results merely allows the user to conclude that within the parameter range covered by the experiments the code is likely to perform well. Whether it will perform equally well in flow regimes which the experiments have not covered remains unclear. Of particular interest in the present context are fire events in very large open-space buildings. Since it is extremely difficult – if possible at all – to downscale such events to laboratory scales, the experimental support for using CFD codes in the fire safety design of such buildings remains scarce.

In summary, the comparison of fire simulations with experimental data is, of course, necessary and useful. But, if it is adopted as the only means of testing, it is insufficient to document the performance of CFD programs. Therefore, a more comprehensive testing strategy is indispensable, and this should include investigations of numerical qualities (e.g., convergence, stability, order of convergence), component-level tests (to avoid spuriously correct results due to incidental error cancellations), and tests of code performance on specific phenomena in simplified set-ups (to test proper interaction of subsets of model components) [14–16].

12.3.1.2 The User's Influence
Even though each successful comparison and check as outlined in the last section increases the reliability of the tested program, such measures cannot *guarantee* the correctness of subsequently computed results. One important reason is that

user-defined data and parameter settings strongly influence the quality and correctness of CFD simulations. Therefore, it is not possible to decouple the discussion of trustable results from the users' knowledge about physical and numerical fundamentals used in the program. To be specific, even if the used program is well verified and free of errors, users may still obtain erroneous results due to incorrect use of the simulation tool.

12.3.2
Examples

In this section we present some examples showing how the user may influence the computed results. Using the well-documented and freely available field model Fire Dynamics Simulator (FDS), we explain some critical numerical aspects of the implemented numerical scheme and discuss their importance for the user. We show, in particular, that trustable results require both well-tested programs and well-trained users.

12.3.2.1 User's Choice of Submodels
CFD program users can often choose different models for the same physical process, and they must define model parameters. Therefore, users need detailed knowledge about the advantages and disadvantages of the available models and parameters in relation to the current problem.

In a benchmark test for the SMARTFIRE program, Grandison, Galea, and Patel [17] compare the results of two different radiation models in a particular test case. In a three-dimensional cubic box (Figure 12.4) with three walls heated at the rate of $1\,W°m^{-2}$, the emissive power along a line parallel to the x-axis is computed (Figure 12.5). There is no fluid flow inside the box, and scattering is neglected. As a reference for their solutions, they use the detailed zone method of Hottel [18, 19]. A six-flux and a multi-ray model are used as options for radiation models in CFD codes. The comparison of the computed data shows that the result of the six-flux model is equal to that from the multi-ray model when it is run with only six rays directed in the coordinate directions, while both results differ from the solution provided by the zone method.

The user of the multi-ray model can specify as many rays and associated directions as desired. For 24 rays, weighted and spread over the 4π steradians, the results tend

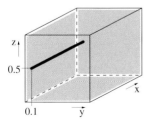

Figure 12.4 Cubic test box with three heated walls.

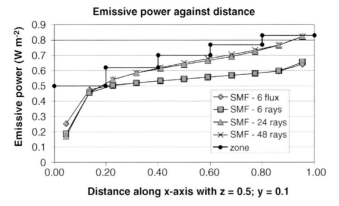

Figure 12.5 Comparison of radiation models [17].

toward the reference solution. Further increase in the number of rays does not further change the multi-ray solution significantly. Thus, the 24-ray solution from the multi-ray model may be considered to be converged (at an empirical level), whereas the six-flux results with either the original six-flux model or with the multi-ray model restricted to only six fluxes are influenced considerably by the particular way the problem has been discretized.

On the other hand, in the multi-ray model each ray direction requires the solution of a linear system of equations, so that computational costs grow rapidly with the number of rays. The model itself, and for the multi-ray model also the number and direction of rays, are at the user's disposal. Therefore, users are faced with a conflict between available compute power and correctness of the solution. Clearly, fundamental insight into the radiation process, the importance of radiation for the current problem, and the properties of the available numerical radiation solvers are required.

12.3.2.2 Influence of a Model's Limits of Applicability

The flow models on which fire simulation codes are based rely on a number of simplifications that are justified in most applications. However, there are practically relevant situations in which these simplifications may fail and lead to erroneous results. This example will demonstrate that the theoretical basis of a simulation code must be accounted for carefully in setting up a fire case for a CFD program.[2] As an example we use the Fire Dynamics Simulator (FDS), but present the theoretical background in more general terms. Specifically, we employ a non-dimensional notation, whereas the FDS program uses a dimensional formulation. The fundamentals are the same, however [21].

Consider the dimensionless governing equations of mass, momentum, and energy for fire-induced fluid flows. To simplify the discussion, all physical processes like radiation, combustion, heat conduction, and so on, are subsumed in general source terms S_ρ, $S_{\rho\vec{v}}$, and $S_{\rho e}$ below. The integral conservation laws for mass,

---

[2] For a detailed description of the theory related to FDS see [15, 20].

momentum, and energy then provide dimensionless governing equations for the density ϱ, velocity vector $\vec{v}$, pressure p, and energy density ϱe as functions of time and space coordinates $(t, \vec{x})$. For arbitrary time-independent control volumes V with boundary $A = \partial V$,

$$\frac{d}{dt}\int_V \varrho \, dV + \int_A \varrho \vec{v} \cdot \vec{n} \, dA = \int_V S_\varrho \, dV$$

$$\frac{d}{dt}\int_V \varrho \vec{v} \, dV + \int_A (\varrho \vec{v} \circ \vec{v}) \cdot \vec{n} \, dA + \frac{1}{M^2}\int_A p \vec{n} \, dA = \int_V S_{\varrho \vec{v}} \, dV \qquad (12.17)$$

$$\frac{d}{dt}\int_V \varrho e \, dV + \int_A [\varrho e + p]\vec{v} \cdot \vec{n} \, dA = \int_V S_{\varrho e} \, dV.$$

Assuming a perfect gas with a constant isentropic coefficient γ, the energy density is

$$\varrho e = \frac{p}{\gamma - 1} + \frac{M^2}{2} \varrho \vec{v} \cdot \vec{v}. \qquad (12.18)$$

Non-dimensionalization, of course, changes neither the mathematical nor the physical content of the equations. However, one advantage of the dimensionless form used here is the occurrence of dimensionless characteristic numbers which weigh the various terms in the equations. In regimes in which the reference quantities chosen for the non-dimensionalization correctly represent the magnitude of the flow variables and the pertinent length and time scales, these characteristic numbers allow us to compare the relative orders of magnitude of the various terms in the equations. Here we consider the reference Mach number M, which is a scale for the importance of compressiblity in a fluid flow

$$M = \frac{v_{ref}}{c_{ref}}. \qquad (12.19)$$

Here v_{ref} and c_{ref} are typical values of the flow velocity and of the speed of sound, respectively. For fire-induced flows we have $v_{ref} \approx 1 \, \mathrm{m \, s^{-1}}$. Of course, this is only an order-of-magnitude estimate, but fluid velocities in fires are in the region of $\mathrm{m \, s^{-1}}$ as a rule. With a speed of sound of $c_{ref} \approx 330 \, \mathrm{m \, s^{-1}}$ the reference Mach number is very small, $M \approx 0.003$.

As a consequence, the pre-factor of the pressure term in the momentum equation in Eq. (12.17) is very large. In the limit of vanishing Mach number, a (mathematical) singularity arises, which signals the physical transition from a compressible to an incompressible flow. Such singularities are very hard to capture accurately in numerical computations. In the present example, the truncation errors associated with the approximation of the pressure gradient become amplified by $1/M^2$ and will destroy the accuracy of the velocity computation as M becomes very small unless the employed computational scheme incorporates suitable measures that remedy the singularity. Numerous solutions to this problem have been proposed, but it is still a matter of active research.

Because for fire-induced flows we are not interested in resolving sound waves, one may completely filter out their influence by considering the asymptotic limit equations that arise as $M \to 0$. The remaining zero-Mach-number equations describe the advection of entropy and vorticity and the influence of the various source terms on these quantities [22–26].

One important result of the related asymptotic analysis is that the pressure may be decomposed into two contributions with very different physical meaning

$$p(\vec{x}, t) = p_0(t) + M^2 p_2(\vec{x}, t). \tag{12.20}$$

The pressure p_0 is constant in space and represents the thermodynamic part of the pressure p, whereas the pressure p_2 represents the hydrodynamic pressure that is responsible for flow acceleration and for maintaining incompressibility.

Using this pressure decomposition, the energy equation in Eq (12.17) reduces to a divergence constraint for the velocity field $\vec{v}$

$$\int_A \vec{v} \cdot \vec{n} dA = \frac{\gamma - 1}{\gamma p_0} \int_V S_{\varrho e} dV - \frac{|V|}{\gamma p_0} \frac{dp_0}{dt}. \tag{12.21}$$

What does this divergence constraint mean in practice?

12.3.2.2.1 Fire in a Closed Room
Consider a closed room (Figure 12.6) with a heat source $S_{\varrho e}$. Because fluid flow across the bounding walls is suppressed, the divergence constraint, Eq. (12.21), reduces to

$$\frac{dp_0}{dt} = \frac{\gamma - 1}{|V|} \int_V S_{\varrho e} dV \tag{12.22}$$

if V is identified with the room's entire volume. p_0 is homogenous in space and can be interpreted as storage or containment pressure.

The divergence constraint, Eq. (12.22), reveals that the volume-averaged heat source $S_{\varrho e}$ directly acts to increase the containment pressure inside the room, ultimately inducing component failure.

12.3.2.2.2 Fire in an Open Room
In the opposite case of an open room with a heat source (Figure 12.7), the ambience and the room are coupled. Because p_0 is constant

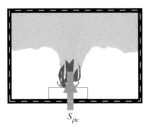

Figure 12.6 Isolated fire room.

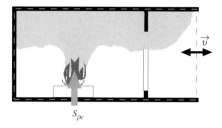

Figure 12.7 Opened fire room.

in space, the ambience determines the thermodynamic pressure. Thus, $p_0 \equiv const$ and the divergence constraint, Eq. (12.21), reduces to

$$\int_A \vec{v}\cdot\vec{n}dA = \frac{\gamma-1}{\gamma p_0}\int_V S_{\varrho e}dV, \qquad (12.23)$$

and the velocity across the open boundary is controlled by the heat source.

12.3.2.2.3 Fire in a 'Leaky' Room Now we consider a 'leaky' room (Figure 12.8) with a closed door, but with a small opening left between the door's lower edge and the floor. Recalling possible grid dependencies, we choose a very fine grid that resolves the small opening. Through the opening, the room is connected to the ambience, so that the thermodynamic pressure p_0 should remain constant according to the theory.

The divergence constraint Eq. (12.21) reduces to Eq. (12.23)

$$\int_A \vec{v}\cdot\vec{n}dA = \frac{\gamma-1}{\gamma p_0}\int_V S_{\varrho e}dV$$

and the gas expansion in the room induced by the heat source $S_{\varrho e}$ results in a large exit velocity across the small open boundary. The smaller the area of the opening the larger will be the induced exit velocity. For sufficiently small openings, the 'zero-Mach-number assumption' that flow velocities remain small in comparison with the speed of sound will become invalid. In this case, the computed solution is outside the range of validity of the underlying flow model of the CFD program, and predictions of flow velocities, pressures, and so on will be erroneous. We conclude that, in order to

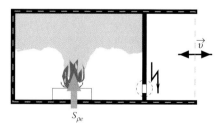

Figure 12.8 'Leaky' fire room.

avoid such inadequate model applications, users have to be aware of the key elements of the theories their programs are built upon.

These examples show that within the theory justifying the simplified equations used, the interpretation of the thermodynamic pressure may differ and depend on the considered case. Based on the open room case, we demonstrate next why the user must know the fundamentals of the theory used in constructing the numerical scheme of the CFD program.

12.3.2.3 The Influence of Grid Dependency

Even if we assume that users cannot choose between different submodels or define model parameters, there are further possibilities of user influence. An important example concerns the grid dependency of computational solutions.

A common way to derive discrete approximations uses Taylor series expansions. The Taylor expansion for $\phi(x_{i+1})$ of a function $\phi(x)$ around $x = x_i$ reads

$$\phi_{i+1} = \phi_i + \left(\frac{\partial \phi}{\partial x}\right)_i \cdot \Delta x + \left(\frac{\partial^2 \phi}{\partial x^2}\right)_i \cdot \frac{\Delta x^2}{2!} + \left(\frac{\partial^3 \phi}{\partial x^3}\right)_i \cdot \frac{\Delta x^3}{3!} + \ldots \quad (12.24)$$

A change in the sequence of terms leads to an approximation for the gradient of ϕ at the point i

$$\left(\frac{\partial \phi}{\partial x}\right)_i \approx \frac{\phi_{i+1} - \phi_i}{\Delta x} \underbrace{- \left(\frac{\partial^2 \phi}{\partial x^2}\right)_i \cdot \frac{\Delta x}{2!} - \left(\frac{\partial^3 \phi}{\partial x^3}\right)_i \cdot \frac{\Delta x^2}{3!} - \ldots}_{\text{truncation error } O(\Delta x)} \quad (12.25)$$

Only a finite number of terms in Eq. (12.25) can be accounted for in approximating the gradient of ϕ. The rest are necessarily neglected and remains as a 'truncation error'. This type of discretization is widely used in CFD programs, and a range of concepts of numerical analysis exist for characterizing the accuracy and correctness of the resulting numerical approximation schemes.

For a successful numerical computation, the truncation error on the right-hand side must be much smaller than the leading term. As the truncation error is determined by the grid size, this is achieved only if the grid size is sufficiently small to ensure that all relevant physical processes are resolved on the grid.

An elucidating example is given by Dreisbach and McGrattan [27], who investigate the grid dependency of the computed plume temperature above a fire source (Figure 12.9) using the Fire Dynamics Simulator (FDS). The measured plume temperature in the vicinity of a fire source in a burning room is compared with three computations based on grid sizes 5, 7.5, and 10 cm. Figure 12.10 shows that only the computation with 5 cm grid size is in good agreement with the experimental data, whereas for the coarser grids large oscillations and deviations from the experimental data occur. Apparently, the coarser grids do not resolve the physical processes in this example, and truncation errors are substantial. This demonstrates that the choice of the compuational grid size is far from being a 'minor issue' but rather requires careful testing and extensive experience.

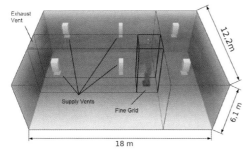

Figure 12.9 Burning room with fire (source Ref. [27]).

The danger here lies in the fact that the fine-grid computations require huge computational resources. In a 3D, time-dependent problem, bisection of the grid size increases run times by a factor of 16 (a factor of 2 for each dimension and for time). Accordingly, Dreisbach and McGrattan's 10 cm grid size computation took hours to run, whereas the 5 cm one took days. Thus it is tempting to choose coarse grids so as to quickly obtain results and be able to move on. Importantly, it is a coincidence that the quick coarse-grid computation resulted in a higher plume temperatures and thus too conservative estimates regarding structural component failure in the current example. Thus, using such solutions is generally a game of chance!

12.3.2.4 Influence of Boundary Conditions

Users without experience in computational fluid dynamics often underestimate the influence of boundary conditions. In fact the discretization algorithms of CFD programs are mostly easy to implement in computational areas far away from the boundary of the domain. However, more or less complicated situations arise in areas

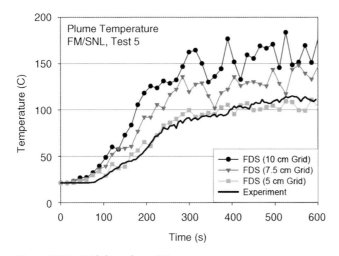

Figure 12.10 Grid dependency [27].

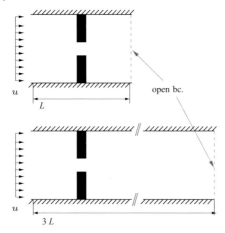

Figure 12.11 Channel with orifice.

near the boundary. For example see the discussion about the closed- and open-room situation in the context of the divergence constraint (Section 12.2.2).

In the example below we have to use an open boundary condition. Open boundaries are numerically ambitious and an important feature of CFD modeling. Open boundaries model the interaction with the ambient outside the computational domain, especially for openings in walls, such as doors, windows, or smoke funnels.

In our test case we consider a constant parallel inflow entering the computational channel domain from the left. Because of the orifice, the flow accelerates and generates vortices in which should get advected through the right boundary. Because there is no analytical solution available, we compare the outflow velocity field of an orifice at the open boundary with the velocity field of another channel simulation at the same position in a longer channel (Figure 12.11).

The comparison between the two simulations in Figure 12.12 shows that some of the vortices issuing from the orifice are reflected instead of passing out of the domain to the right.

Similar well-chosen and sufficiently simple tests enable users to assess some of the uncertainties associated with compuational simulations in their own fields of application.

12.4
Conclusions

In this paper we have highlighted three important limitations of computational fluid dynamics in the context of safety-related simulations:

i) It is not possible in everyday routine simulations to resolve all details of a flow field. As a consequence, some ways of representing the net effect of processes acting on scales smaller than the computational grid size need to be introduced.

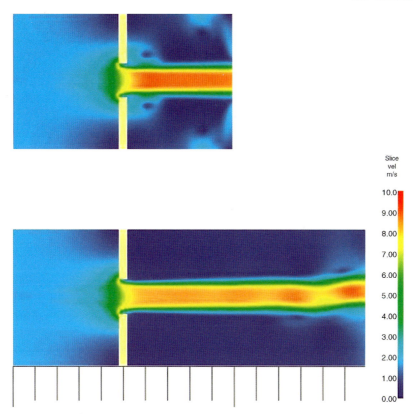

Figure 12.12 Velocity field.

Such model modifications are introduced in ways that are either not compatible with the fundamental design principles of the numerical methods used to solve the flow equations in the first place (LES), or they are compatible but only if combined with very sophisticated dynamically adaptive numerical procedures (RANS). Therefore, as useful as CFD simulations have turned out to be in engineering practice in many instances, the CFD user cannot claim to be on proven, mathematically sound grounds.

ii) In assessing a given situation, safety engineers are faced with much larger uncertainties than are engineers working on some typical design or optimization problem. In fact, the space of possible accident scenarios is typically infinite in, say, fire safety engineering. It is important to notice that any single CFD simulation, for example, for the spread of smoke in a building, can always only provide detailed information about just one such scenario. Therefore, a pressing issue today is the development of systematic strategies for the design of simulation ensembles that provide maximal safety-relevant information from the least amount of computational work. The recently developed computational methods of uncertainty quantification may prove to be a promising starting point.

iii) Computational fluid dynamics codes are complex, and they incorporate a broad range of knowledge, from fundamental fluid mechanics via numerical mathematics to computer science. The model equations used as a basis of a code, the numerical schemes adopted, and the particular algorithms employed to implement these schemes all come with limitations and caveats. Furthermore, it is not possible today to prove or even accurately assess the correctness of a CFD simulation in a mathematically rigorous sense. As a consequence, any CFD code is only applicable within the range of flow scenarios for which is has been thoroughly tested and validated. For safety-critical applications it is thus of the utmost importance that detailed documentation and test programs or (automated) testing procedures are made available which allow the user to reproduce and check out the claimed regime of validity of a CFD code. Moreover, users who are not sufficiently well trained in matters of the theory and practice of CFD simulations will generally not be aware of a code's limitations. Such users may easily end up applying the code to scenarios for which it is not designed and tested and for which it therefore may or may not produce correct results.

The three problem areas outlined here will not be entirely resolved in the near future. Fundamental developments on the issues raised in items (i) and (ii) are needed and are being pursued in many research projects worldwide, however, and progress will be made. Even within its current limitations, CFD has had many useful applications in safety engineering already, particularly in the context of the design and construction of safety technology, where the above-mentioned limitations are less of an issue than in the context of safety assessments for given installations, as explained in this chapter. As the results of ongoing research become incorporated into state-of-the-art CFD tools, the range of applications will further increase. Thus we do see a promising future for CFD in safety technology.

Such promising developments can be endangered, however, if the capabilities of the available simulation tools are over-stated and their limitations are not properly and very openly highlighted at any time. If they are not, it is very likely that CFD techniques will be applied under erroneous circumstances, and one way or another disaster will sooner or later occur. An important part of this is appropriate training of both potential CFD users and of members of regulatory institutions. As we have emphasized under item (iii) above, we can be on the safe side with such applications only if the CFD users at both the producing and the receiving end of the simulation process are acutely aware of both the capabilities and the limitations of CFD tools.

References

1 Braess, D. (2010) *Finite Elemente: Theorie, schnelle Löser und Anwendungen in der Elastizitätstheorie*, 4th edn, Springer Verlag.
2 Kröner, D. (1996) *Numerical Schemes for Conservation Laws*, Wiley and Teubner.
3 LeVeque, R.J. (2002) *Finite Volume Methods for Hyperbolic Problems*, Cambridge University Press, Basel.
4 Pope, S.B. (2000) *Turbulent Flows*, Cambridge Univ. Press.

5 von Larcher, T. and Klein, R. (2010) DFG Schwerpunktprogramm 1276: Multiple Skalen in Meteorologie und Strömungsmechanik. URL http://metstroem.mi.fu-berlin.de.

6 Sagaut, P. (2002) *Large Eddy Simulation for Incompressible Flow*, 2nd edn, Scientific Computing, Springer.

7 Peters, N. (2000) *Turbulent Combustion*, Cambridge Univ. Press.

8 Adams, N., Hickel, S., and Franz, S. (2004) Implicit subgrid-scale modeling by adaptive deconvolution. *Journal of Computational Physics*, **200**, 412–431.

9 Grinstein, F.F., Margolin, L.G. and Rider, W.J. (eds) (2007) *Implicit Large Eddy Simulation: Computing Turbulent Fluid Dynamics*, Cambridge Univ. Press.

10 Gravemeier, V. (2006) The variational multiscale method for laminar and turbulent flow. *Archives of Computational Methods in Engineering*, **13**, 249–324.

11 Gravemeier, V., Gee, M.W., Kronbichler, M., and Wall, W.A. (2010) An algebraic variational multiscale–multigrid method for large eddy simulation of turbulent flow. *Computer Methods in Applied Mechanics and Engineering*, **199**, 853–864.

12 LeMaitre, O.P. and Knio, O.M. (2010) *Spectral Methods for Uncertainty Quantification with Applications to Computational Fluid Dynamics*, 1st edn, Scientific Computation, Springer Verlag.

13 Nist fire dynamic simulator (FDS) and smokeview. URL http://fire.nist.gov/fds.

14 Münch, M. and Klein, R. (2008) Critical numerical aspects for field model applications. *EUSAS-Journal*, **4**, 41–54.

15 Kilian, S. and Münch, M. (2009) A new generalized domain decomposition strategy for the efficient parallel solution of the FDS-pressure equation, Part I: Theory, Concept and Implementation, Tech. Rep. ZR-09-19, Konrad-Zuse-Zentrum für Informationstechnik Berlin. ISSN 1438-0064.

16 Münch,, M. and Kilian,, S. (2009) A new generalized domain decomposition strategy for the efficient parallel solution of the FDS-pressure equation, Part II: Verification and Validation, Tech. Rep. ZR-09-20, in preparation, Konrad-Zuse-Zentrum für Informationstechnik Berlin. ISSN 1438-0064.

17 Grandison, A., Galea, E., and Patel, M. (2001). Fire Modelling Standards/Benchmark, Report on SMARTFIRE Phase 2 Simulations, Tech. Rep., Fire Safety Engineering Group, University of Greenwich, London.

18 Fiveland, W. (1988) Three-dimensional radiative heat-transfer solutions by the discrete-ordinates method. *Journal of Thermophysics*, **2** (4), 309–316.

19 Larsen, M. (1983) Exchange Factor Method and Alternative Zonal Formulation for Analysis of Radiating Enclosures Containing Participating Media. Ph.D. thesis, University of Texas, Austin.

20 Münch, M. (2006) Kritische numerische aspekte bei der anwendung von feldmodellen, in 55. Jahresfachtagung der Vereinigung zur Förderung des Deutschen Brandschutzes e.V., Tagungsband, Salzburg, vfdb, Vereinigung zur Förderung des Deutschen Brandschutzes e.V., Postfach 1231, 48338 Altenberge, pp. 255–289.

21 National Institute of Standards and Technology, Building and Fire Research Laboratory, current version available at, Fire Dynamics Simulator Technical Reference Guide, Volume 1–4. http://www.fire.nist.gov/fds/documentation.html.

22 Klainerman, S. and Majda, A. (1982) Compressible and incompressible fluids. *Communications on Pure & Applied Mathematics*, **35**, 629–653.

23 Klein, R. (1995) Semi-implicit extension of a Godunov-type scheme based on low Mach number asymptotics. I: one-dimensional flow. *Journal of Computational Physics*, **121**, 213–237.

24 Klein, R., Botta, N., Schneider, T., Munz, C.D., Roller, S., Meister, A., Hoffmann, L., and Sonar, T. (2001) Asymptotic adaptive methods for multi-scale problems in fluid mechanics. *Journal of Engineering Mathematics*, **39**, 261–343.

25 Rehm, R.G. and Baum, H.R. (1978) The equations of motion for thermally driven buoyant flows. *Journal of Research*, **83** (3), 297–308.

26 Schochet, S. (2005) The mathematical theory of low Mach number flows. *M2AN*, **39** (3), 441–458. doi: 10.1051/m2an:2005017

27 Dreisbach, J. and McGrattan, K. (2006) Verification and validation of selected fire models for nuclear power plant applications, Volume 6: Fire dynamic simulator (FDS), Tech. Rep., U.S. Nuclear Regulatory Commission, Office of Nuclear Regulatory Research (RES), Rockville, MD: 2005 and Electric Power Research Institute (EPRI), Palo Alto, CA. NUREG-1824 and EPRI 1011999.

13
Validation of CFD Models for the Prediction of Gas Dispersion in Urban and Industrial Environments

Michael Schatzmann and Bernd Leitl

13.1
Introduction

CFD (computational fluid dynamics) models have become the preferred tool for predicting flow and dispersion within the urban canopy layer. The increasing use of these models is, however, paralleled by a growing awareness that the results obtained with their aid are to a large degree dependent on both the particular model chosen and the person using it (see, e.g., Ketzel *et al.* [1]). Since the differences in model output are usually large, this is a severe issue, above all when the results form the basis for decisions with profound economic or political consequences.

The European Union responded to the situation by launching 'COST action 732,' which was tasked to improve the situation. In line with quality assurance initiatives in other fields of application, the action prepared several documents [2–6] (which can be downloaded from http://www.mi.uni-hamburg.de/Home.484.0.html) and suggested an evaluation procedure consisting of 6 steps:

- Model Description: this should be a brief description of the characteristics of the model, the intended range of applicability, the theoretical background on which the model development was based, the software and hardware requirements, and so on.
- Database Description: a complete description of the database that has been employed for the evaluation of the model, including the reasons why this specific database was chosen. An estimation of the data variability is required.
- Scientific Evaluation: this is a description of the equations employed to describe the physical and chemical processes that the model has been designed to include. If appropriate it should justify the choice of the numerical modeling procedures and should clearly state the limits with respect to the intended applications.
- (Code) Verification: this process is to verify that the model produces results that are in accordance with the actual physics and mathematics that have been employed. This is to identify, quantify, and reduce errors in the transcription of the mathematical model into a computational model and the solution (analytical or numerical) of the model.

Process and Plant Safety: Applying Computational Fluid Dynamics, First Edition. Edited by J. Schmidt.
© 2012 Wiley-VCH Verlag GmbH & Co. KGaA. Published 2012 by Wiley-VCH Verlag GmbH & Co. KGaA.

- Model Validation: this is a structured comparison of model predictions with experimental data and is based on statistical analyses of selected variables. It seeks to identify and quantify the difference between the model predictions and the evaluation datasets; it provides evidence as to how well the model approximates to reality. A quantification of the uncertainty of the model predictions should be produced.
- User-Oriented Assessment: this is intended to find a readable, comprehensive documentation of the code including technical description, user manual, and evaluation documentation. The range of applicability of the model, the computing requirements, installation procedures, and troubleshooting advice should be available.

Whereas five of the six steps of the evaluation procedure are relatively straightforward, step 5, the model validation step, is not. It will be given more attention here. Focus will be on the provision of suitable validation data.

13.2
Types of CFD Models

Tools which have the potential to accommodate obstacles in a reasonable way are micro-scale meteorological CFD models of the prognostic type, these being based on the Navier–Stokes equation. To directly solve the equation in a turbulent flow requires a very fine grid to capture all the relevant scales, down to the so-called Kolmogorov scale (in the atmosphere usually less than a millimeter). Furthermore, a time-dependent solution over a sufficiently long period is needed to yield stable time averages of the flow variables. This approach is called direct numerical simulation (DNS). As its computational demand is too high for the Reynolds numbers typically encountered in atmospheric boundary layers, DNS is not applicable here.

The computational demand can be substantially reduced if the time-dependent equations are solved on a grid that is fine enough (less than a few m in the case of an industrial plant or a city quarter) to resolve the larger atmospheric eddies. This approach is called large eddy simulation (LES). The small scales are formally removed from the flow variables by spatially filtering the Navier–Stokes equations. The influence of the small scales then appears as sub-filter stresses in the momentum equation. Since the large eddies are always unsteady, LES models require input conditions which are time dependent as well. While being less demanding than DNS, LES still requires significant computer resources which go beyond the capabilities many users presently have.

In view of this, the still most widespread method used for the computation of turbulent atmospheric flows is the Reynolds-Averaged Navier–Stokes (RANS) approach. In this approach, the equations are averaged in time over all turbulent scales to directly yield the statistically steady solution of the mean and turbulent flow variables. Like LES, this averaging leads to additional terms, known as the Reynolds stresses, in the momentum equation. They represent the effects of the turbulent fluctuations on the averaged flow and have to be parametrized. This is the task of turbulence closure models. In most models presently in use, the Reynolds stresses are assumed to depend linearly on the strain rate, as do the molecular stresses. The eddy viscosity appears as a proportionality factor that can be calculated using additional differential equations for the various order moments. Many modelers regard the two-equations turbulence closure schemes, which solve differential equations for the turbulent kinetic energy, k, and the dissipation rate, ε, as a good compromise between universal validity and operating expense. In particular, the standard k-ε model is widely used in engineering and micro-meteorological applications, despite the fact that it produces too much turbulent kinetic energy in regions of stagnant flow and so on. Several modifications have been proposed which ease this problem, but mostly these are at the expense of the quality of other flow property predictions.

13.3
Validation Data

The validation of mathematical flow and dispersion models relies markedly on the availability of experimental data for model testing and for the evaluation of model quality. There has been no clear definition of what should be called 'validation data' and which quality requirements should be met by validation data. Consequently, various sources of atmospheric flow and dispersion data have been used, and – in many cases – misused for validation purposes. Since the quality and completeness of the test data significantly affects the outcome of the validation process, it must be stated clearly that the result of any model validation activity will critically depend on the quality of the data which were used as reference.

13.3.1
Validation Data Requirements

The validation data requirements are essentially based on the type of problem being considered and the type of model being used to simulate the problem. The model types, as introduced in Section 13.2, need validation data which match the complexity of the models.

COST action 732 dealt mainly with Urban RANS CFD models, since these are presently the models most widely used in the COST member states. This model type applies steady boundary conditions, that is, one solution for one specific meteorological situation is generated. The variability of atmospheric situations is indirectly allowed for by repeating the runs for different wind directions, stabilities, and so on, with each of these runs being steady-state again.

Steady-state models require validation data from experiments which were taken under steady-state atmospheric conditions, and here the problem begins. The atmosphere is intrinsically time dependent and never steady-state. The weather is continuously changing, due to both the atmospheric circulation and the diurnal cycle. The fluctuations and gusts in the incoming wind interact with equally important vortex shedding from buildings to control the local flow fields and the dispersion of pollutants. To overcome this problem quasi-steady-state situations are commonly defined which are composed of mean values averaged over for example, 20 or 30 min. However, as is later shown in the example of a detailed analysis of data from urban field experiments, the time scale of the naturally occurring turbulent structures significantly exceeds such short averaging periods. As a consequence, the commonly determined short-time mean values measured inside the urban canopy layer have the character of random samples out of an ensemble of values with significant scatter.

A second problem caused by the naturally occurring atmospheric variability concerns the determination of representative input conditions for the model runs. In most field experiments only one reference station exists that can be used to characterize the meteorological situation and the model input. Fortunately, however, there are a few exceptions, most prominent among them being the Joint Urban 2003 experiment in Oklahoma City [7]. In this experiment the wind vector was measured with high temporal resolution at multiple locations in parallel. Two of the instruments with which the wind was measured were mounted on poles well above the rooftops of buildings located outside the Business District and about 1 km apart. The data from these two sites (called PNNL and PWIDS) are often selected to determine the inflow profiles for numerical model runs. At least for southerly and westerly wind directions the flow approaching the instruments is regarded as being undisturbed.

Hertwig [8] compared, for several periods, the wind velocities and directions, which were measured simultaneously at the two stations. A typical result is shown in Figure 13.1. The data are presented as they were measured, that is, no corrections

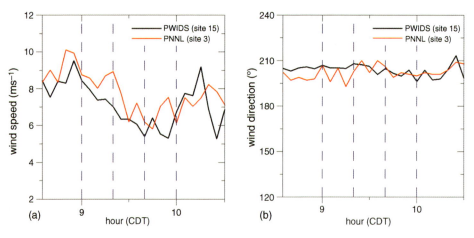

Figure 13.1 Five-minute averaged wind speeds (a) and directions (b) measured at the stations PWIDS 15 and PNNL 3 during the Joint Urban 2003 campaign IOP3. The dashed lines mark 20-min intervals (from [8]).

were made for the different measurement heights (PWIDS = 49 m and PNLL = 37 m above ground). In order to reduce the scatter, averages over 5 min have been calculated.

As can be seen, there is significant variability in the wind velocities, whereas the wind direction is comparatively constant. This was not always the case; other examples showed just the opposite or a variation of both. Not very surprisingly, the curves in Figure 13.1 differ from each other, and a correction of the data for different measurement heights does not diminish but increases the differences. Although not the raw data, 5-min mean values are presented. The data do not change in synchrony, which indicates that there are structures with time scales much longer than the averaging time that are superimposed on the mean flow.

If 20-min averages were used to determine the input for a model run, it becomes very obvious that the models would predict quite distinct flow and, above all, concentration fields, depending on which measurement site the input data were chosen from. In the Oklahoma field experiment, more than 100 anemometers, Sodars (sonic detection and ranging systems), for example, were deployed and measured in parallel. The data of many of these would likewise be suitable to be used as reference velocities. All of them deliver different model inputs, and the differences are much larger than the instrument uncertainty. A model user who knows what the model should predict could select the input which leads to the best fit. To regard such a test as a serious proof for model quality would certainly not be justified.

The ambiguity in the right choice of input data for model simulations concerns every field data set, not only those in which the model input can be selected from several meteorological towers. In experiments with one reference station only it simply remains unknown how representative this tower measurement really is. The large turbulent elements embedded in the mean flow create changes from location to location and from averaging interval to averaging interval which could be determined only if the mean weather remained constant and a sufficiently large ensemble of short-time means could be collected and statistically analyzed.

A third difficulty which has been encountered in the search for reliable validation data sets originates from the fact that resources are limited and field measurements usually are made at a few selected points only. A CFD model, on the other hand, requires at the inflow boundary of the domain a complete data array, and this not only for the mean velocity but also for more complex properties like the turbulent kinetic energy or the dissipation rate. Provided the measurement was made with a high-resolution instrument that delivers a velocity–time series (which is usually not the case in older field data sets) this information is available – but for the measurement position only. It needs to be extrapolated laterally and vertically, which requires additional assumptions to be made (e.g., assessment of a roughness length, a logarithmic wind profile in the surface layer, lateral homogeneity, etc.). In view of what has already been explained, none of these assumptions is likely to be justifiable over short averaging intervals. A CFD model presumes the existence of long periods of constant weather, the existence of a boundary layer in equilibrium with the underlying surface, and so on. As a consequence, the model is fed with input which most likely does not sufficiently correspond to the flow or concentration measurements which are used in the validation process to check the accuracy of the model output.

13.3.2
Analysis of Data from an Urban Monitoring Station

It is commonly believed that field data are best suited for the purpose of model calibration and validation since they represent 'reality'. Using the example of data from an urban monitoring station, it will subsequently be demonstrated that this belief may well be a fiction. As will be shown, data sampled within the urban canopy exhibit a large inherent variability, which makes direct comparisons of data with model results a bold venture [9, 10].

One of the most pertinent pollutant sources within the urban canopy layer is traffic. Since the vehicles are moving it is usually assumed that their emissions can be represented by a line source, since it is not yet feasible to simulate a large number of individually moving cars.

For momentum-free line sources and under conditions as subsequently specified, it is to be expected that the concentration C [g m^{-3}] (in excess above background) at any point in the vicinity of the source will be proportional to the source strength Q/L [g s^{-1} m^{-1})], with L the length of the source, and inversely proportional to the wind velocity u [m s^{-1}] measured at a reference height well above the buildings. Dimensional reasoning suggests the introduction of a normalized concentration c^*, which, under steady state ambient conditions, depends on the following non-dimensional variables:

$$c^* = \frac{C \cdot u \cdot H}{(Q/L)} = f\left(DD, \frac{l_i}{H}, Re, \frac{L_M}{H}, TIT\right) \tag{13.1}$$

with the additional parameters:

H characteristic building height (in m),
DD wind direction (in degrees),
l_i/H multiple length scales (normalized by H) describing all details of the urban geometry,
Re Reynolds number (Re = H • u/v),
L_M/H Monin-Obukhov length L_M (normalized by H) which characterizes the density stratification of the wind flow,
TIT an appropriately defined dimensionless parameter describing the traffic-induced turbulence.

Chemical reactions are excluded from the analysis, that is, only passive tracer dispersion is considered here. If the wind speed is sufficiently high, the Reynolds number takes on an above-critical value and the turbulence within the canopy layer is dominated by shear turbulence. This means that effects of stratification or traffic-induced turbulence should be of minor importance for the pollutant transport within the UCL. Then, for a given urban landscape (H and all l_i/H values are fixed), c^* is a function of the wind direction alone

$$c^* = \frac{C \cdot u \cdot H}{(Q/L)} = f(DD) \tag{13.2}$$

For more than a decade, the Lower Saxony State Agency for Ecology (NLÖ) has operated a monitoring station at the pedestrian walkway in an urban street canyon (GoettingerStrasse in Hanover, Germany [11]). The GoettingerStrasse is a busy inner-city street with four traffic lanes. The traffic load is about 30 000 vehicles/day, about 15% of which are trucks. Automated traffic counts provide the vehicle flow rate for each lane separately. In combination with general information on the composition of the German vehicle fleet, reasonable estimates of pollutant emission rates can be obtained. The width of the street and the height of the buildings alongside the street are both approximately 25 m. Most of the buildings are joined, thereby forming a street canyon with an aspect ratio (height to width) of approx. 1. The orientation of the street canyon is virtually perpendicular to the prevailing westerly wind directions (Figure 13.2).

The meteorological measurements cover wind direction and wind velocity within the canyon and above-roof (10 m above the roof level of the highest building). The measurements are made with ultrasonic anemometers, which provide the 3 components of the mean wind and the turbulent wind velocity fluctuations. Pressure, temperature, global radiation, and precipitation are also measured (only at the above-roof station).

In addition to the meteorological parameters, the GoettingerStrasse monitoring station is tasked to collect air quality data. The pollutants measured at street level comprise NO, NO_2, CO, benzene, toluene, xylene, and soot. The above-roof station provides the corresponding background measurements plus SO_2, O_3 and PM10 values. At the street monitoring station, samples are routinely taken at 'nose height'

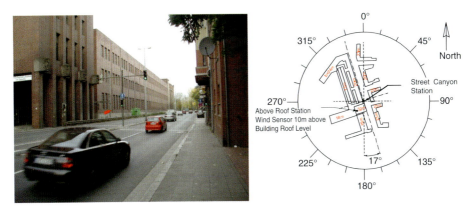

Figure 13.2 Photography and orientation of the field test site 'GoettingerStraße' in Hanover (from [11]).

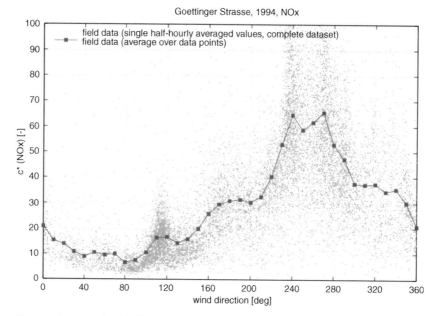

Figure 13.3 Normalized half-hourly mean concentration values as a function of wind direction measured over a period of one year at the street monitoring station 'GoettingerStrasse' in Hanover/Germany.

(1.5 m above ground), but facilities are available to collect samples at other heights (up to 10 m) as well. More information on the site can be found in Refs. [11, 12].

A permanent monitoring station like that operated by NLÖ in Hanover provides the opportunity to analyze data collected over long periods of time. This creates the possibility that particular dispersion situations are encountered several times and that the statistical significance of measured results can be assessed.

Figure 13.3 shows c^* concentrations of NO_x for an entire year (1994), presented in accordance with Eq. (13.2) as a function of wind direction only [13]. NO_x, calculated from measured NO and NO_2, can be regarded as a passive tracer for the short dispersion time periods of interest here. Each of the approximately 12 500 dots in Figure 13.4 represents a half-hourly mean value, and the curve the ensemble average over all 30-min values grouped according to the wind direction. As can be seen, the scatter of data points in Figure 13.3 is striking.

In order to decrease the spread, the data points were subsequently re-analyzed, and all those points which did not properly meet the conditions underlying Eq. (13.2) were eliminated. This was done in a step-by-step procedure by scrutinizing the individual parameters that enter the equation.

In the definition of c^*, C is the concentration excess above ambient, which means that the background concentration C_b needs to be subtracted from the value C_m measured at the monitoring station. In a city environment with numerous sources and large local concentration differences it is not easy to determine a meaningful

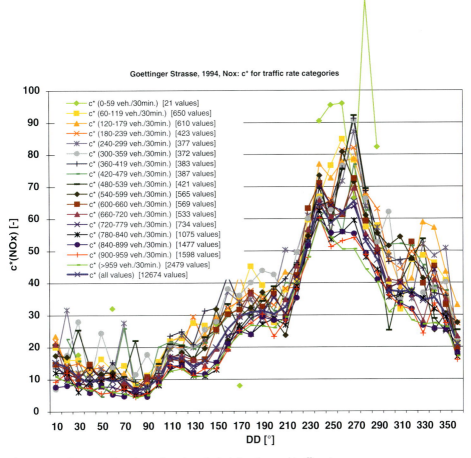

Figure 13.4 Concentration c^* as a function of wind direction and traffic rate.

background. In case of the GoettingerStrasse, the background C_b is measured at the top of the highest (NLÖ) building (left hand side in Figure 13.2) about 30 m above street level. Although for certain wind directions the background is likely to be increased by pollutants flushed out of the street canyon, this cannot be quantified. Thus, C_b as measured is assumed to be correct.

The wind vector is also measured at the top of the NLÖ building at the top of a mast 10 m above the highest elevation of the building complex and 42 m above street level. The wind directions used in Figure 13.3 are those which were directly measured. From the velocity u_{42}, a hopefully 'representative' free-stream reference wind speed u_{100} (100 m above ground) was calculated assuming the existence of a power law wind profile above the urban canopy with an exponent of $n = 0.3$. Possible wind directional changes between 42 m and 100 m height were neglected. The present practice is similar to that frequently applied in CFD modeling. The velocity $u\ (= u_{100})$ might correspond to the wind speed at the top of the numerical domain. Although the velocity is measured 10 m above roof level, free stream conditions are certainly not yet met. It remains unknown to what degree the flow is disturbed by surrounding obstacles or the NLÖ building itself. If such disturbances occur, they are surely different for different wind directions.

The c^*-equations (Eqs (13.1) and (13.2)) are valid only for line sources. The question arises of what traffic rate must be exceeded before the line source approximation holds. To find an answer, all data were grouped according to the traffic rate and plotted according to Eq. (13.2). As Figure 13.4 shows, there is large scatter between the different curves. However, with the exception of the lowest traffic category (< 60 vehicles/30 min), all curves have a similar shape, which indicates a certain consistency of source conditions. In the subsequent analysis, not only the lowest but also the second lowest class will be neglected, that is, only half-hourly values with 120 or more vehicles/min will be taken into account. The elimination of low traffic data points corresponds to the elimination of some of the night time measurements.

The determination of the emission rate per unit length Q/L is usually something of a problem. To obtain an estimate as reliable as possible, automated traffic counters are used at the Hanover site. These counters register not only the number of vehicles per time interval and per lane, they also discriminate between passenger cars and light and heavy trucks. In combination with knowledge of the composition of the German vehicle fleet in the year 1994, the prevailing driving pattern alongside the monitoring station, and emission factors for specific vehicle types, Q/L values can be estimated. The present study used the emission model MOBILEV, which is recommended by the German Environmental Agency. It should be noted, however, that there are several sets of emission factors presently in use. Depending on which one is chosen, the whole ensemble of measured points in Figure 13.3 moves up or down by about 50 %; the shape of the curves and the degree of scatter is not affected.

In the derivation of Eq. (13.1) it was made clear that the c^*-concept is not applicable to low-wind situations. Only if the wind speed rises above a certain minimum value can it be assumed that the critical Reynolds number is exceeded, that stability effects inside the canopy layer are negligible, and that the dispersion is governed by wind-generated rather than by traffic-induced turbulence.

In order to determine the minimum wind speed, the data were split into 9 velocity classes. As Figure 13.5 shows, at low wind speeds c^* appears to be fairly independent of the wind direction. This suggests that traffic-induced turbulence is the major mixing mechanism, since there is no preferred direction for pollutant transport perpendicular to the traffic lanes. With increasing velocity, the wind seems to form a secondary flow inside the canyon with the consequence of higher concentrations for westerly than for easterly winds. The street canyon has an approximate north-south orientation (Figure 13.2). Winds from 77° or 257° would be exactly perpendicular to the canyon. The monitoring station is positioned at the walkway west of the traffic lanes. The c^* values show a maximum for westerly winds which is in line with expectation. It appears that the curves take on a similar form for wind velocities $u_{100} > 3.9 \text{ m s}^{-1}$, which corresponds to wind speeds in unobstructed terrain upstream of the city and at the standard anemometer height of $u_{10} \approx 2\text{--}3 \text{ m s}^{-1}$.

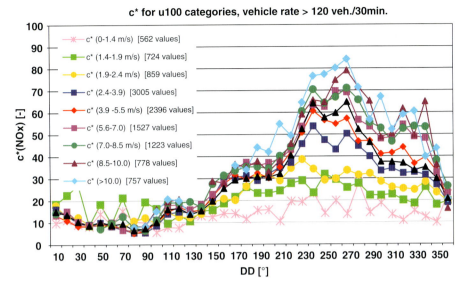

Figure 13.5 Concentration c^* as a function of wind direction and wind velocity interval.

Figure 13.6 replicates the data presentation of Figure 13.3 but includes, out of the original 12 424 values, only those 6 562 values that survived the filtering process explained before. The ensemble averaged values increase by about 10% (different for different directions). This is small compared to data sets from other monitoring stations for which a similar analysis was made [14].

Compared to Figure 13.3 the spread of data points is somewhat reduced but still significant. Disturbing is the fact that in spite of the filtering, the data not only scatters but shows a clear trend. The c^*-values increase with increasing wind velocity (Figure 13.5). This is contrary to expectation and seems to undermine the c^*-concept with its underlying assumption that C and u are inversely proportional to each other. It has been speculated that C might have a more complex relationship to u or that traffic-induced turbulence might be important even under strong wind conditions [1, 15]. However, if that were the case, the strong dependence of c^* on wind direction (for higher wind speeds) could not be properly explained.

The real explanation is probably much simpler and has to do with the unsteady behavior of plumes dispersing inside urban canopy layers. As can be made visible in smoke experiments, plumes meander inside street canyons and city quarters. Even for fairly steady above-roof winds, instantaneous pictures show a plume that moves to one canyon wall, remains there for a while, and then, in a random manner, flaps to the opposite wall and back again before the material moves into one or the other neighboring street or is flushed out of the canopy. The presence of an organized secondary flow inside the canyons, an assumption on which many simple models are based, is only observable in long-time averages of the velocity or concentration fields. It appears logical to assume that high-resolution concentration measurements at receptor points located at the pedestrian walk way would reveal a highly intermittent

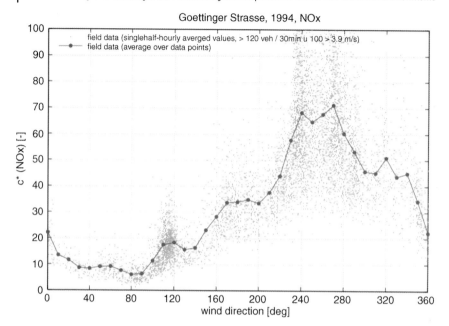

Figure 13.6 Normalized half-hourly mean concentration as a function of wind direction. Data points not in line with the assumptions underlying Eq. (13.2) were eliminated.

signal, that is, periods of low or even zero concentrations (in excess above ambient) are interspersed with high, fluctuating concentrations.

The field concentrations sampled at the GoettingerStrasse monitoring station are mean values based on 30 min averaging intervals. However, absolute averaging times are not entirely meaningful in physical contexts [16]. Dimensionless averaging or sampling times $t^* = t/(L_{ref}/u_{ref})$ should be used, with t the absolute averaging time, L_{ref} a properly chosen length scale (e.g., the reference height) and u_{ref} again the reference wind speed. Strictly, mean values are only comparable with each other if averaged over the same dimensionless sampling (averaging) time t^*. With respect to field measurements this would require adjusting the absolute sampling time to the actual above-roof wind velocity. This is not common practice. Automated systems are used that always sample over the same time interval. For two cases with above-critical Reynolds numbers that differ in wind speed by a factor of 2 and are otherwise identical, the eddy velocities would be twice as fast for the strong wind case. The concentration-versus-time trace would contain roughly twice as many 'events' within 30 min. Although the picture is not yet clear, it might be speculated that the increase of c^* with u in the higher velocity range (Figure 13.5) could have something to do with the averaging procedure.

Another notable point is that concentrations cannot be negative and that at most receptor points mean concentrations are small compared to concentration fluctuations. The probability density function of concentration fluctuations is skewed. A 'mean' value obtained under those conditions is distinct from that of a normal

Figure 13.7 Sketch of the artificial line source located on the median strip between the traffic lanes (from Ref. [17]).

Gaussian distribution. The experimentalist who samples probes over fixed periods of time or arithmetically averages over time series measured online with low frequency response instrumentation is often not familiar with those problems.

When these results were presented to experts in the field it was frequently argued that the scatter observed in the data would reflect the many uncertainties in connection with the assumption of the traffic-generated 'line source' rather than the urban dispersion conditions. A few years later the chance arose to confute this hypothesis [17, 18]. An artificial line source of about 100 m length was positioned at the median strip between the four traffic lanes of the GoettingerStrasse (Figure 13.7).

A carefully controlled flux of a real passive tracer (SF_6) was continuously released, and during several campaigns half-hourly SF_6 mean concentrations were measured next to the monitoring station. During the time of the SF_6 experiments there was normal traffic operation. The result, shown in Figure 13.8, exhibits a scatter similar to that found previously.

The wide scatter of data points (Figures 13.6 and 13.8) makes it clear that the 'period' of concentration fluctuations measured within the urban canopy is not small compared to the sampling time. Therefore, the common 30 min mean concentrations measured inside the urban canopy are members of an ensemble of values which can have significant scatter. Depending on the wind direction, the variability between seemingly identical cases can be large. To simply increase the sampling time would

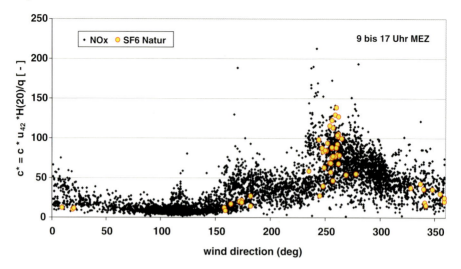

Figure 13.8 Normalized half-hourly mean concentration as a function of wind direction. Black data points represent measurements of traffic-generated NO_x immissions, yellow data points represent corresponding SF_6 immissions originating from the artificial line source shown in Figure 13.7 (from Ref. [17]).

not solve but worsen the problem since over periods longer than 30 min a systematic trend in meteorological conditions due to the diurnal cycle has to be expected. These figures furthermore indicate that single measurements cannot be representative of locations exposed to highly fluctuating and intermittent concentrations. Each individual data point in this figure is likewise justifiable. The usefulness of data generated in episodic field campaigns for model validation purposes is, therefore, clearly limited. They cannot safely be regarded as of the reliable standard a numerical model should meet.It must be concluded that with respect to the generation of validation data only long-duration measurement campaigns within which similar dispersion episodes occur several times are meaningful. Such experiments allow representative ensemble mean values to be determined (curve in Figure 13.6). But even if such long-term data are available, they must be handled with care. As was shown, there are numerous and mostly somewhat arbitrary decisions to be made when processing the raw data. If these decisions are not documented and communicated to the user of such data, he or she will be misled with respect to data reliability. Our example shows that, although reasons were given for the elimination of parts of the measurements or for the selection of a particular set of emission factors, other similarly plausible choices could have been made, with the consequence of different values for the processed data.

The widespread belief that the uncertainty of field data sets is primarily related to the inaccuracy of the instruments is surely fiction. Since it is unfair to blame a model for 'flaws' contained in the data and which originate from the inherent variability of the atmosphere, and since all models which were tested by the action were run in steady-state mode, action COST 732 came, after lengthy discussions, to the conclusion that the validation work should be predominantly based on wind tunnel data. In order to provide the opportunity to compare the wind tunnel data with field data, it was decided to accept only data from laboratory experiments in which a real field situation was replicated. Furthermore, it was required that the detailed wind tunnel boundary layer characteristics were known and documented in sufficient detail.

13.4
Wind Tunnel Experiments

Section 13.3 demonstrated that data from field experiments carried out within the urban canopy layer must be handled with care. The uncertainty – or better – variability of such data is usually unknown. Cases as reported in Section 13.3.2 indicate that data variability can be greater than a factor of 2, depending on the particular situation. A numerical model developer who needs to test the quality of his model would prefer to have more accurate data.

An alternative to field data is data from scaled laboratory experiments. Such data have their own limitations but can be generated under carefully controlled ambient conditions. The experiments for the GoettingerStrasse in Hanover were carried out in the 'WOTAN' wind tunnel of Hamburg University. This tunnel is more than 25 m long and has a cross section 13.4 m wide and 2.75–3.25 m high. It is equipped with an adjustable ceiling and two turntables. Probe positioning is carried out by a computer-controlled traverse system with a positioning accuracy of better than 0.1 mm on all three axes. An extensive custom-made software package has been developed for automated and semi-automated measurements, mass flow controller calibration and probe positioning, online data visualization, data reduction, and data validation.

The generation of a wind tunnel boundary layer which replicates an urban boundary layer at a given model scale is not an easy task. In order to achieve a laboratory boundary layer which properly covers the lower part of its atmospheric counterpart, it is common to use a combination of vortex generators and roughness elements at the tunnel floor (Figure 13.9). Which particular combination is used plays a decisive role for the distribution of mean and turbulent flow characteristics upstream from the urban model.

Since it is well known to both numerical and physical modelers that the model results are usually quite sensitive to the conditions at the domain entrance, target values for the flow properties are needed which should be based on pertinent field measurements. These target values were derived from field measurements in an urban boundary layer taken at the 300 m high NDR radio transmitter tower in Hamburg. This tower is equipped with modern meteorological instrumentation (ultrasonic anemometers etc.). Time series of the wind velocity were measured at several heights above ground, and properties like the vertical momentum flux, power spectra, macro scales of turbulence, and wind directional variations were determined [19]. The vortex generators and roughness elements were varied until the wind tunnel boundary layer matched the corresponding field data in the scale 1: 250 (which was the scale of the physical model). We restricted ourselves to only modeling the

Figure 13.9 Physical model in the scale 1 : 250 of the GoettingerStrasse site in Hamburg University's large boundary layer wind tunnel 'WOTAN'. In the background the combination of vortex generators and roughness elements can be seen.

lower part of the boundary layer and neutral stability cases. Wind speeds were chosen sufficiently high to ensure above-critical Reynolds numbers. Since the Hanover project focused on the generation of validation data for RANS CFD-models, only cases with steady mean boundary layer conditions were investigated. In contrast to field work, it was possible to control the 'weather' in the wind tunnel and keep it constant as long as necessary.

Wind tunnel experiments not only have a geometrical scale but also a time scale. When an experiment in the scale 1 : 250 is carried out in the wind tunnel, then all processes take place 250 times faster than in reality, provided the reference wind speed in model and prototype is the same and the Reynolds number in both cases is above its critical value. For example, a time series of about 3 min in the tunnel corresponds to a time series of 12.5 h in reality, but taken under steady mean atmospheric conditions for the whole period of time. This long time series can now be broken into 30 min (real time) sequences. Figure 13.10 shows the two 30 min sequences with the biggest and smallest mean concentration for a reference wind speed of 4 m s^{-1} from the south-west [20].

The figure fully corroborates the hypothesis introduced in Section 13.3.2, that for urban dispersion processes as discussed there 30 min mean values show great variability. Depending on the details of the particular case (wind direction etc.), the uncertainty of the data can be in the region of a factor of 2 or more, this already being under steady, carefully controlled laboratory conditions. In reality, the uncertainty is certainly greater, since not all the conditions presumed in the derivation of Eq. (13.2) will be effectively fulfilled.

The scatter of the field data shown in Figure 13.6 is replicated in the wind tunnel experiments as well. Depending on the complexity of the flow at the measurement point, which in a given building geometry corresponds to the wind direction, 30 min (full scale time) intervals with high- and low-concentration means were determined

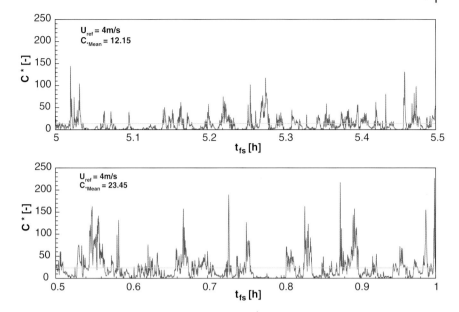

Figure 13.10 Two 30-min intervals selected from different sections of the same long-time series which was measured at the position of the GoettingerStrasse street monitoring station in a corresponding wind tunnel experiment (from Ref. [20]).

for all wind directions. If the scatter bars (based on minimum and maximum values) determined in the wind tunnel at the location of the field observation station are attached to the ensemble mean curve based on the field measurements, reasonable agreement is found (see Figure 13.11).

Although the laboratory experiments were carried out in a 4 m wide wind tunnel which at the given scale was able to accommodate turbulent eddies in the approach flow with length scales of up to about 1 km, the largest atmospheric structures have most likely not yet been matched. Thus we generally observe more variability in the field than in the wind tunnel. The variability quantified in the wind tunnel experiments should thus be regarded as a lower estimate of the variability which is to be expected in reality.

13.5
Summary

This chapter deals with the well-known fact that data are always to some degree uncertain. Taking the specific example of measurements taken within an urban environment, this uncertainty (better: variability) is investigated and quantified. It happens that the uncertainty of field data is only marginally caused by instrumental error. The measurement systems nowadays in use usually give quite accurate results, provided they are properly calibrated and applied. The major source of uncertainty

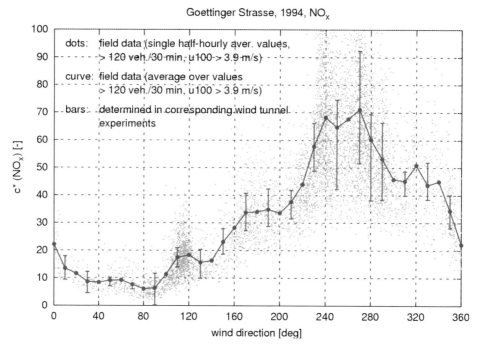

Figure 13.11 Normalized half-hourly mean concentration values as a function of wind direction measured over the period of one year at the street monitoring station GoettingerStrasse in Hanover/Germany. Each individual dot represents a 30-min mean value. The data set was filtered as in Figure 13.6. The 'variability-bars' (see text) were determined in corresponding wind tunnel experiments.

derives from the question of measurement representativeness. With respect to time, this lack of representativeness is caused by the fact that the time scales of atmospheric turbulence are not small compared to the duration of individual measurements, with common averaging intervals of 10 min or 30 min. As has been shown, measurements taken under apparently identical ambient conditions can differ by a factor of 2 or more. Data which show such a large inherent variability are not suited for CFD validation purposes.

It is usually impossible to determine the variability of field measurements by field measurements alone, although observations repeated over long periods of time may be an exception. In order to achieve data of known quality/uncertainty, as is necessary for model validation purposes, it is therefore advisable to combine field measurements with corresponding laboratory measurements, as was decided in COST 732 [21]. Boundary layer wind tunnels equipped with modern instrumentation are able to properly replicate urban flow and transport processes. The fact that most tunnels are restricted to neutral stratification is not a serious drawback for many applications of theoretical or practical concern.

The results of physical simulations are likewise sensitive to the proper choice of inflow boundary conditions, as are the results of numerical models. It appears to be vital for the quality of wind tunnel experiments to generate an approach flow boundary layer which matches the field conditions with respect to both mean and turbulent properties in the scale of the model. For typical urban model scales in the range of 1 : 100 to 1 : 400 this usually requires large tunnels with widths and heights of a few meters.

Acknowledgements

The authors are grateful for financial support from the German Ministry for Education and Research (BMBF) through the Atmospheric Research Program AFO2000.

General ideas on the applicability of field and laboratory data for validation purposes have been developed based on very constructive discussions with many colleagues participating in COST Action 732, which was funded by the European Science Foundation.

Finally, support by the German Science Foundation through the Cluster of Excellence 'Integrated Climate System Analysis and Prediction (CliSAP)' and the Priority Program 1276 'MetStroem' is gratefully acknowledged.

References

1 Ketzel, M., Louka, P., Sahm, P., Guilloteau, E., Sini, J.F., and Moussiopoulos, N. (2001) Inter-Comparison of Numerical Urban Dispersion Models. Proceedings, 3rd Int. Conf. on Urban Air Quality, Loutraki, Greece.

2 Schatzmann, M. and Britter, R. (eds) (2005) *Proceedings from the International Workshop on 'Quality Assurance of Microscale Meteorological Models'*, European Science Foundation, ISBN 3-00-018312-4.

3 Britter, R. and Schatzmann, M. (eds) (2007) *Model Evaluation Guidance and Protocol Document*, COST Office Brussels, ISBN 3-00-018312-4.

4 Britter, R. and Schatzmann, M. (eds) (2007) *Background and Justification Document to Support the Model Evaluation Guidance and Protocol Document*, COST Office Brussels, ISBN 3-00-018312-4.

5 Franke, J., Hellsten, A., Schlünzen, H. and Carissimo, B. (eds) (2007) *Best Practice Guideline for the CFD Simulation of Flows in the Urban Environment*, COST Office Brussels, ISBN 3-00-018312-4.

6 Schatzmann, M., Olesen, H., and Franke, J. (eds) (2010) *COST 732 Model Evaluation Case Studies: Approach and Results*, COST Office Brussels, ISBN 3-00-018312-4.

7 Allwine, K.J. (2004) Overview of JOINT URBAN 2003 - An atmospheric dispersion study in Oklahoma City. Proceedings of the AMS Symposium on Planning, Nowcasting, and Forecasting in the Urban Zone, January 11–15, Seattle, Washington, USA.

8 Hertwig, D. (2008) Dispersion in an urban environment with a focus on puff releases. Study Project, University of Hamburg, Meteorological Institute.

9 Boris, J., Patnaik, G., Lee, M.Y., Young, T., Leitl, B., Harms, F., and Schatzmann, M. (Jan. 5–8 2009) Validation of an LES Urban aerodynamics model for homeland security. Proceedings, 47th AIAA Aerospace Sciences Meeting, Orlando, USA.

10 Klein, P., Leitl, B., and Schatzmann, M. (2009) Driving Physical Mechanisms of

Flow and Dispersion in Urban Canopies. *International Journal of Climatology,* **27**, 1887–1907.

11 Müller, W.J., Heits, B., and Schatzmann, M.(October 14–17 2002) A prototype station for the collection of urban meteorological data. Proceedings, 8th International Conference on Harmonisation within Atmospheric Dispersion Modelling for Regulatory Purposes, Sofia, Bulgaria.

12 Lufthygienisches, N.L.Ö. (1995) Überwachungssystem Niedersachsen – Standortbeschreibung der NLÖ Stationen. Bericht, Niedersächsisches Landesamt für Ökologie, Göttinger Str. 14, 30449 Hannover (in German).

13 Schatzmann, M., Grawe, D., Leitl, B., and Müller, W.J.(May 26–30 2003) Data from an urban street monitoring station and its application in model validation. Proceedings, 26th International Technical Meeting on Air Pollution Modelling and its Application. Istanbul, Turkey.

14 Schatzmann, M., Frantz, H., Grawe, D., Leitl, B., and Müller, W.J. (March 19–23 2001) Do field data represent the truth? Proceedings, 3rd Int. Conf. On Urban Air Quality, Loutraki, Greece.

15 Kastner-Klein, P., Berkowicz, R., and Fedorovich, E. (March 19–23 2001) Evaluation of scaling concepts for traffic produced turbulence based on laboratory and full scale concentration measurements in street canyons. Proceedings, 3rd Int. Conf. On Urban Air Quality, Loutraki, Greece.

16 Schatzmann, M. and Leitl, B. (2002) Validation and application of obstacle resolving dispersion models. *Atmospheric Environment,* **36**, 4811–4821.

17 Bächlin, W., Theurer, W., Müller, W.J., and Frantz, H. (2004) Tracermessungen in einer Straßenschlucht – neue Daten zur Validierung mikroskaliger modelle. *Gefahrstoffe – Reinhaltung der Luft,* **64**, 393–398. (in German).

18 Schatzmann, M., Bächlin, W., Emeis, S., Kühlwein, J., Leitl, B., Müller, W.J., Schäfer, K., and Schlünzen, H. (2006) Development and validation of tools for the implementation of european air quality policy in Germany (Project VALIUM). *Atmospheric Chemistry and Physics,* **6**, 3077–3083.

19 Pascheke, F., Leitl, B., and Schatzmann, M. (2001) COST 715 - workshop on urban boundary layer parameterisations. Office for Official Publications of the European Communities, ISBN 92-894-4143-7, 73–83.

20 Pascheke, F., Systematische Untersuchung von mikroskaligen Strömungs- und Transportprozessen in städtischer Bebauung. PhD Thesis, University of Hamburg (in German, under preparation).

21 Leitl, B. and Schatzmann, M. (2010) Validation data for urban flow and dispersion models - Are wind tunnel data qualified? The Fifth International Symposium on Computational Wind Engineering (CWE2010), Chapel Hill, North Carolina, USA, May 23–27.

14
CFD Methods in Safety Technology – Useful Tools or Useless Toys?
Henning Bockhorn

14.1
Introduction

Safety technology reflects the scientific/technical background of 'System Safety Engineering,' which may be regarded as a subset of the applied science field 'Systems Engineering.' System safety engineering assures that a safety-critical system behaves as needed even when pieces fail or run out of control. In general, in system safety engineering two strategies are pursued. The traditional safety strategies rely on control of conditions and causes of system failure or runaway based either on the epidemiological analysis or as a result of investigation of individual past accidents [1]. In contrast to that, the system safety concept calls for a risk management strategy based on identification and analysis of failures and hazards and application of remedial controls using a systems-based approach [2]. The systems-based approach to safety requires the application of scientific, technical, and managerial skills to failure identification, failure analysis, and elimination, control, or management of hazards throughout the life-cycle of a system [2].

A system in this sense is defined as a set or group of interacting, interrelated, or interdependent elements or parts that are organized and integrated to form a collective unity or a unified whole to achieve a common objective [3, 4]. This definition lays emphasis on the interactions between the parts of a system and the external environment and includes not only the product or the process but also the influences that the surrounding environment including human interactions may have on the safety performance of the product or process. Conversely, system safety also takes into account the effects of the system on its surrounding environment. Broader definitions of a system are the hardware, software, human systems integration, procedures, and training. Therefore system safety as part of the systems engineering process should systematically address all of these domains and areas in engineering and operations in a concerted fashion to prevent, eliminate, and control failures and hazards.

Referring to the above definitions, safety technology comprises a vast plurality of methods according to the diversity of systems and tremendous variety of parts of systems in technical and everyday life. A subset of methods is given by 'computational fluid dynamics' (CFD) being applied to parts of a system which involve fluid flows. Such parts of systems prevail, for example, in chemical and process industry, refineries, steel production and treatment, energy conversion, utilities and disposal systems, and so on. Examples are chemical reactors with chemical reactions between flammable components, plants where potentially hazardous materials are treated, and chemical reactors with simply exothermic reactions and hence the potential of thermal runaway. CFD is applied to predict the behavior of these parts of systems, to identify and analyze failures of these devices, to develop remedial controls, and to prevent hazardous events, even at the design stage of the devices. Thus, CFD forms a scientific basis of system safety engineering.

Again, according to the broad field of occurrence of devices where fluid flow is involved, CFD addresses a vast variety of questions, problems, and applications. The common approach in CFD to these various tasks is via the solution of the governing equations for the description of fluid flow (generally with chemical reactions). Normally this is composed of solving – in most cases numerically – the Navier–Stokes equations in combination with balance equations for the energy and the mass of chemical species including equations of state, constitutive equations, and transport laws for numerous and vastly different initial and boundary conditions.

It is obvious from the above notes that covering the entire field of applications of CFD in system safety engineering is beyond the scope of this article, and therefore only a limited number of applications can be brought into focus due to space limitations. These applications concern characteristic properties of combustion systems such as ignition temperature, ignition delay time, laminar and turbulent flame velocity, and so on, and the importance of these properties in designing combustion devices with regard to the above-mentioned tasks of identifying and analyzing failures of these devices, developing remedial controls, and preventing hazardous events.

14.2
Characteristic Properties of Combustion Systems

14.2.1
Ignition of Flammable Mixtures

Generally, in process and chemical industries, refineries, and energy conversion processes flammable substances are on hand and have to be treated. In the above discussed definition of system safety engineering they form a part of the entire system. Flammable substances form explosive mixtures with oxidants (oxygen or air) within the explosion limits. For system safety engineering, explosion limits, ignition behavior, and ignition delay times are basic characteristic properties of combustion systems. For the determination of these characteristic properties, standard procedures have been developed in the past, for example, DIN 51794 [5] for the determination of the self-ignition temperature and others. The experimental procedure

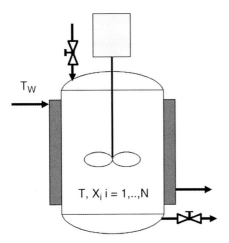

Figure 14.1 Representation of the procedure for investigation of ignition behavior by means of a well-stirred, isobaric chemical reactor.

according to DIN 51794 can be represented by a well-stirred, isobaric chemical reactor with defined temperature T_w (Figure 14.1).

The well-stirred reactor is conditioned at an initial temperature, which may be controlled appropriately by the wall temperature T_w, and filled with the flammable mixture under consideration. Ignition behavior then may be observed by temperature measurement or visually.

Modeling. Physically the ignition behavior can be traced by the non-stationary evolution of the temperature T and the mass fraction w_i or mole fractions X_i of the N chemical species of the flammable mixture inside the well-stirred reactor. Accordingly, description of this adiabatic, isobaric system can be performed by the energy balance and the mass balances for the N chemical species, which read with some simplifying assumptions (see, for example, Ref. [6]):

$$\varrho \frac{dw_i}{dt} = M_i r_i, \quad i = 1, \ldots, N \tag{14.1}$$

$$\varrho c_p \frac{dT}{dt} = -\sum_{i=1}^{N} h_i r_i \tag{14.2}$$

Equations (14.1) and (14.2) form a system of ordinary differential equations, coupled via the mixture density $\varrho = f(T, w_i)$ and the reaction rates of the chemical species $r_i = f(T, w_i)$. Also the mixture specific heat capacity is a function of temperature and composition, $c_p = f(T, w_i)$. The initial conditions are given by $T_0 = T_w$ and w_{i0}, $i = 1, \ldots, N$. For the solution of this initial value problem a number of software packages are available (see, for example, Refs [7, 8]), including the computation of mass specific heat capacities c_p and molar enthalpies h_i. M_i is the molar mass of species i.

Example: Ignition behavior of Di-tert-butyl peroxide (DTBP). As an example, the ignition behavior of di-*tert*-butyl peroxide is given here, the ignition of this compound

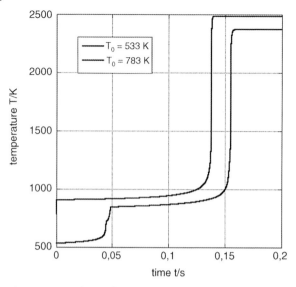

Figure 14.2 Evolution of temperature during the ignition of stoichiometric mixtures of DTBP in air at different initial temperatures.

being typical of that of organic peroxides. DTBP is used in the chemical industry as a radical initiator for the polymerization of olefins or cross linking of silicone rubber. Figure 14.2 gives the temperature evolution of a stoichiometric mixture of DTBP with air at initial temperatures of 533 and 783 K, respectively, as calculated by the solution of Eqs (14.1) and (14.2). As can be seen from the figure, the ignition of DTBP occurs in a two-stage process where, in a first step, a comparatively small amount of heat is released which is not sufficient to raise the temperature to a level for initiating the main oxidation of the intermediates formed in this first step. At higher initial temperatures this first step occurs more rapidly and the temperature rise is sufficient to induce the oxidation of the intermediates.

The evolution of the mole fractions of some involved chemical species during the ignition of stoichiometric mixtures of DTBP with air is given in Figure 14.3. As can be derived from this figure, the first step at 533 K consists of a relatively slow and complete decomposition of DTBP to acetone, which is then completely burned in the second step. Accordingly, the first step exhibits only slight oxygen consumption and formation of only a little water and carbon dioxide. At higher initial temperatures the first step occurs almost instantaneously, favoring the initial conditions for the second ignition step. This kind of two-step ignition process has been confirmed by experiments with self-ignition temperatures of 453 and 758 K [9].

The reaction mechanism used for the above simulations consists of 667 chemical reactions between 121 chemical species and is based on a C_4-hydrocarbon reaction mechanism [10] including a number of reactions for the decomposition and oxidation of DTBP [11].

Challenges and perspectives. The mathematical background and numerical techniques for the solution of the initial value problem given by Eqs (14.1) and (14.2) have

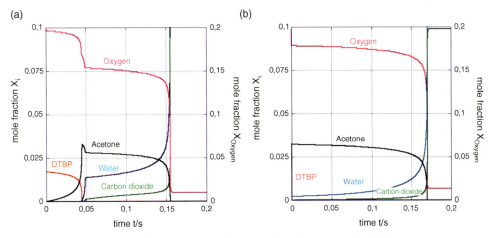

Figure 14.3 Evolution of mole fractions of chemical species during the ignition of stoichiometric mixtures of DTBP in air at different initial temperatures; (a) 533 K, (b) 783 K.

now been developed to a mature state and are available via commercial or free-access program packages. Handling of these software packages is possible on desktop computers with moderate computational power. The main deficiencies when applying the discussed approach lie in the lack of knowledge of chemical reaction mechanisms for complex molecules. Here, some progress has been achieved in the recent past by applying quantum chemistry methods combined with statistical reaction path analysis [12].

14.2.2
Ignition Delay Times

Another characteristic property of combustion systems is ignition delay time. For the characterization of flammable mixtures as part of a more complicated total system, ignition delay times are important properties in system safety engineering. This property may be defined by the threshold for the temperature increase during the ignition of combustible mixtures or the maximum slope of $T = f(t)$.

Modeling. The mathematical problem is the same as for the ignition behavior of combustible mixtures. The ignition delay times can be traced by the non-stationary evolution of the temperature T and the mass fraction w_i or mole fractions X_i of the N chemical species of the flammable mixture under homogeneous conditions. Accordingly, the energy balance and the mass balances for the N chemical species, Eqs (14.1) and (14.2), under adiabatic, isobaric conditions have to be solved for specified initial conditions, particularly for varying initial temperatures.

Examples: Ignition delay times of hydrogen, methane, and methane/n-butane mixtures with air. Figure 14.4 displays the ignition delay time of stoichiometric mixtures of hydrogen, methane, and methane/5% n-butane mixtures with air at 0.1 MPa pressure in dependence on the initial temperature T_0.

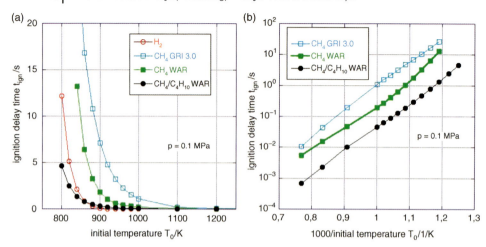

Figure 14.4 Ignition delay times for stoichiometric mixtures of hydrogen, methane and methane/5% n-butane in air in dependence on initial temperature T_0 (a) and Arrhenius-plot of the ignition delay times (b).

The calculations displayed in Figure 14.4 have been performed based on a reaction mechanism for hydrogen combustion according to Ref. [13], with 40 chemical reactions between 9 chemical species, for methane combustion based on the Gas Research Institute (GRI) mechanism 3.0 (Ref. [14]) and, alternatively, based on a reaction mechanism for the combustion of hydrocarbons up to C_4 according to Ref. [15]. The results of these calculations clearly demonstrate that the ignition delay times significantly and non-linearly depend on the initial temperature. Extrapolating the curves to ignition delay times longer than 300 s would result in self-ignition temperatures as defined by DIN 51794. For hydrogen, one obtains about 522 °C, which is somewhat lower than the recommended value (560 °C [16]), and for methane, 597 °C can be derived from Figure 14.4 using the results based on the mechanism from Ref. [15]. This is in good agreement with the recommended value (595 °C [16]). The GRI-mechanism results in somewhat higher self-ignition temperatures. The findings also clearly show that the addition of small amounts of higher hydrocarbons to methane considerably decreases the ignition delay times. The 5% addition of n-butane lowers the ignition delay time by about one order of magnitude, and consequently the self-ignition temperature is also lowered by about 100 °C. The more readily ignitable hydrocarbon n-butane (T_{ign} = 365 °C [16]) dominates the ignition behavior of the mixture. Putting it the other way round, methane may be used for making the more reactive hydrocarbon more inert. Finally, the figure clarifies that the ignition delay times as well as the self-ignition temperatures are very sensitive to the reaction mechanism used for the calculations. Nevertheless, using the above method for the determination of characteristic properties of combustible mixtures enables the estimation of the dependence of these properties on, for example, mixture composition, initial temperatures, and pressure. As an example Figure 14.5 gives the ignition delay times of methane/5% n-butane mixtures under stoichiometric conditions in

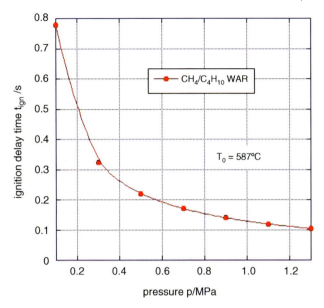

Figure 14.5 Ignition delay times for stoichiometric mixtures of methane/5% n-butane in air in dependence on pressure.

dependence on pressure. A clear decrease of the ignition delay time by one order of magnitude for a pressure change of about 1 MPa is obvious from that figure.

Challenges and perspectives. As shown above, the mathematical background and numerical techniques for the solution of the initial value problem given by Eqs (14.1) and (14.2) appear not to be the main problem in computing the properties of combustible mixtures. The main deficiencies in applying the discussed approach lie in the lack of knowledge of chemical reaction mechanisms for complex molecules and the corresponding reaction rate coefficients. Generally, these reaction mechanisms are based only partially on elaborated experiments and more on theoretical methods or pure estimates. Consequently, these mechanisms have to be validated against measured properties of combustible mixtures, and hence are not able to predict reliably other properties than the ones used for validation.

14.2.3
Laminar Flame Velocities

The laminar flame velocity describes the propagation of a flame through a combustible mixture. This property characterizes the conversion rate of combustible mixtures by the flame and its knowledge is necessary, for example, to estimate stabilization of flames at burner outlets, flame heights, non-stationary flame propagation, and transition of deflagrations to detonations. Again, for the characterization of flammable mixtures as part of a more complicated total system, flame velocities are important properties in system safety engineering. For a stationary flame

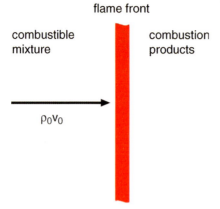

Figure 14.6 Definition of the laminar flame velocity of combustible mixtures.

(see Figure 14.6), the laminar flame velocity S_l is defined by the component of the mass flux $\varrho_o v_o$ of the combustible mixture perpendicular to the flame front.

$$\varrho_0 S_l = \varrho_0 v_0 \tag{14.3}$$

The laminar flame velocity is a fuel-specific property and depends on mixture composition, initial temperature, and pressure.

Modeling. The laminar flame velocity can be calculated by means of the energy and mass balances for the arrangement sketched in Figure 14.6. For the direction z perpendicular to the flame front and with some simplifying assumptions these balances read [6]:

$$0 = \frac{\partial(\varrho v)}{\partial z}, \text{ or } \varrho v = \varrho_0 v_0 = \text{const.} \tag{14.4}$$

$$\varrho c_p \frac{\partial T}{\partial t} = \frac{\partial}{\partial z}\left(\lambda \frac{\partial T}{\partial z}\right) - \left(\varrho v c_p + \sum_{i=1}^{N} j_i c_{p,i}\right)\frac{\partial T}{\partial z} - \sum_{i=1}^{N} h_i r_i \tag{14.5}$$

$$\varrho \frac{\partial w_i}{\partial t} = \frac{\partial}{\partial z}\left(D_i^M \varrho \frac{\partial w_i}{\partial z}\right) - \varrho v \frac{\partial w_i}{\partial z} + M_i r_i, \quad i = 1, \ldots, N \tag{14.6}$$

In Eqs (14.4) to (14.6) λ means the mixture heat conductivity; D_i^M are the species diffusion coefficients and j_i the diffusion fluxes. The term containing j_i vanishes if the component's specific heats $c_{p,i}$ are the same. Equations (14.4) to (14.6) contain the muss flux ϱv as eigenvalue, and the solution of this set of differential equations with the appropriate boundary conditions enables the calculation of the laminar flame velocities. Again, for the solution of this boundary value problem software packages are available (see, for example, Refs [7, 8]) that can be used conveniently for the calculation of laminar flame velocities.

Examples: Laminar flame velocities of hydrogen and methane mixtures with air or oxygen. Figure 14.7 gives examples of the calculation of the laminar flame velocities of

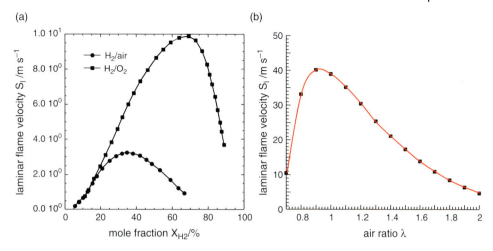

Figure 14.7 Calculated laminar flame velocities of hydrogen/air and hydrogen/oxygen (a) and methane/air mixtures (b); $T_0 = 298$ K, $p_0 = 0.1$ MPa.

hydrogen/air and hydrogen/oxygen mixtures as well as methane/air mixtures. The reaction mechanism for hydrogen combustion is from Ref. [13], and the one for methane combustion is the GRI-mechanism 3.0. As can be seen from the figure, the dependence of the flame velocities on the content of the fuel in the mixture (air ratio λ) is predicted very well, the maximum value of the flame velocities lying at slightly fuel-rich conditions. The combustion of hydrogen with pure oxygen enhances the maximum burning velocity by about a factor of 3.

Equations (14.4) to (14.6) allow also the calculation of the dependence of the laminar flame velocities on initial pressure and temperature. Some results for methane/air mixtures are given in Figure 14.8. The flame velocity of methane/air mixtures scales with pressure by $S_l \sim p^{-0.5}$. From the simple thermal theory of flame velocities, the pressure dependence is given by Eq. (14.7), so that the overall reaction order n for the combustion of methane appears to be approximately 1.

$$S_l \propto p_0^{\frac{n-2}{2}} \tag{14.7}$$

The scaling of the laminar flame velocity with initial temperature is more complicated, even when applying the simple thermal theory of flame velocities. Often this can be done by a power series in temperature, and from the results in Figure 14.8 one can derive:

$$S_l = 10 + 3.71 \cdot 10^{-4} (T_0/K)^2 \tag{14.8}$$

Challenges and perspectives. For the solution of the boundary value problem given by Eqs (14.4) and (14.6) at present numerous software packages are available by which calculations of flame velocities can conveniently be performed. As for the properties of combustible mixtures discussed in Sections 2.1 and 2.2 the main deficiencies in

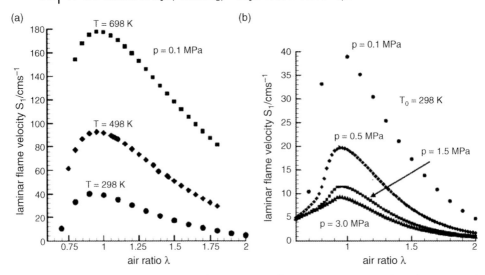

Figure 14.8 Calculated laminar flame velocities of methane/air mixtures in dependence on initial temperature (a) ($p_0 = 0.1$ MPa) and pressure (b) ($T_0 = 298$ K).

applying the discussed approach lie in the lack of knowledge of chemical reaction mechanisms for complex fuel molecules, the corresponding reaction rate coefficients, and the extension of these mechanisms to high pressure, unusual combustion conditions (very lean conditions, high dilution of oxidizer, etc.) and, for example, bio fuels.

14.2.4
Turbulent Flame Velocities

In technical devices or in flames or fires evolving from hazardous events or malfunction of plants, flow is mostly turbulent. Therefore, explosion and combustion in the context discussed here occur generally under turbulent flow conditions. Turbulent flow contains eddies of different sizes that enhance momentum, mass, and heat transport. These enhanced fluxes of momentum, mass, and heat increase the burning velocities compared with laminar flows. The simplest approach to combustion under turbulent conditions assumes wrinkling of the laminar flame front due to turbulent eddies [17] (see Figure 14.9).

The turbulent burning velocity then is given by

$$S_t = S_l \frac{A_t}{A} \tag{14.9}$$

where A_t is the area of the quasi-laminar flame front wrinkled by the turbulent eddies and A the area of the mean turbulent flame front. The problem for calculating the turbulent burning velocities is to find models for the area of the wrinkled quasi-laminar flame front. The simplest approach for modeling the area of the wrinkled flame front is given by Ref. [17] as

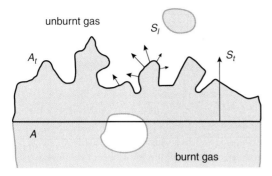

Figure 14.9 Explanation of the turbulent flame velocity of combustible mixtures according to Ref. [17].

$$\frac{A_t}{A} = 1 + \frac{\sqrt{u'^2}}{S_l} \qquad (14.10)$$

where u' are the velocity fluctuations of the turbulent flow. The simple approach according to Ref. [17] has been shown not to sufficiently represent the turbulent burning behavior (see Figure 14.10) and some other models, for example, Ref. [18], have been developed, where the interaction of turbulent and chemical time scales is considered in more detail:

$$\frac{S_t}{S_l} = 1 + c_S \frac{\sqrt{u'^2}}{S_l}\left(\frac{Da_t^2}{1+Da_t^2}\right)^{0.25} \qquad (14.11)$$

In Eq. (14.11) Da_t is the turbulent Damköhler number, which is given as the ratio of a turbulent time scale to a chemical time scale. Figure 14.10 gives a correlation of the turbulent flame velocity with the development of the area of the wrinkled flame front

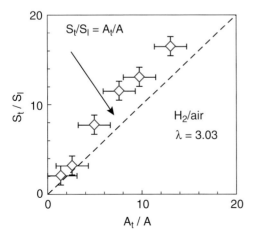

Figure 14.10 Correlation of turbulent flame velocity with wrinkled laminar flame front [19]; experimental results are indicated by symbols.

from experiments [19] with very lean hydrogen/air mixtures, and it is obvious that the simple approach cannot be applied without further refinement. For the development of correlations for the turbulent flame velocities, direct numerical simulation (DNS) can be applied to turbulent combustion scenarios. In DNS of turbulent flames the wrinkled flame front and the effects of turbulent fluctuations on the propagation of the flame can be directly calculated.

Modeling. For the direct numerical simulation of flame propagation under turbulent conditions, the Navier–Stokes equations in combination with the energy and mass balances for a computational set-up have to be solved numerically. In DNS, no filtering is applied to the set of equations, that is, all length scales of the turbulence spectrum are to be resolved in the calculations. The set of equations, again including some simplifications in component notation, is given as:

$$\frac{\partial \varrho}{\partial t} = -\frac{\partial}{\partial x_i}(\varrho u_i) \qquad (14.12)$$

$$\varrho c_p \frac{\partial T}{\partial t} = \frac{\partial}{\partial x_i}\left(\lambda \frac{\partial T}{\partial x_i}\right) - \left(\varrho u_i c_p + \sum_{i=1}^{N} j_i c_{p,j}\right)\frac{\partial T}{\partial x_i} - \sum_{j=1}^{N} h_j r_j \qquad (14.13)$$

$$\varrho \frac{\partial w_j}{\partial t} = \frac{\partial w_j}{\partial x_i}\left(D_j^M \varrho \frac{\partial w_j}{\partial x_i}\right) - \varrho u_i \frac{\partial w_j}{\partial x_i} + M_j r_j, \quad j = 1, \ldots, N \qquad (14.14)$$

$$\varrho \frac{\partial u_i}{\partial t} = -\varrho u_j \frac{\partial u_i}{\partial x_j} - \frac{\partial}{\partial x_i}\bar{\bar{p}} + \varrho g_i \qquad (14.15)$$

In Eq. (14.15) $\bar{\bar{p}}$ is the pressure tensor containing the hydrostatic pressure and the viscous shear pressure. When solving this set of equations for turbulent conditions all scales of the turbulent fluctuations have to be resolved. This restricts the application of DNS to moderate turbulent Reynolds numbers and requires a very large number of grid points for the discretization of the computational domain. The source terms of the N species equations are obtained by Arrhenius-type reaction rate expressions, which contain an exponential dependence on temperature. This gives rise to a very wide set of characteristic chemical time scales, and results in a stiff system of equations which can severely limit the integration time step, thus increasing the computational costs. Furthermore, the properties of the gaseous mixture change with composition and temperature. This has to be accounted for as well, which also is more demanding than in standard CFD applications.

Examples: Turbulent burning velocities of hydrogen mixtures with air. For the case of simulating the turbulent burning velocities of hydrogen/air mixtures the following set-up has been applied.

Thermochemistry according to Ref. [13] has been used (9 chemical species, 40 chemical reactions). For the initial conditions the propagation of premixed planar flames in 2-dimensional configurations has been considered. The initial conditions of the planar flames have been constructed from a 1-dimensional steady-state solution obtained using the method described in Section 2.3. This solution is extended to a 2-dimensional laminar plane flame.

The turbulent flow field is obtained from a 2-dimensional turbulent kinetic energy spectrum with turbulence forcing. For this, isotropic turbulence is generated in Fourier space with a von Kármán energy spectrum with Pao correction, while the phase of the flow, which determines the shape of turbulent structures, is generated with random numbers [20]. The kinetic energy spectrum, initiated with $k^{-5/3}$, decreases toward a value close to k^{-4}. To maintain the energy level and to obtain a flow with physical structures, turbulence forcing was applied by starting from values of velocity fluctuations u' and v' in physical space with a corresponding kinetic energy spectrum. From this an energy spectrum in Fourier space $E(k)$ is computed and compared with a reference kinetic energy spectrum $E_{REF}(k)$ that possesses the desired properties (i.e., u', the rms value of velocity fluctuations, and Λ, the turbulent integral length scale). Then, by using inverse Fourier transform, new values of u' and v' are generated in physical space and used in the code. The forcing procedure is sketched in Figure 14.11. From the frozen periodic fields, columns of data are extracted at each time step and prescribed as inlet conditions. Turbulence is then injected at U_{inlet} equal to the space-averaged fuel consumption rate (turbulent burning velocity) that is computed at each time step.

In the simulations a computational domain of up to 4×4 cm has been considered discretized with an equidistant grid in both directions by 600×600 points. Periodic boundary conditions are used normal to the direction the planar flame front on inflow and outflow boundaries, while non-reflecting boundary conditions are used in the direction parallel to the flame front. The calculations are initialized with reactants on

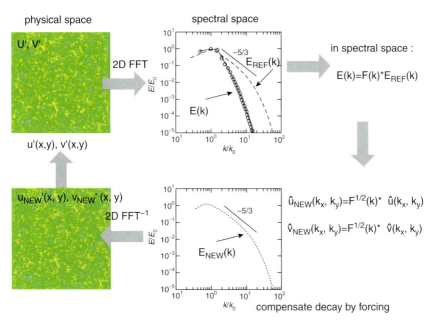

Figure 14.11 Turbulence forcing principle. During a run, the turbulence level is maintained in order to conserve the properties of the experimental flow which is to be simulated.

one side of the computational domain and products on the other side; they are separated by a laminar premixed flame.

The direct numerical simulation (DNS) comprises solving the Navier–Stokes equations (Eqs (14.12) to (14.15)) in their complete form, without any averaging or filtering accounted for by a model. The code employed for this task is the code *PARCOMB*, developed by Thévenin and co-workers [21]. Detailed diffusion of species and heat is included. High-order discretization is used in order to reduce numerical dissipation. In space, this is a sixth-order central finite-difference scheme along with a third-order differencing at the boundaries. In time, a fourth-order Runge-Kutta scheme is employed.

Figure 14.12 gives some snapshots of the developing turbulent flame according to the conditions for case 4 from Table 14.1 for successive times. As can be seen from the figures, the flame front, indicated in the figures by tracking the position of maximum heat release, is wrinkled, and pockets of fresh mixture move into the burned gases, changing the local conditions for burning the mixture. The flame front, which is resolved in the computations with at least 25 grid points, can be tracked, and the local conditions and their influence on the quasi-laminar displacement velocity of the flame front can be isolated. The simplest approach is the evaluation of the area and correlation according to Eq. (14.10). This is given in Figure 14.13, where the computational results are compared with the experiments and the simple correlation.

The results, displayed in Figure 14.13, show that the simple correlation of wrinkled laminar flame fronts with turbulent burning velocities corresponding to Eq. (14.10) is not adequate to reproduce the experimental findings. On the other hand, the agreement between the numerical results and the experiments is quite good, so that this method is suitable (i) to calculate turbulent burning velocities and (ii) to enable modeling of the turbulent burning velocities in dependence on the different physical effects in turbulent flame propagation.

Challenges and perspectives. The main challenge of DNS of turbulent flames is the development of efficient computational codes coping with the tremendous requirements for the solution of the set of equations discretized for the computational domain and the necessarily high resolution in time and space. For the case demonstrated here the computing times amount to several weeks per case on an high performance computing (HPC) cluster, though the computations have been 2-dimensional only. This restricts the consideration of realistic geometries in DNS and

Figure 14.12 Snapshots of flame front from DNS of turbulent hydrogen/air flames, case 4 from Table 14.1.

Table 14.1 Conditions of the four cases for DNS of hydrogen/air flames; air ratio (λ) in all cases 3.03.

Case 1	Case 2	Case 3	Case 4
$U_{rms} = 0.27$ m s^{-1}	$U_{rms} = 0.60$ m s^{-1}	$U_{rms} = 1.60$ m s^{-1}	$U_{rms} = 2.00$ m s^{-1}
Re = 65	Re = 143	Re = 382	Re = 478
$\tau_t = \Lambda/u' = 14.5$ ms	$\tau_t = \Lambda/u' = 6.5$ ms	$\tau_t = \Lambda/u' = 2.4$ ms	$\tau_t = \Lambda/u' = 1.95$ ms
$\tau_F = \delta/S_l = 1.84$ ms	$\tau_F = \delta/S_l = 1.84$ ms	$\tau_F = \delta/S_l = 1.84$ ms	$\tau_F = \delta/S_l = 1.84$ ms
Da = $\tau_t/\tau_F = 7.9$	Da = $\tau_t/\tau_F = 3.5$	Da = $\tau_t/\tau_F = 1.3$	Da = $\tau_t/\tau_F = 1.06$

the solution of technical problems. However, DNS opens up the possibility of developing models for very complex phenomena such as combustion under turbulent conditions and offers information that cannot be gathered by experiments.

14.3
Practical Problems

14.3.1
Mixing of Fuels with Air in Jet-In-Cross-Flow Set-ups

In modern technical combustion systems generally, homogeneous conditions are required to avoid high temperature and species concentration gradients and, connected with this, high formation rates of, for example, NO$_x$. Under these conditions, premixing of fuel and oxidant is necessary as far as possible. In premixed mixtures of fuel and oxidizer, self-ignition may occur (see Sections 2.1 and 2.2) if the mixture is heated up to temperatures above the self-ignition temperature. Heating of the mixture may occur at hot parts of the combustion system.

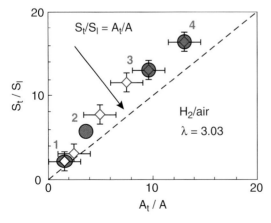

Figure 14.13 Correlation of turbulent flame velocity with wrinkled laminar flame front; comparison of experimental results with DNS (filled dots); numbers correspond to cases from Table 14.1.

Figure 14.14 Jet-in-cross-flow arrangement for mixing gas and air in a gas turbine.

For mixing of fuel and air, for example, in modern gas turbines, often jet-in-cross-flow setups are applied (see Figure 14.14). From the system safety aspect this kind of mixing device can be considered as part of a more complicated total system. These arrangements are advantageous for mixing flows with quite different flow rates, as in the combustion of methane, where the air flow is more than ten times that of the gas flow. This kind of mixer simply consists of multiple channels for the air, and the gas is injected through small holes within the channel wall (see Figure 14.14, right). Somewhere downstream of the mixer the flame is stabilized, and from the system safety aspect the problem of preventing flash-back or ignition in the mixing region has to be solved. As a first approximation, the residence time and residence time distribution in the mixing region must be estimated, and ignition may occur if the residence times are in the same order of magnitude as the ignition delay times (see Section 2.2). This problem may be solved by CFD methods.

Modeling. For the above-sketched problem of mixing fuel and oxidizer in a jet-in-cross-flow arrangement, turbulent conditions generally prevail. As pointed out in Section 2.4, the vast spectrum of length scales that has to be covered is prohibitive for direct numerical simulation of this problem. However, instead of DNS, large eddy simulation (LES) may be applied where direct numerical simulation in principle is applied but the numerical grid is not extended to the smallest scales. Instead, fewer grid points are applied, so that the larger scales are computed by DNS and the smaller, unresolved ones (subgrid scales) are modeled using appropriate turbulence models.

The equations to be solved are developed from Eqs (14.12) to (14.15) by filtering with the unresolved scales, which results in balance equations for the filtered quantities $\tilde{\phi}$ and some unclosed terms which have to be modeled. The filtered momentum equations then are given by

$$\frac{\partial \bar{\varrho} \tilde{u}_i}{\partial t} = -\frac{\partial \bar{\varrho} \tilde{u}_i \tilde{u}_j}{\partial x_j} - \frac{\partial \tilde{p}}{\partial x_i} + \frac{\partial}{\partial x_j}\left(2\mu \tilde{S}_{ij}\right) - \frac{\partial \bar{\varrho} \tau_{ij}^{sgs}}{\partial x_j} \qquad (14.16)$$

where

$$\tau_{ij}^{sgs} = -2\nu_t \tilde{S}_{ij} \qquad (14.17)$$

$$\nu_{t\,sgs} = C_\mu \overline{\Delta^2 |\tilde{S}|} \qquad (14.18)$$

$$\tilde{S}_{kl} = \frac{1}{2}\left(\frac{\partial u_k}{\partial x_l} + \frac{\partial u_l}{\partial x_k}\right) \qquad (14.19)$$

$$|\tilde{S}| = \sqrt{2\tilde{S}_{kl}\tilde{S}_{kl}}$$
(14.20)

C_μ is a locally varying factor which is determined dynamically, and Δ is the grid distance. In the mass balances for the chemical species and the energy balance, similar unclosed terms appear upon filtering which are closed analogously, for example

$$\overline{\varrho w_i u_j} - \bar{\varrho}\tilde{w}_i\tilde{u}_j = -\frac{\mu_t}{Sc}\frac{\partial w_i}{\partial x_j}$$
(14.21)

where Sc is a turbulent Schmidt number.

Examples: Mixing in a jet-in-cross-flow arrangement. For a jet-in-cross-flow arrangement according to that given in Figure 14.14, this approach has been applied employing the program package LESSOC [22]. Figure 14.15 gives the computed fields for isothermal conditions of averaged jet species concentrations $<S_j>$ and averaged velocities indicated by stream lines. Calculations have been performed with a 3-dimensional block-structured grid of blocks with different grid sizes and resolutions. The figure indicates a rather complex flow pattern in this arrangement, with counter-acting vortices upstream of the jet and a recirculation zone in its wake. This flow pattern is the source of a relatively broad residence time distribution of the jet fluid in the mixing region. More details of the numerical simulations can be found in Ref. [23].

For analyzing the residence time distribution, stream line tracking for 100 stream lines has been performed based on the results displayed in Figure 14.15 as indicated in Figure 14.16. The volume for this analysis is indicated in the figure by solid lines and comprises the main mixing region and recirculation zone of the jet.

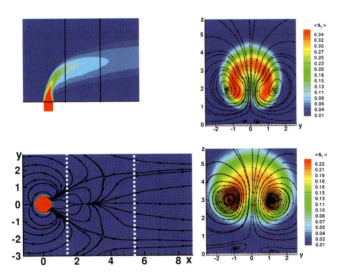

Figure 14.15 Flow and concentration fields of a jet-in-cross-flow problem. Given are the fields of the averaged concentration of the jet fluid and the stream lines for the arrangement, $u_{jet}/u_{cross} = 3$; upper left: section in x-z- direction; upper right and lower right: cross sections as indicated in upper left; lower left: cross section in y-x- direction; all distances are normalized to the jet diameter at the nozzle outlet.

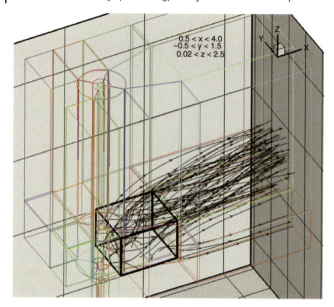

Figure 14.16 Stream line tracking for 100 streamlines for the jet-in-cross-flow from figure 14.15. The probing volume is surrounded by solid lines.

The results for the residence time distribution given in Figure 14.17 clearly indicate that there is a comparatively broad residence time distribution ranging over more than one order of magnitude. If this order of magnitude lies within the ignition delay times of the fuel (see Section 2.2) the mixture may ignite within the mixing region and may damage the mixing arrangement. By varying the flow conditions and the geometry of the jet-in-cross-flow arrangement the critical

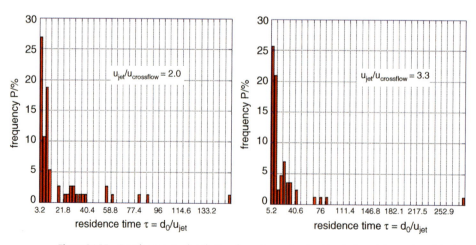

Figure 14.17 Residence time distribution from stream line tracking in the mixing region of the jet-in-cross-flow arrangement given in Figure 14.14

condition can be identified with the help of LES and avoided even in the design state of the device.

Challenges and perspectives. Similarly to DNS of turbulent flames, the main challenge in LES of turbulent reacting flows is the development of efficient computational codes to cope with the tremendous requirements for the solution of the LES equations for the computational domain and the necessary high resolution in time and space. Additionally, the development of turbulence models (sub-grid scale models) is challenging and at present an active field of research. Carrying on the analysis beyond the status given above (isothermal flow fields) brings sub-grid scale modeling of chemical reactions into play, which also is an active field of research. However, in its present state, LES is a helpful tool for successfully approaching system safety problems, and, similarly to DNS, LES opens up the possibility of describing and developing models of very complex phenomena such as combustion under turbulent conditions and offers information that cannot be gathered by experiments.

14.3.2
Chemical Reactors for High-Temperature Reactions

In the chemical industry, numerous processes are conducted at high temperatures which are provided by combustion processes. One example is the production of sulfur dioxide by combustion of sulfur. A typical reactor for that process is given schematically in Figure 14.18. The figure exhibits the discretization of the reactor for numerical simulation.

The reactor consists of a horizontal cylinder. At the left hand side, liquid sulfur is injected with the help of two pressure atomizers into the reactor and mixed with the combustion air that enters tangentially at two inlets. Between the two air ducts a restriction is installed for homogenization of the flow and mixing field. On the right the products leave the reactor in a tangential direction and enters a downstream heat exchanger. Problems can occur if the liquid sulfur is not completely burned, and

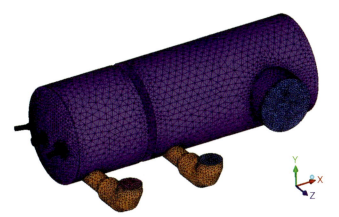

Figure 14.18 Discretized reactor for synthesis of sulfur dioxide.

droplets of the liquid phase can impinge on the wall. This would be more likely to happen in the region of the restriction in the reactor, and the deposited sulfur could ignite and lead to a destructive fire in the refractory lining. CFD calculations, in this case at the design stage, are helpful for preventing these events.

Modeling. For this case, the large dimensions of the apparatus and the highly turbulent flow is prohibitive for the application of the methods discussed in the previous sections (DNS or LES). For this kind of applications the Navier–Stokes equations and mass and energy balances are used in Reynolds-averaged form (RANS). In Reynolds-averaging, the filter width is expanded to a size such that all turbulent fluctuations are cut out. Formally, this can be achieved by introducing the decomposition of the physical variables into a mean value and a fluctuating value

$$\phi = \bar{\phi} + \phi' \tag{14.22}$$

and averaging over time. With density-weighted averages $\tilde{\phi}$ (Favre averages), for example, the momentum equation reads with some simplifications

$$\frac{\partial \bar{\varrho} \tilde{u}_i}{\partial t} = -\frac{\partial \bar{\varrho} \tilde{u}_i \tilde{u}_j}{\partial x_j} - \frac{\partial \tilde{p}}{\partial x_i} - \frac{\partial}{\partial x_j}\left(\bar{\varrho}\tau_{tij}\right) \tag{14.23}$$

The problem of this approach is to model the turbulent stresses or fluxes, the latter appearing from averaging the mass and energy balances:

$$\bar{\varrho}\tau_{ij} = \overline{\varrho u''_i u''_j}, \overline{\varrho u''_i \phi''} \tag{14.24}$$

Modeling of these unclosed terms is generally performed by gradient transport assumptions

$$\overline{\varrho u''_i \phi''} = -\bar{\varrho}\nu_t \frac{\partial \tilde{\phi}}{\partial x_j} \quad \text{with} \quad \nu_t = C_\nu \tilde{k}^2 \tilde{\varepsilon} \tag{14.25}$$

In Eq. (14.25) k is the turbulence energy and ε is the rate of dissipation of turbulence energy, ϕ being the mass fraction of chemical species or the energy the turbulent viscosity has to be scaled with the inverse of a turbulent Schmidt or Prandtl number.

A second closure problem appears using the above approach when averaging the equations for the species mass fractions. Here the chemical reaction rates have to be averaged, and, due to the non linear functions $\bar{r}_j = \overline{f(T, w_i)} \neq f(\bar{T}, \bar{w}_i)$, so that modeling of the averaged reaction rates is necessary. For modeling the averaged reaction rate, a reasonable assumption is that the chemical reaction rate cannot be faster than the break-up of the turbulent eddies. The averaged reaction rate is assumed to be proportional to the eddy turnover frequency, which can be calculated from the turbulence energy and the dissipation rate of turbulence energy.

$$\bar{r}_j \propto \frac{\tilde{\varepsilon}}{k} \tag{14.26}$$

For the numerical solution of RANS, software packages, for example, Ref. [24], are available which can be used conveniently and can be adapted to a variety of problems.

Example: Reactor for producing sulfur dioxide. The reactor for the production of sulfur dioxide schematically sketched in Figure 14.18 has been numerically

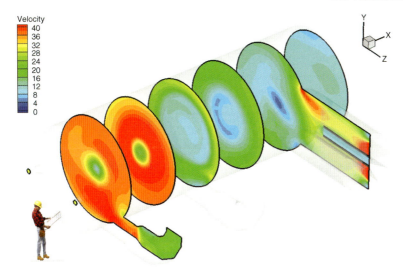

Figure 14.19 Velocity fields at six cross sections in the reactor for synthesis of sulfur dioxide.

simulated using the approach given above. The general boundary conditions have been employed for this case, and the computational domain has been discretized by about one million elements.

Figures 14.19–14.21 display the velocities, temperatures, and sulfur dioxide mass fraction in the reactor. As can be seen from the figure the restriction within the reactor homogenizes the velocities somewhat and thereby temperatures and mass fraction

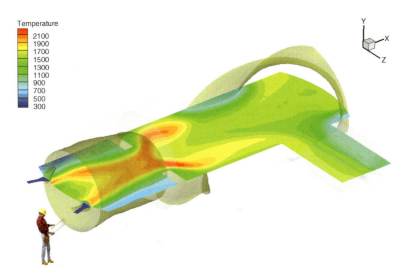

Figure 14.20 Temperature field in the reactor for synthesis of sulfur dioxide; brown surface represents the isotherm of 800 K.

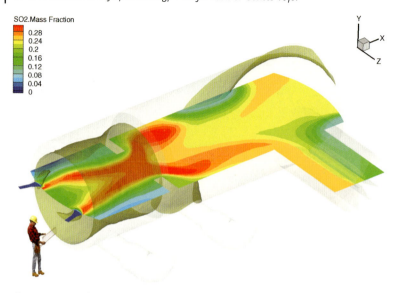

Figure 14.21 Sulfur dioxide mass fraction field in the reactor for synthesis of sulfur dioxide; brown surface represents the isosurface with mass fraction 0.08.

profiles. For the calculations it has been assumed that the vaporization of sulfur is not rate limiting, so that the temperature field is determined by the mixing of the vaporized fuel with the air. This is sufficient to get a first estimate of the fluid flow and temperatures in the reactor.

For a more exact estimate, the vaporization of the liquid sulfur droplets has been treated in more detail. For this purpose a number of streamlines were selected at random and the movement and vaporization of the droplets were calculated along these trajectories. The initial droplet size distribution was obtained from experiments for that particular atomizer. The equation for the movement of the droplets in the fluid flow is given according to Basset, Boussinesq, and Oseen as

$$m_p \frac{dv_p}{dt} = \frac{1}{8}\pi \varrho_f d^2 C_D |v_f - v_p|(v_f - v_p) + \frac{\pi d^3 \varrho_f}{6} \frac{dv_f}{dt} + \frac{\pi d^3 \varrho_f}{12}\left(\frac{dv_f}{dt} - \frac{dv_p}{dt}\right)$$

$$+ F_b + \frac{3}{2}d^2 \sqrt{\pi \varrho_f \mu} \int_{t_0}^{t} \left(\frac{dv_f}{dt'} - \frac{dv_p}{dt'}\right)(t-t')^{-0.5} dt'$$

$$- \frac{\pi d^3}{6}(\varrho_p - \varrho_f)\omega \times (\omega \times \vec{R}) - \frac{\pi d^3 \varrho_p}{3}\omega \times v_p + F_U$$

(14.27)

Having calculated the droplet trajectories, the traveling time along the trajectories can be calculated and compared to the vaporization time. The vaporization time may be obtained from the d^2-law for droplet vaporization which is given by

$$t_b = \frac{d_0^2}{k_b} \tag{14.28}$$

where

$$k_b = \frac{8\varrho_f a_f}{\varrho_p} \cdot \ln(B+1) \tag{14.29}$$

with the temperature conductivity of the fluid $a_f = \lambda_f/(\varrho_f c_{p,f})$ and the mass transfer coefficient B:

$$B = \frac{\Delta H_R \cdot f \cdot Y_{Ox,\infty} + c_{p,f}(T_\infty - T_b)}{L + c_{p,p}(T_b - T_S)} \tag{14.30}$$

If the vaporization time for the single droplets is smaller than the traveling time along the trajectories from nozzle exit to the wall the sulfur is totally vaporized and no liquid sulfur impinges at the wall. Figure 14.22 displays the trajectories of a droplet ensemble injected into the reactor. As can be seen from the figure, a considerable number of droplets are hitting the wall, and this leads to deposition of liquid sulfur at these points.

Analyzing the trajectories and counting the number of droplets hitting the wall, a probability of wall hitting can be estimated in dependence on the droplet size. This

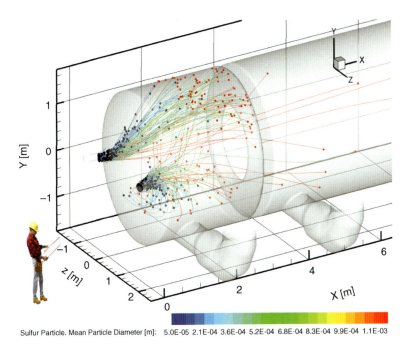

Sulfur Particle. Mean Particle Diameter [m]: 5.0E-05 2.1E-04 3.6E-04 5.2E-04 6.8E-04 8.3E-04 9.9E-04 1.1E-03

Figure 14.22 Droplet trajectories in the reactor for synthesis of sulfur dioxide; droplet trajectories with red end points hit the wall, droplet trajectories with black end points vaporize completely, trajectories are colored with droplet sizes.

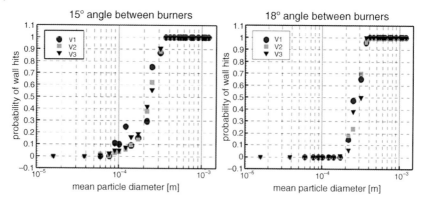

Figure 14.23 Probability of wall hitting in dependence of droplet sizes. The different colors refer to different initial droplet size distributions. The two cases refer to two configurations with different angles between the atomizers.

analysis is illustrated in Figure 14.23 where the probability of hitting the wall in dependence of the droplet sizes is plotted for two configurations of two atomizers.

As is obvious from the figure, slight changes in the orientation of the two atomizers reduce the probability of wall hitting, particularly for the smaller droplets. Using this methodology, the deposition of liquid sulfur onto the reactor walls can be prevented by a proper design of the atomizer geometry, circumventing fires, burn-out of the brick lining, and malfunctions of the system.

Challenges and perspectives. Similarly to DNS and LES, RANS simulation of turbulent reacting flows requires efficient computational codes coping with the requirements for the solution of the RANS equations for complex geometries. Additionally, the development of turbulence models and models of other physical and chemical processes occurring in reactive flow for RANS simulations is challenging and at present an active field of research. Nevertheless, RANS simulations have been developed in the recent past into a powerful tool that is helpful and convenient to use even at the design stage of safety-sensitive systems.

14.4
Outlook

The above discussion has demonstrated on the one hand that CFD methods are useful tools for predicting the behavior of safety-crucial parts of systems, enabling system failures and possible critical states to be identified and analyzed. Remedial controls can be developed and strategies for the prevention of hazardous events designed. Particularly at the design stage of systems, CFD methods are used, being a scientific basis of system safety engineering. On the other hand, CFD methods need to be developed further to a state where new and extended questions can be raised. For this purpose numerical methods need to be further developed so that LES and, in the long term, DNS can be applied for complex systems. Most importantly, the physicochemical background of the applied models in CFD tools needs to be advanced.

References

1 Rasmussen, J., Pejtersen, A.M., and Goodstein, L.P. (1994) *Cognitive Systems Engineering*, John Wiley & Sons, Hoboken.
2 Roland, H.E. and Moriarty, B. (1990) *System Safety Engineering and Management*, John Wiley & Sons, Hoboken.
3 Kossiakoff, A. and Sweet, W.N. (2003) *System Engineering Principles and Practice*, John Wiley & Sons, Hoboken.
4 Wasson, C.S. (2006) *System Analysis, Design and Development*, John Wiley & Sons, Hoboken.
5 (2005) DIN 51794 Prüfung von Mineralölkohlenwassesrtoffen - Bestimmung der Zündtemperatur, Beuth Verlag, Berlin.
6 Warnatz, J., Maas, U., and Dibble, R.W. (1996) *Combustion*, Springer Verlag, Berlin, Heidelberg, New York.
7 Kee, R.J., Rupley, F.M., and Miller, J.A. (1991) Chemkin II: A Fortran Chemical Kinetics Package for the Analysis of Gas-Phase Chemical Kinetics. SANDIA Report SAND89-8009B, Albuquerque.
8 COSILAB® Collection (2007) Rotexo-Softpredict-Cosilab, Bad Zwischenahn.
9 Wehrstedt, K. (2009) Personal communication.
10 Curran, H.J., Pitz, W.J., Westbrook, C.K., Dagaut, P., Boettner, J.-C., and Cathonnet, M. (1998) *International Journal of Chemical Kinetics*, **30**, 229.
11 Griffith, J.F., Jiao, Q., Kordylewski, W., Schreiber, M., Meyer, J., and Knoche, K.F. (1993) *Combustion and Flame*, **93**, 303.
12 Gardiner, W.C. Jr. (ed.) (2000) *Gas-Phase Combustion Chemistry*, Springer Verlag, New York.
13 Maas, U. and Warnatz, J. (1988) *Combustion and Flame*, **74**, 53.
14 Smith, G.P., Golden, D.M., Frenklach, M., Moriarty, N.W., Eiteneer, B., Goldenberg, M., Bowman, C.T., Hanson, R.K., Song, S., Gardiner, W.C. Jr., Lissianski, V.V., and Qin, Z., http://www.me.berkeley.edu/gri_mech/.
15 Baulch, D.L., Bowman, C.T., Cobos, C.J., Cox, R.A., Just, T., Kerr, J.A., Pilling, M.J., Stocker, D., Troe, J., Tsang, W., Walker, R.W., and Warnatz, J. (2005) *Journal of Physical and Chemical Reference Data*, **34**, 757–1397.
16 CHEMSAFE ORS 2010.1 (2010) DECHEMA, Frankfurt.
17 Damköhler, G. (1940) *Zeitschrift für Elektrochemie*, **46**, 601.
18 Schmid, H.-P., Habisreuther, P., and Leuckel, W. (1998) *Combustion and Flame*, **113**, 79.
19 Weiß, M., Zarzalis, N., and Suntz, R. (2008) *Combustion and Flame*, **154**, 671.
20 Lecanu, M., Mehravaran, K., Fröhlich, J., Bockhorn, H., and Thévenin, D. (2008) Computations of premixed turbulent flames, in *High Performance Computing in Science and Engineering '07* (eds W.E. Nagel, W. Jäger, and M. Resch), Springer Verlag, Berlin, pp. 229–239.
21 Thévenin, D., Behrendt, F., Maas, U., Przywara, B., and Warnatz, J. (1996) *Computers and Fluids*, **25**, 485–496.
22 Hinterberger, C. (2004) Dreidimensionale und tiefengemittelte Large-Eddy-Simulation von Flachwasserströmungen. PhD thesis, University of Karlsruhe, Karlsruhe.
23 Denev, J., Fröhlich, J., and Bockhorn, H. (2009) *Physics of Fluids*, **21**, 015101.
24 ANSYS CFX, CFX Berlin Software GmbH CFD Software und Dienstleistungen.

Part Five
Dynamic Systems – Are 1D Models Sufficient?

15
Dynamic Modeling of Disturbances in Distillation Columns

Daniel Staak, Aristides Morillo, and Günter Wozny

15.1
Introduction

The distillation process for the thermal separation of substances is of outstanding importance within the chemical industry. Humphrey [1] estimates the number of installed distillation columns, within the USA only, to be about 40 000, with a total investment value of more than 8 billion US $. Because of their contents, often consisting of large amounts of flammable and/or toxic liquids, distillation columns have a high potential safety risk. Any breakdown of plant components can cause the process to diverge from normal operation, ultimately leading to an emergency situation in which the intervention of safety measures becomes necessary. The commonest safety precaution is to install pressure relief valves (PRVs) for an efficient and fast pressure reduction.

However, the design, positioning and sizing of PRVs is crucial. Also the dynamic behavior of a distillation column after the breakdown of a plant component (e.g., condenser failure) is not well known and, due to complex interactions, hard to predict. For example, the time to reach the maximum allowable working pressure and thus the set pressure of the relief valve after an operation failure and a pressure increase is not easy to estimate. Therefore, the sizing and layout of PRVs or other relief devices is still done today by using short-cut methods, heuristics or rules of thumb based on stationary assumptions. Common short-cut methods are to adjust the relief capacity to the reboiler or the overhead vapor rate. Another very common heuristic is to adjust the relief area of the safety relief valve to the cross-sectional area of the overhead vapor pipe (often used: 40%). These methods are quick, but they involve a high level of uncertainty. Furthermore, important aspects of the plant dynamics and the interaction of different plant components are completely neglected. A frequently used method based on dynamic simulation is a reduction of the distillation column to a simple pressure tank with no internals and no peripheral devices and assuming an external heat input. The commoner reactor simulation models

---

Process and Plant Safety: Applying Computational Fluid Dynamics, First Edition. Edited by J. Schmidt.
© 2012 Wiley-VCH Verlag GmbH & Co. KGaA. Published 2012 by Wiley-VCH Verlag GmbH & Co. KGaA.

for emergency pressure reliefs then can be applied. However, this method has critical disadvantages. The fluid dynamics and the thermo-physical interaction happening inside the column are not considered adequately, and therefore the quality and the reliability of the simulation results are limited. The reason for this simplification is that still today appropriate dynamic simulation models for the description of distillation columns during an emergency situation including a pressure relief do not exist.

In this chapter a dynamic simulation model is presented which is especially designed to describe the behavior of a distillation column in the course of such events. Applying the model allows the engineer to recheck the design specification of safety measures and to get an inside view of the ongoing dynamics after an operation failure.

15.2
Dynamic Simulation Model

In defining the model assumptions, a reasonable trade-off between accuracy on the one hand and computing time and convergence behavior on the other had to be achieved. Nowadays, CFD (computational fluid dynamics) calculations are frequently used for a detailed investigation of the fluid dynamics in subsystems of distillation columns, such as the flow in a PRV or the liquid–vapor distribution on column internals like trays or structured packings [2], or the condensation in a condenser. However CFD modeling of a whole distillation column including reboiler, condenser, separation stage, and PRV is, if possible at all, still today very time-consuming and not justified for the desired task. The CFD model has to be adapted to the specific geometry of the column and therefore cannot easily be applied to different column systems. Therefore the model finally chosen is a 'hybrid model.' It consists of an equilibrium stage model (column, reboiler, and condenser) and the flow model according to Diener and Schmidt [3, 4] for the PRV. The model validation in the following section shows the suitability of the model within the scope of the desired application. The model is described in detail below.

The dynamic simulation model was implemented in gProms®. The different units are connected to a composite model by incoming or outgoing predefined streams. The streams contain the mass flow, composition, temperature, pressure, and enthalpy transferred between the elements. Regarding the desired model application for safety analysis in any distillation column, the design of the composite model is kept flexible. Not only can the column parameters such as diameter, weir height, number of stages and so on be changed, but so also can the number and connection points of the feed flows and the number and connection points of relief devices. A possible column design is shown in Figure 15.1. In this case one feed stream and one relief device on top of the column are applied.

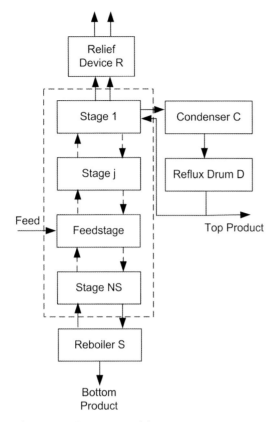

Figure 15.1 Composite model.

15.2.1
Column Stage

The model presented here is only applicable for tray columns. The following assumptions were made:

- Equilibrium stage model with Murphree efficiency is assumed.
- The liquid can flow over the weir and also through the opening holes of the tray ('weeping').
- Liquid entrainment caused by the vapor flow is possible.
- The liquid and vapor hold-up are considered.

The schematic of a column stage is shown in Figure 15.2.

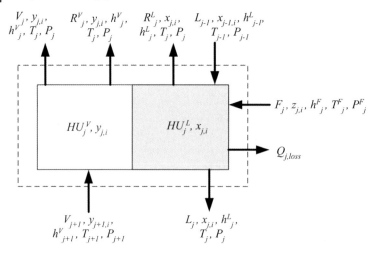

Figure 15.2 Column stage model.

15.2.1.1 Balance Equations

Each column stage is modeled completely dynamically including mass, energy, and component balances and the summation relation. The vapor and liquid hold-up are considered. The energy hold-up of the column shell and the column internals are also included in the energy balance. Eqs. (15.1)–(15.7) hold for $i = 1$ to nc–1.

Mass Balance

$$\frac{d\left(HU_j^L + HU_j^V\right)}{dt} = L_{j-1} - L_j + V_{j+1} - V_j + F_j - R_j^L - R_j^V \tag{15.1}$$

Energy Balance

$$\frac{d\left(HU_j^L u_j^L + HU_j^V u_j^V + M_{St} C_{p,St}(T_j - T_A)\right)}{dt} = L_{j-1} h_{j-1}^L - L_j h_j^L + V_{j+1} h_{j+1}^V - V_j h_j^V$$
$$+ F_j h_j^F - R_j^L h_j^L - R_j^V h_j^V - Q_{j,loss} \tag{15.2}$$

Component Balance

$$\frac{d\left(HU_j^L x_{j,i}^L + HU_j^V y_{j,i}^V\right)}{dt} = L_{j-1} x_{j-1,i} - L_j x_{j,i} + V_{j+1} y_{j+1,i} - V_j y_{j,i}$$
$$+ F_j z_{j,i} - R_j^L x_{j,i} - R_j^V y_{j,i} \quad \text{for } i = 1 \text{ to nc}-1 \tag{15.3}$$

Summation Equation

$$\sum_{i=1}^{nc} x_{j,i} = 1; \tag{15.4}$$

$$\sum_{i=1}^{nc} y_{j,i} = 1 \tag{15.5}$$

15.2.1.2 Phase Equilibrium

The non-ideal behavior of the liquid phase is considered by including the activity coefficient $\gamma_{j,i}$. The non-ideality of the vapor phase is represented by the fugacity coefficient $\varphi_{j,i}$. The Pointing Correction $\pi_{j,i}$ can be neglected at operating pressures below 10 bars.

$$\gamma_{j,i} x_{j,i} p_{j,i}^{LV} \pi_{j,i} = \varphi_{j,i} y'_{j,i} p_j \tag{15.6}$$

The deviation from the real $y_{j,i}$ composition compared to the equilibrium composition $y'_{j,i}$ is described by using the Murphree efficiency $\eta_{j,i}$.

$$y_{j,i} = \eta_{j,i,\text{Murphree}} \left(y'_{j,i} - y_{j+1,i} \right) + y_{j+1,i} \tag{15.7}$$

15.2.1.3 Incoming Vapor Flow

The actual vapor flow is given then by the vapor velocity (15.8), which results from the pressure prevailing under the considered tray j. It can be calculated from the pressure drop caused by the vapor flow and hydrostatic pressure of the liquid layer (15.9).

$$V_j = w_{j+1}^V A_j \frac{1}{v_{j+1}^V} \tag{15.8}$$

$$\Delta p_j = \Delta p_{j,\text{dry}} + \Delta p_{j,\text{hydro}} = \varsigma_j \frac{\rho_{j+1}^V}{2} \left(w_{j+1}^V \right)^2 + H_j^L \rho_j^l g \tag{15.9}$$

15.2.1.4 Outgoing Liquid Flow

Liquid can leave the stage over the weir or through the holes of the tray. The flow over the weir $L_{j,\text{weir}}$ is calculated by the Francis weir formula.

$$L_{j,\text{weir}} = l_{j,\text{weir}} C_{j,fw} \left(H_j^L - H_{j,\text{weir}} \right)^{1.5} \frac{1}{v_j^L} \tag{15.10}$$

The liquid flowing through the holes of the tray is described by adapting the Torricelli equation. Ω_j is described according to [5] as a weeping factor, which is between 0 and 1.

$$L_{j,\text{weep}} = \Omega_j A_{\text{hole}} \sqrt{2g H_j^L} \tag{15.11}$$

For the description of the weeping factor a correlation was developed and validated experimentally in a test apparatus describing the weeping factor as a function of the tray free area and the F-factor. Using this correlation, a description of column states where weeping occurs becomes possible. The parameters k_1^{ent}, k_2^{ent}, k_3^{ent} were investigated by experimental investigation in a test facility and are given in the Nomenclature Parameter list.

$$\Omega_j = \text{Min}\left[1; k_3^{\text{weep}} \exp\left(-k_2^{\text{weep}} \frac{F_j}{\varphi_{\text{free},j}} + k_1^{\text{weep}}\right)\right] \tag{15.12}$$

The complete outgoing liquid flow is calculated by adding the weir overflow and the weep flow.

$$L_j = L_{j,\text{weir}} + L_{j,\text{weep}} \tag{15.13}$$

15.2.1.5 Additional Equations

The hold-up of the vapor phase (HU^V) is calculated via the stage hold-up (Vol) and the liquid hold-up (HU^L)

$$\text{Vol}_j = HU_j^V \frac{\tilde{M}_j^V}{\rho_j^V} + HU_j^L \frac{\tilde{M}_j^L}{\rho_j^L} \tag{15.14}$$

The reboiler and the condenser models are not given here, but are modeled accordingly.

15.2.2
Relief Device

In contrast to the models of the other column units it is necessary to describe not only the thermodynamics and the flow model, but also the dynamic behavior of the valve itself. The state of the valve is changing over time (open ⇔ closed) and therefore cannot be neglected. Thus the model is divided into two sub-models: the flow model and the opening characteristic model.

Flow model: The schematic of a Relief Device Flow model is shown in Figure 15.3. The flow model describes the thermodynamic change of state when the fluid is flowing through the safety valve. To be able to describe a two-phase relief flow, two-phase flow models were also implemented. It is generally assumed that the safety valve does not possess a hold-up. Therefore the mass balance becomes:

$$0 = R_{\text{in}}^V + R_{\text{in}}^L - R_{\text{out}}^V - R_{\text{out}}^L \tag{15.15}$$

The description of a one-phase vapor relief flow is generally approved. An isentropic expansion of the fluid in the safety valve is assumed.

$$R_{\text{in}}^V = \frac{\dot{M}}{\tilde{M}_R^V} = \alpha_V A_R \Psi_{OP} \frac{1}{\tilde{M}_R^V} \sqrt{\frac{2p_{R,\text{in}}}{v_{R,\text{in}}^V}} \tag{15.16}$$

Ψ_{OP} is a function of the pressure ratio and the isentropic exponent and is calculated according to Diener and Schmidt [4]. To be able to calculate a two-phase mass flow the stagnation quality at the inlet of the relief device and thus the volumetric vapor fraction ε and the vapor mass fraction $\dot{x}$ must be known. These

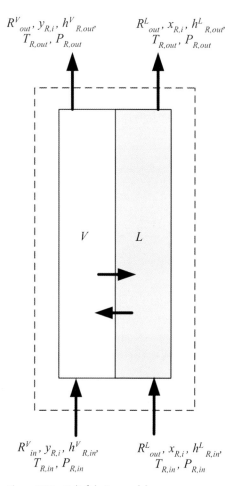

Figure 15.3 Relief device model.

are determined by an entrainment correlation based on experimental analysis in a test stand. The given correlation is valid for the spray regime.

$$E_{\text{spray}} = \frac{\dot{V}^L_{\text{in}}}{\dot{V}^V_{\text{in}}} = k_1^{\text{ent}}(F/F_{\text{max}})^2 + k_2^{\text{ent}}(F/F_{\text{max}}) + k_3^{\text{ent}} \tag{15.17}$$

$$\varepsilon = \frac{1}{E_{\text{spray}} + 1} \tag{15.18}$$

$$\dot{x} = \frac{\varepsilon v^L_{R,\text{in}}}{(1-\varepsilon)v^V_{R,\text{in}} + \varepsilon v^L_{R,\text{in}}} \tag{15.19}$$

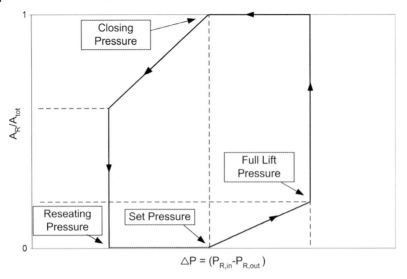

Figure 15.4 Opening and closing cycle.

Opening characteristic: The opening and closing characteristic displays the influence of the operational behavior of the relief device, for example, the properties of the spring and the shape of the disc of a spring-loaded safety valve and is given by the valve manufacturer. The free area A_R of the valve is defined as a function of the pressure drop over the relief device.

$$\frac{A_R}{A_{tot}} = f(p_{R,out}, p_{R,in}) \qquad (15.20)$$

Four distinct pressures are defined Figure 15.4: Set Pressure, Full Lift Pressure, Closing Pressure, and Reseating Pressure. By setting these pressures, the opening and closing characteristic of the relief device is determined. The valve is moving along this cycle. The dynamic behavior of different types of safety valves (Normal, Proportional, Full Lift) and also bursting disks can be covered with the implementation shown. The coupling of the two sub-models is done via the valve free area A_R.

15.3
Case Study

The model presented in the previous section is validated by experimental analysis of a complete failure scenario in a pilot scale column. The distillation column used for this purpose is especially designed for the investigation of failure scenarios and in particular for the analysis of pressure relief events. The column has a height of 4 m and a diameter of 100 mm. It is fully automated and equipped with 16 sieve trays and temperature and pressure sensors evenly distributed over the whole column. A specially designed column head allows a quick and easy exchange of the PRV and therefore the investigation of the impact of different PRVs on the dynamics of the plant. A blow-down system especially developed for this purpose allows the analysis of the relief flow (mass flow and phase composition). A flowsheet of the pilot plant is shown in Figure 15.5.

For the experiment shown, the column is operated with the system methanol/water. A condenser failure is simulated by simply stopping the cooling water flow. The course of events and the impact on the column head pressure after the condenser failure is shown in Figure 15.6. The dotted line represents the simulation results.

The head pressure of the column increases rapidly. The pressure increase is accelerated by weeping from the light boiler from the column head toward the sump. This effect is very well reproduced by the simulation model. The effect of an accelerated pressure increase caused by weeping is also mentioned by different authors (e.g., Ref. [6]). After reaching the set pressure of the PRV the first pressure relief event occurs and the pressure decreases again. The PRV closes as soon as the closing pressure is reached. The pressure increases again and the second relief cycle occurs. The total mass discharged during the two relief events is shown in Figure 15.7. The predicted total mass is slightly lower than the actual measured mass flow. This is mainly due to the fact that the real PRV remains open slightly longer than predicted. The mass flow, however, which is represented by the slope of the graph during the pressure relief event, is in excellent accordance with the simulated results. Even though the simulation model is able to predict and calculate two-phase relief flows, both events remain single-phase under the given operating conditions. However two-phase relief flows are possible in distillation columns and should be considered [7].

Once the valve closes for the second time, the intervention of an operator or the process control system is simulated by a reboiler shut-down. However, as predicted by the simulation and also proven by the experimental results, a further pressure increase is not safely prevented by this means. This effect was also observed by different authors (e.g., Ref. [6]) and is discussed in detail by Staak [8].

This example shows that the prediction of a complete failure scenario in a distillation column, including a pressure relief event, weeping, entrainment, and so on, becomes possible using the simulation model. Access to all relevant plant parameters (like head pressure, differential pressure, relief mass flow, time to relief event, etc.) is available. The executing engineer obtains a number of additional items of information which allow a more efficient conceptual design and layout of safety measures. However, if a more detailed analysis of flow dynamics is needed, different models with a higher resolution of the integral volume should be applied.

15.4
CFD- The Future of Safety Technology?

Most methods applied nowadays for the safety investigation of distillation columns are based on short-cut calculations or heuristics and are used only for the design of the PRV according to the 'worst-case scenario' or for qualitative statements about the effects of different operation failures. These methods consider different aspects of the plant operation separately and in a stationary state. However, the time-dependent

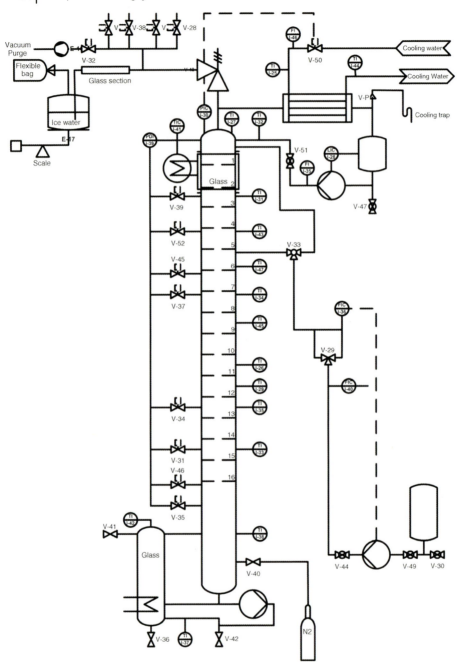

Figure 15.5 Pilot plant flowsheet.

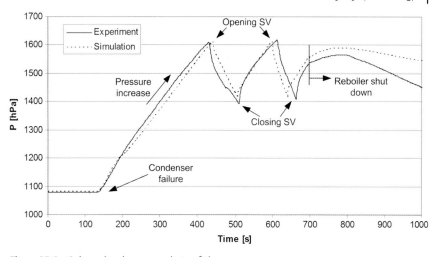

Figure 15.6 Column head pressure during failure scenario.

course of such an event, the interaction of different plant components, and the impact of the different influences on the whole plant remain uncertain.

The method proposed in this chapter enables the investigation of the dynamics of a distillation column after a number of different operation failures. It allows the design engineer to predict, to detect, and to evaluate the impact of different disturbances or operation failures on the complete plant; it therefore represents a more holistic approach.

The dynamic simulation model used for this purpose is a 'hybrid' model. It consists of an equilibrium stage model for the column and the reboiler. The flow through the PRV is calculated according to Diener and Schmidt [3]. The case study in the previous section shows the capability of this approach to achieve the desired purpose. Different aspects of the plant dynamics can be easily investigated. The accuracy of the model is adequate for the desired application.

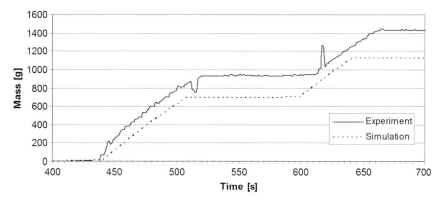

Figure 15.7 Discharged mass during pressure relief.

CFD modeling is today already a powerful tool for the investigation of flow regimes in parts of distillation columns (e.g., trays or packing) or in PRVs. But to bring the different pieces together and to use a pure CFD model for the complete plant is still today beyond the scope of actual applications. However, due to the enormous advances in CFD modeling and the continuous increase in computer power, a more intensive incorporation of CFD simulation into dynamic simulations for safety analysis is certain to occur in the near future. In a first step, this will result in the modeling of multi-scale systems (combination of CFD modeling and integral-balance modeling) and should finally lead to pure CFD safety analysis of complete plants.

15.5
Nomenclature

Latin Symbols	Description	Unit
A	Cross-sectional area	m^2
C_{fw}	Francis weir parameter	—
C_p	Heat capacity	$kJ\,kg^{-1}\,K^{-1}$
dt	Time step	s
E_{spray}	Entrainment in spray regime	m^3/m^3
F	Feed flow	$mol\,s^{-1}$
F	F Factor	$Pa^{0.5}$
g	Acceleration due to gravity	$m\,s^{-2}$
H	Height	m
h	Molar enthalpy	$J\,mol^{-1}$
HU	Molar holdup	mol
l_{weir}	Weir length	m
L	Liquid flow	$mol\,s^{-1}$
M	Mass	g
$\tilde{M}$	Molar weight	$g\,mol^{-1}$
$\dot{m}$	Mass flow density	$g\,s^{-1}\,m^{-2}$
$\dot{M}$	Mass flow	$g\,s^{-1}$
p	Pressure	Pa
Q	Heat flow	$J\,s^{-1}$
R	Relief flow	$mol\,s^{-1}$
T	Temperature	K
t	Time	s
u	Molar inner energy	$J\,mol^{-1}$
v	Molar volume	$m^3\,mol^{-1}$
V	Vapor Flow	$mol\,s^{-1}$
Vol	Volume	m^3
$\dot{V}$	Volumetric Flow	$m^3\,s^{-1}$
W	Velocity	$m\,s^{-1}$
$\dot{x}$	Vapor Mass Fraction (stagnation quality)	—
x	Liquid Mole Fraction	—

y	Vapor Mole Fraction	—
z	Feed Flow Mole Fraction	—

Greek Symbols	Description	Unit
Φ	Fugacity Coefficient	—
φ_{free}	Ratio of free area to total area	—
Π	Pointing Correction	—
π	Pressure ratio	—
$\eta_{Murphree}$	Murphree Efficiency	—
Ω	Weeping Factor	—
γ	Activity Coefficient	—
α	Discharge Coefficient	—
ς	Pressure Drop Coefficient	—
ρ	Density	kg/m^3
Ψ	Outflow function (safety valve)	—
ε	Volumetric vapor fraction (void fraction)	—

Indices	Description (referring to)
A	Ambience
dry	Dry pressure drop
F	Feed
hole	Tray holes
hydro	Hydrodynamic pressure drop
i	Component
j	Stage
L	Liquid phase
loss	Heat loss
LV	Liquid–vapor equilibrium
OP	One-phase flow
R	Relief
spray	Referring to spray regime on a tray
St	Steel
tot	Total
TP	Two-phase flow
V	Vapor phase
weep	Weeping
ent	Entrainment

Abbreviation	Description
nc	Number of Components
ns	Number of stages
PCS	Process control system
PRV	Pressure relief valve

Parameter	
k_1^{weep}	2.2
k_2^{weep}	0.69
k_3^{weep}	0.85
k_1^{ent}	$9.47 \cdot 10^{-3} \exp(-7.4 \cdot H_{stage})$
k_2^{ent}	$-3.24 \cdot 10^{-3} \exp(-6.8 \cdot H_{stage})$
k_3^{ent}	$0.27 \cdot 10^{-3} \exp(-5.8 \cdot H_{stage})$

References

1 Humphrey, J.L. (1995) Separation processes: Playing a critical role. *Chemical Engineering Progress*, **91** (10), 31–32.
2 Kenig, E.Y. and Shilkin, A. (2010) Anwendungsspezifisches design von strukturpackungen. *Chemie Ingenieur Technik*, **82** (9), 1365.
3 Diener, R. and Schmidt, J. (2004) Sizing of throttling device for gas/liquid two-phase flow part 1: safety valves. *Process Safety Progress*, **23** (4), 335–345.
4 Diener, R. and Schmidt, J. (2004) Sizing of throttling device for gas/liquid two-phase flow part 2: control valves. *Process Safety Progress*, **24** (1), 29–37.
5 Wijn, E.F. (1998) On the lower operating range of sieve and valve trays. *Chemical Engineering Journal*, **70**, 143–155.
6 Hallenberger, K. and Vetter, M. (2002) Plate damage as a result of delayed boiling. International Conference on Distillation and Absorption, VDI- Gesellschaft (Hrsg.), Baden-Baden.
7 Staak, D., Repke, J.-U., and Wozny, G. (2007) Outflow conditions and strains on internals during pressure relief events in distillation columns – experimental investigation and dynamic simulation. *Chemical and Process Engineering*, **28**, 33–45.
8 Staak, D. (2010) Absicherung von Destillationskolonnen im nichtbestimmungsgemäßen Betrieb. Dissertation TU-Berlin.

16
Dynamic Process Simulation for the Evaluation of Upset Conditions in Chemical Plants in the Process Industry

16.1
Introduction

The computational steady-state simulation of chemical processes emerged in the 1970s as computational power became more affordable and more abundant. Initially only specific parts of a chemical unit were simulated. Today the simulation of whole process plants is common practice.

The steady-state 1-D models represent a closed mass and energy balance of a process in an equilibrium state. The steady-state simulation is therefore considered to be a reliable tool for defining the appropriate operating conditions for continuous chemical processes.

As changes over time are ignored, the steady-state simulation is *per se* not able to predict the behavior of a plant leaving one state of equilibrium 'A' and moving to another state of equilibrium 'B' in a finite time period (see Figure 16.1).

By including time dependence in the models via derivative terms, the calculation of non-equilibrium states is also possible, that is, the model is able to accumulate mass and energy. To predict real life as well as steady-state behavior, the interacting environment must be incorporated into the model as well, specifically:

- equipment size and design properties influencing heat and mass transfer between media and equipment,
- dynamic reaction behavior,
- physical properties of the system,
- control properties and control parameters.

It is evident that a dynamic process simulation is significantly more complex and needs more input and much more computational power compared to a steady-state simulation, depending on the model complexity.

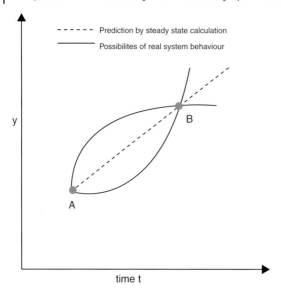

Figure 16.1 Changeover characteristics.

16.1.1
Dynamic Process Simulation for Process Safety

Knowledge of the behavior of a system in the course of upsets is essential for a safety engineer to define appropriate precautions to ensure the integrity of a plant. A dynamic simulation of unplanned occurrences, predicting the transition between two states 'A' and 'B' with sufficient accuracy, leads to better process understanding and to a more accurate process safety design of the plant.

It should be noted that the simulation approach is stated to be an adequate method in ISO 23 251 (API 521).

In general, a simulation is based on a simplified model – an approximation and simplification of the reality using mathematical methods. The reliability of the simulation model is crucial, particularly for safety-relevant applications.

Of course, real plant data are not available to a designer of a plant yet to be built. Therefore, to utilize the expected advantages of the dynamic process simulation, the quality and safety of the dynamic process simulation models must be ensured. The following principles should be followed for the application of a dynamic simulation model:

- Split processes into separate, smaller, independent units (e.g., rectification unit, cooling chain). One task – one unit concept.
- Verify the dynamic model by comparing the results with measured data from an operating process unit.

- Utilize the model for comparable process units and scenarios.
- Verify the dynamic process model/template after the plant has been constructed and started up.

Model building is understood as the design of re-usable, dynamic unit templates. The unit-specific template is built such that the template incorporates the generic attributes of the unit and is scalable. A typical example is the template for a rectification unit with a steam-operated reboiler. This reboiler, together with its dynamic behavior, is considered generic, whereas the equipment size, process control, and the rectification column properties are considered specific.

In the verification phase, the results of the dynamic model are compared with real operating data, and the significant parameters are determined. The verification should ensure that the model limits are not exceeded in the simulation of the unwanted occurrences. Therefore, real operating data must represent these conditions as closely as possible. It is a challenge to provide real operating data of sufficient quality and accuracy.

After finishing the verification phase the model can be used to simulate comparable units. It is good engineering practice to re-evaluate the model as soon as relevant process data are available.

In the following we will present the Linde approach for the model phase, the verification phase, and the design phase using two actual applications. One example will reflect our approach for the dynamic simulation of hydrocarbon rectification units. For these units, a dynamic process model has been developed and is used as template for the process safety design of new rectification systems. In the second example, the verification phase of the dynamic process model for hydrogen plants will be introduced and discussed.

16.2
Application of Dynamic Process Simulation

16.2.1
Rectification Systems

16.2.1.1 General

Rectification systems are essential process units in petrochemical plants. For example, hydrocarbon mixtures are separated into different fractions by many distillation stages in series inside a column. To achieve this, several mass flows and heat fluxes must be fed and withdrawn to and from the system. Consequently, numerous upsets are possible within such systems which may lead to an unacceptable rise in pressure or temperature. An unexpected rise in pressure above the mechanical design of the system is usually protected by means of safety valves, which release excess gases to a safe location (in petrochemical plants to the flare system). The correct design of these safety valves, as well as of the downstream disposal system, is very important for the integrity of the process plant.

16.2.1.2 Verification of the Dynamic Process Simulation

Major process parameters such as pressure, temperature, and composition may change significantly in the course of an upset. But also the mass and heat inventories of the equipment may change. The principle of closed mass and energy balances applied for steady-state calculations is no longer valid. Moreover, the system has the ability to store and release mass and heat.

The conditions prior to an upset and – in some cases – the steady-state conditions after the upset can be represented by a steady-state simulation. To determine the behavior of the system during the transition from state 'A' to state 'B,' a dynamic process simulation is mandatory (see Figure 16.1).

As described above, the dynamic simulation model must be verified before it can be used for the design of safety systems. An adequate method for this verification is the comparison of measured plant data of a real rectification system with the results of a dynamic simulation model of this system under the same constraints. If there is good correlation between the results of the dynamic simulation and the behavior of the real plant, it can be concluded that the results of the simulation model are reliable, even under different constraints. As a secondary benefit, the knowledge which was gained during this verification phase can be used for future dynamic simulation models.

In cooperation with a plant operating company, a C3/C4 rectification system was chosen for verification (see Figure 16.2). This is a typical rectification system, in which a C3/C4-hydrocarbon mixture with traces of C5–C7 is separated into a C3– overhead product stream and a C4+ bottom product stream. The reboiler is

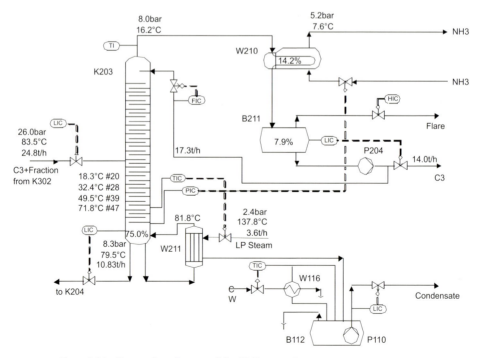

Figure 16.2 Process flow diagram of the C3/C-separation system.

operated with low pressure steam. The overhead condenser is designed as a kettle type using ammonia as refrigerant. The process data of the steady-state prior to the upset are given in Figure 16.2. These values are measured plant data used for monitoring and controlling of the process. Furthermore, samples were taken from the feed as well as from the top and bottom products and analyzed.

On this basis, a steady-state model was developed by using the process simulator UniSim Process. This model was applied for the determination of process mass and heat flows which were not covered by the measured data. Also, the efficiency of the installed trays was determined, that is, the actual installed 50 practical trays were modeled as 29 theoretical trays. For the calculation of the physical properties the Peng-Robinson equations in combination with the UniSim internal data base were used. Figure 16.3 shows the relatively simple main flow sheet with the calculated steady-state plant data.

The dynamic components were then added to the UniSim steady-state model. These were, among others:

- Control valves and controllers, including their specific parameters such as Kv value, characteristics (e.g., linear, equal percentage), PID control parameters, and so on
- Pressure losses of the single components
- Geometry of the equipment
- Models of the column trays
- Modeling of the reboiler as two steam heat exchangers   Modeling of the condenser as kettle type.

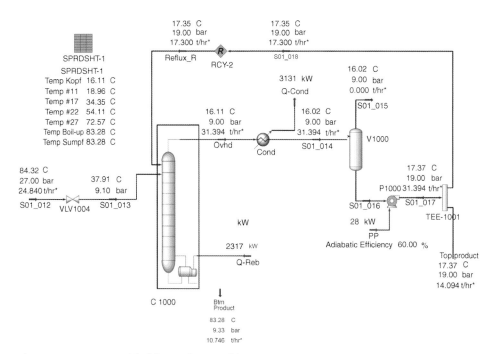

Figure 16.3 UniSim model of the steady state of the C3/C4 separation system.

16 Dynamic Process Simulation for the Evaluation of Upset Conditions

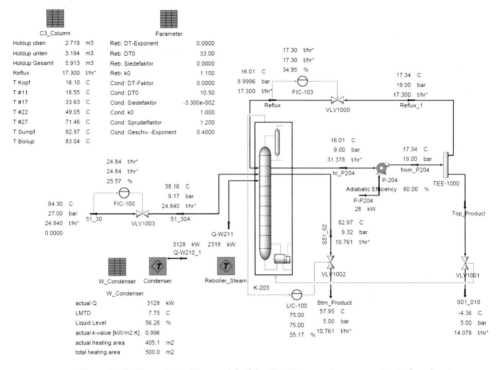

Figure 16.4 Dynamic UniSim model of the C3/C4 separation system (main flowsheet).

Furthermore, the dynamic behavior of the heat transfer within the reboiler and the condenser was investigated, depending of the fluid composition, velocity, and so on. The results were incorporated into the dynamic model.

Figures 16.4 and 16.5 show the updated dynamic UniSim model. It can be seen that the dynamic model is much more complex and detailed and therefore very time-consuming and more difficult than a simple steady-state model.

The data given in Figures 16.4 and 16.5 reflect the steady-state condition of the system.

For the verification of the dynamic model a defined upset was triggered on both the real plant and the dynamic model.

The upset chosen for verification should be preferably well-defined, that is, only one plant parameter should be changed, and the subsequent behavior of the system should be observed without any further interference. As the parameter for the upset, the steam supply to the reboiler was chosen. In the course of the upset, the set point of the steam flow controller was ramped up from 3.6 th^{-1} to 8.0 th^{-1} within 5 min.

The assessment of the simulation results was done on the basis of significant system parameters:

- Amount of steam to the reboiler
- Overhead pressure of the column
- Temperature profile over the column
- Control output, indicating the opening of the ammonia control valve.

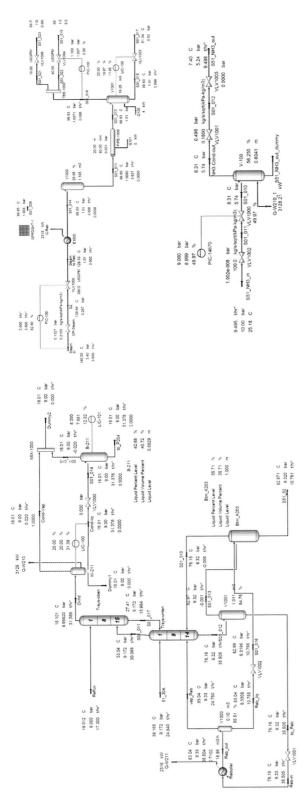

Figure 16.5 Dynamic UniSim model of the C3/C4 separation system (sub-flowsheets).

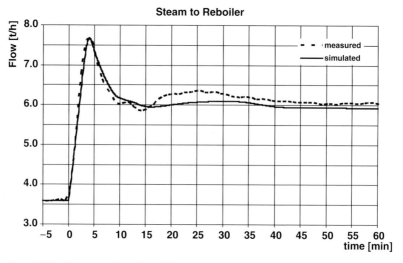

Figure 16.6 Steam supply to the reboiler.

The comparison of the simulation results with the measured plant data is given in Figures 16.6–16.9

Figures 16.6–16.9 show a very good correlation of the simulation results with the measured plant data during the first 10 min. Over the full 60 min the behavior of the system was reproduced, not only on a qualitative but also on a quantitative basis.

In summary, it can be stated that the dynamic behavior of the selected rectification system in the course of the upset could be reproduced sufficiently accurately by means of a dynamic simulation. Based on this, the conclusion can be drawn that the

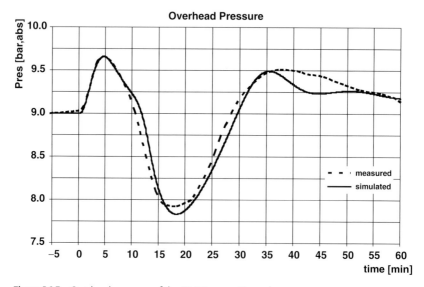

Figure 16.7 Overhead pressure of the C3/C4 separation column.

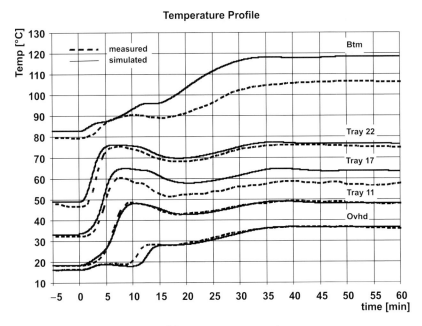

Figure 16.8 Temperature profile of the C3/C4 separation column.

dynamic simulation can be used as a reliable tool for the process safety-related design and protection of similar rectification systems.

It should be pointed out that the generation of the dynamic simulation model as shown here was a long development process, during which we gained valuable

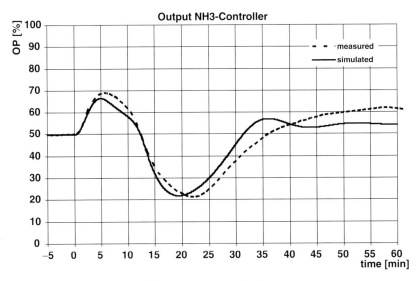

Figure 16.9 Output/opening of the ammonia control valve.

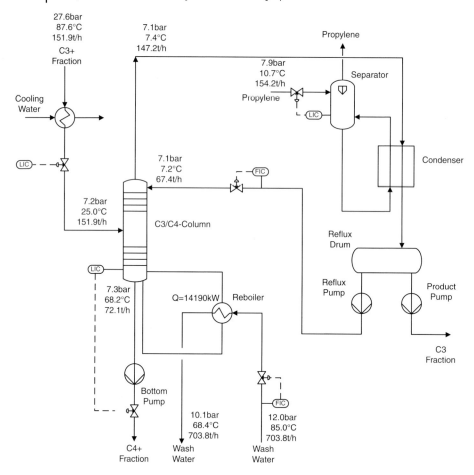

Figure 16.10 Process flow diagram of the C3/C4 separation column in an olefin plant.

experience with regard to both the modeling of a dynamic simulation model and the priority of system parameters. This experience will be used in the development of future dynamic models.

16.2.1.3 Process Safety-Related Application of a Dynamic Process Simulator

Based on the expertise gained during the verification phase, we were able to use the dynamic simulation model for related applications in comparable rectification systems. In the following, the influence of the results of a dynamic simulation on the process safety design of a system is demonstrated, using the example of a C3/C4 rectification system such as is usually found in olefin plants (see Figure 16.10).

Basically the process design of this C3/C4 separation system is similar to that used in the verification phase with the following exceptions:

- Reboiler is operated with hot water (Wash Water) instead of steam. The inlet temperature of the hot water is considered constant at 85 °C. The overhead condenser is designed as a plate fin heat exchanger with phase separator to ensure natural circulation, as instead of ammonia refrigerant propylene is used.
- The C3 + feed to the rectification column is subcooled to 40 °C by means of cooling water.

After the customization of the dynamic model as well as the re-adjustment of all system-specific parameters, the steady-state conditions which represent the normal operating conditions were established. Starting from these conditions, several upsets can be simulated. For the design of the safety systems the following upsets are typically relevant:

- Instantaneous loss of the overhead condenser due to a sudden trip of the refrigerant compressor
- Instantaneous increase of the Wash Water flow to the reboiler due to an inadvertent opening of the Wash Water control valve
- Gas–breakthrough from the upstream system. Instead of a liquid C3+ mixture a gaseous mixture of lighter hydrocarbons is fed to the rectification system.

Besides these so-called single upsets, major plant-wide upsets like 'total electric power failure' or 'cooling water failure' must also be considered.

In the following, the results of the dynamic simulation for the upset 'cooling water failure' will be presented and discussed.

In case of a plant-wide 'cooling water failure' the propylene refrigerant compressor will trip automatically and the condenser duty is reduced to zero immediately. In addition, the cooling water to the feed cooler fails.

It is a golden rule in process safety engineering that worst-case conditions must always to be considered. Therefore, the following extended constraints must apply:

- The heat transfer area of the reboiler is clean (no fouling to be considered). Accordingly, the heat transfer coefficient and hence the transferable heat could be significantly higher than during normal operating conditions.
- The Wash Water control valve is highly overdesigned – the maximum Wash Water flow is limited by the dynamic pressure losses within the system only.

Figures 16.11–16.13 show the simulated dynamic behavior of the system under the above-mentioned constraints.

Due to the loss of condensation in the overhead condenser and the loss of cooling water to the feed cooler, the pressure in the whole C3/C4 separation system rises until it reaches the set pressure of the safety valves (after approx. 2 min). Subsequently, the safety valve opens and releases the excess gas to the flare system at an initial rate of 170 th^{-1}. After approx. 10 min the flow via the safety valve drops to 105 th^{-1} and a new steady state is established (see Figure 16.11).

The feed flow to the column stays constant at 152 th^{-1} over the observed period. Due to the loss of the overhead condenser the level in the reflux drum decreases. The reflux to the column remains constant for approx. 3 min due to the hold-up of the reflux drum. Afterwards the reflux completely fails as the reflux pump is stopped (see Figure 16.11).

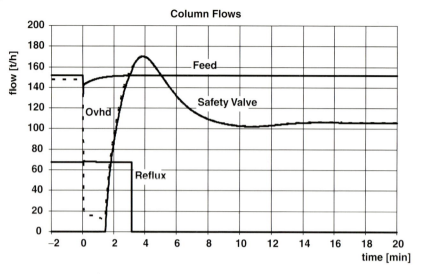

Figure 16.11 Mass flows to and from the C3/C4 separation system on 'cooling water failure'.

As a result of the increased pressure, the temperature in the bottom of the column also rises. The temperature difference between the Wash Water and the bottom product is reduced, in conjunction the transferred heat (see Figure 16.12). The heat duty control tries to keep the set point of 14.2 MW and opens the Wash Water control valve to 100% within 2 min.

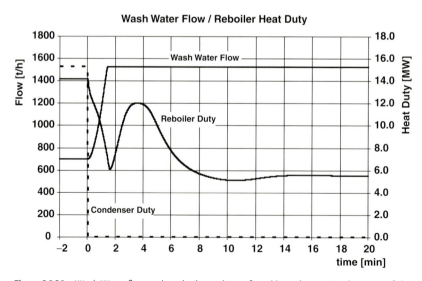

Figure 16.12 Wash Water flow to the reboiler and transferred heat duty on 'cooling water failure'.

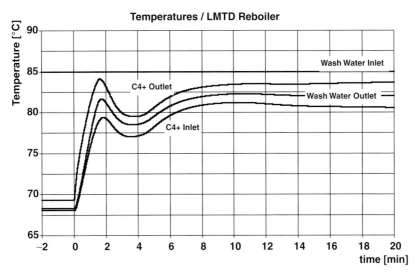

Figure 16.13 Temperature profile in the reboiler on 'cooling water failure'.

Due to the sharply reduced heat transfer in the reboiler, the counter flow inside the column is reduced as well. For a short period of time an increased liquid flow from the trays to the bottom of the column is observed. Thus the composition in the bottom changes slightly in the direction of lighter components. Therefore the bubble temperature of the bottom product decreases again. This leads to a restart of the reboiler and hence to an increased heat transfer. Consequently, the relief load via the safety valve increases as well. In the further course of the upset, the temperature in the bottom rises again and the heat transfer is reduced accordingly, until a new steady state is established after approx. 10 min (see Figure 16.13).

As a result of the dynamic simulation, the safety valve in the overhead system of the column as well as the connected flare system must be designed for a maximum flow rate of 170 t h^{-1}.

If the required relief load for the safety valve is calculated by means of a common steady-state simulation, additional constraints must be defined to achieve closed mass and heat balances:

- The lower part of the column (below the feed tray) is acting as a stripper column, as there is no more reflux.
- In the upper part of the column (above the feed tray) there is no mass and heat transfer, and this can therefore be neglected.

Taking into account these additional constraints, the steady-state simulation results in a relieving requirement for the safety valve of 105 t h^{-1}. This corresponds to the conditions of the new steady state of the dynamic simulation after 10 min.

Thus, the use of a steady-state simulation would lead to a critical undersizing of the concerned safety valve by roughly 60% for the scenario considered.

16.2.2
Hydrogen Plant

16.2.2.1 General

Hydrogen plants produce hydrogen with a purity of 99.99 vol.% via a steam-reforming process. Typically, natural gas is used as the feedstock. The main endothermic reaction of natural gas and steam takes place in direct-fired tubes filled with catalyst, the so-called Steam Reformer. Afterwards, the hydrogen-rich process gas stream needs to be cooled down, the steam condensed, and the water removed. For this purpose, a cooling train consisting of various heat exchangers and separators is connected to the outlet of the Steam Reformer. In total, the hydrogen content after the water removal is above 70 vol.%. The further purification of the hydrogen up to 99.99 vol.% is done in a pressure swing adsorption unit (PSA). Figure 16.14 shows a simplified process flow diagram from the natural gas feed down to the inlet of the PSA unit.

In the case of process upsets, a safe shutdown of the process must be ensured. Due to the heat capacity of the Steam Reformer, the reaction tubes inside the Steam Reformer must be cooled and purged, even while the process is being shut down. Therefore, steam is still fed to the Steam Reformer. This steam is heated up and is supplied to the cooling train. Depending on the kind of upset, the cooling of this steam in the cooling train can also be affected by the upset. The worst case is usually a general electrical power failure, when all the major equipment fails simultaneously. To ensure the integrity of the system, the maximum temperature which could occur

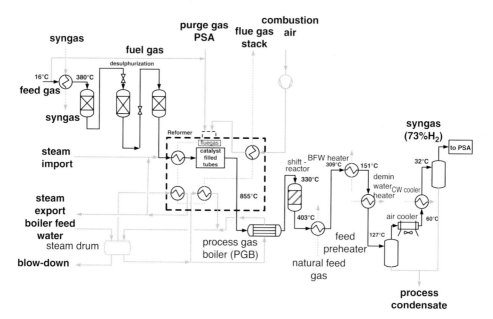

Figure 16.14 Simplified process flow diagram of the hydrogen plant.

under such circumstances must be considered in the mechanical design of the concerned equipment.

The cooling train is a complex system which is highly integrated within the whole process. Knowledge of the dynamic behavior of this system and the expected maximum temperatures is very important. Due to the kind of upset and the consequences to the process, a steady-state simulation can only be performed under simplified constraints and will not lead to realistic results. Thus a dynamic process simulation seems to be mandatory for a realistic assessment of the consequences.

16.2.2.2 Model Building and Verification of the Dynamic Process Simulation

The first step of the verification phase was the development of a steady-state model against which the measured process data could be verified. The dynamic features and parameters were then added to the model. Among others, the following parameters have been considered and implemented in the dynamic model:

- Inertias of the system.
- Volumes and liquid hold-ups.
- Model of the steam system.
- Variable heat transfer coefficients of heat exchangers, depending on the kind of media (e.g., synthesis gas and steam).
- Behavior of the air cooler during fan trip (e.g., in the case of electric power failure). Due to natural draft, a certain amount of air still flows across the air cooler, and hence a certain heat duty is still removed from the process gas.
- Control parameters for PID-controllers.
- Different control sequences for the simulation of upsets.

Figure 16.15 shows the dynamic simulation model of the hydrogen plant. The data given in Figure 16.15 represent the steady state prior to the upsets. By activation of predefined control sequences different upsets can be simulated.

For the verification of the dynamic model the behavior of a real plant in the case of an upset is needed. To achieve comparable conditions between the real plant and the dynamic simulation model, a defined upset is required. The most common upset in a hydrogen plant is the emergency shutdown of the Steam Reformer. Measured data of an existing Steam Reformer for an emergency shutdown at 100% plant load were available. The changing feed, steam, and air flows to the Steam Reformer were implemented to the control sequence for the simulation of this upset. Figure 16.16 shows the measured and simulated flows.

After the implementation of these mass flows into the sequence control of the dynamic simulation model (see Figure 16.16), the dynamic simulation could be started and the results could the compared with the corresponding measured data of the real plant.

The most important parameter is the temperature profile along the cooling train. Figure 16.17 shows the comparison of the results of the dynamic simulation with the measured data. The results correspond satisfactorily. Specifically, the behavior of the reformer outlet temperature (one of the most important parameters) corresponds very well throughout the considered time period of 60 min.

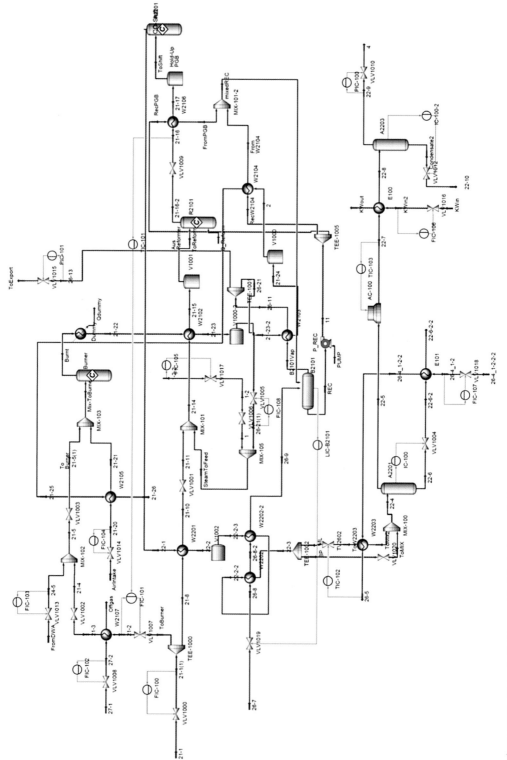

Figure 16.15 Dynamic UniSim model of the hydrogen plant.

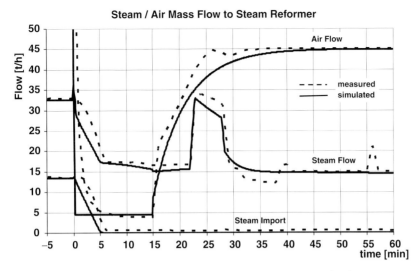

Figure 16.16 Steam and air flows to the reformer during an emergency shutdown.

During the course of the upset, the pressure levels of the cooling train as well as of the steam system are reduced in a defined way. These pressures have a significant influence on the overall behavior of the system. Figure 16.18 shows the comparison between the results of the dynamic simulation with the measured data for these pressures.

The simulated pressure curves match the real plant data quite well. The pressure curve of the steam drum shows basically the same trend. The larger changes at certain simulation times are a consequence of changes in mass flows in the plant.

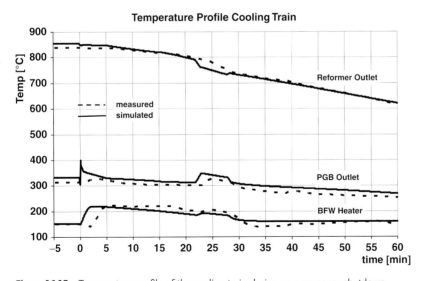

Figure 16.17 Temperature profile of the cooling train during an emergency shutdown.

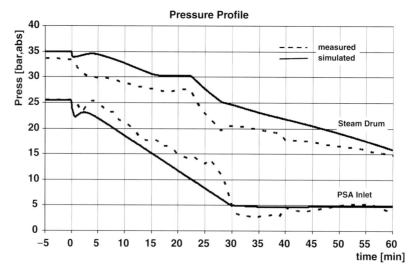

Figure 16.18 Pressure in cooling train and steam drum during an emergency shutdown.

All in all it can be stated that the results of the dynamic simulation of an emergency shutdown of the Steam Reformer are reflected with sufficient accuracy. Thus, the dynamic model is fit for further investigations for other upsets.

With the verified dynamic simulation model, the consequences of other upsets like a 'total electric power failure' can be investigated. A 'total electric power failure' has the following consequences on the Steam Reformer and the cooling train:

- Emergency shutdown of the Steam Reformer.
- Boiler feed water (BFW) supply to the BFW heater fails.

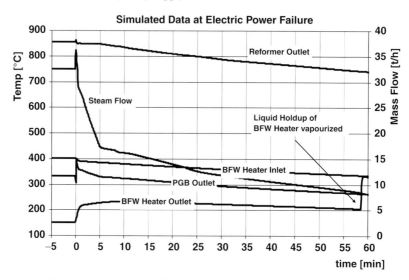

Figure 16.19 Simulation results of the upset electric power failure.

- Demin(eralized) water supply to the demin water heater fails.
- Air cooler fan stops.
- Cooling water (CW) to the CW cooler fails.
- PSA unit is tripped.

Figure 16.19 shows the results of the dynamic simulation bearing in mind the above-mentioned constraints.

The main result is that the temperature downstream of the BFW heater increases after approx. 60 min from 210 °C to about 320 °C. After that time the liquid hold-up in the BFW heater is vaporized and the cooling of the process gas is lost. This effect must be considered and adequately protected in the process safety design of the system.

16.3
Conclusion

The 1-D dynamic process simulation model is a powerful tool for process safety design in the hands of skilled experts.

The equipment models typically provided by the simulation software may not fully satisfy the specific requirements of the dynamic simulation and need to be extended by the model designer. Dynamic process model design is, compared to steady-state process simulation design, more time-consuming and must be performed by engineers experienced in both process safety and simulation. Specifically, the improvement of the calculation models for process conditions deviating from those associated with normal operation are challenging tasks and should not be underestimated.

Summing up the results of both studies, the capabilities of the 1-D dynamic simulation software for the selected process units and scenarios are adequate – the calculated results correspond well with the measured data from existing plants.

From the safety design point of view it has become evident that the common short-cut methods for the designing of safety systems can fail under certain circumstances. Dynamic process simulation can deliver more accurate and reliable results. Utilization of these results will lead to a better understanding of plant behavior during upsets and therefore to a safer plant.

It can therefore be stated that dynamic process simulation is not an everyday design tool. It is a tool for the experienced designer. Properly used it helps in gaining a profound understanding of a plant and allows a more accurate process safety design.

16.4
Dynamic Process Simulation – The Future of Safety Technology?

In the next decade, we expect that dynamic process simulation in process safety will become a standard expert tool, and will emerge from its current niche.

The application of dynamic simulation in the process industry will make a leap forward, and, in effect, the engineering knowledge and availability of dynamic simulation models, in combination with the expected increase in available computer power, will increase. We expect the momentum gained by dynamic simulation in the process industry will fruitfully and significantly influence its utilization in process safety.

17
The Process Safety Toolbox – The Importance of Method Selection for Safety-Relevant Calculations

J. Andrew Jones

17.1
Introduction – The Process Safety Toolbox

One of the first decisions that an engineer faces when confronted with a process safety problem is the selection of an appropriate analysis technique. Just as a carpenter has an array of tools available for use, the engineer has a wide range of methods at his disposal. Unfortunately, the choice that the engineer must make is not as easy as the decision between a hammer and a screwdriver. This paper will discuss the nature of the decision process and will provide numerous examples in which proper or improper decisions were made.

In general, the tools available to the process safety engineer can be classified based on their level of sophistication (ranging from simple to complex) and on their level of rigor (ranging from approximate to rigorous). For example, a pressure relief problem may be approached with techniques ranging from a simple set of equations for single-phase vapor relief to complex dynamic simulations involving two-phase relief with partial vapor-liquid disengagement. A dispersion analysis may be conducted utilizing Gaussian dispersion models for substances diluted in air, whereas a computational fluid dynamics model may be necessary for heavier-than-air mixtures encountering complex terrain or surroundings. With such a wide range of tools, how is a process safety practitioner to determine which tool is best for a given application?

For a process safety design problem, the optimal tool will give the lowest-cost design while ensuring that all assumptions are sufficiently conservative and that all constraints are satisfied. It should be noted that the total cost of the design also includes the cost of the engineering effort. Therefore, a more rigorous technique requiring higher engineering costs may not be worthwhile if the cost reduction in the resulting process design is minimal.

In other problems, an engineer must determine whether an existing installation sufficiently satisfies process safety requirements. For these situations, the optimal tool is the lowest-cost technique yielding a definitive answer on the acceptability of the existing installation. For example, conservative approximations may indicate that a design is insufficient. However, additional rigor or more accurate assumptions may reveal that the design is in fact acceptable. If the marginal costs of this additional rigor are less than the costs required to modify the existing installation, utilization of the more rigorous methodology was worthwhile.

The selection of the most appropriate analysis methodology is not always simple. With experience, a process engineer will be more capable of determining which technique should be utilized first for a given problem. In addition, it is not unusual for an engineer to analyze the same problem with multiple techniques before arriving at a final answer or final design. The following sections will further illustrate the importance of understanding the strengths and weaknesses of the available methodologies.

17.2
Flow through Nitrogen Piping During Distillation Column Pressurization

17.2.1
Initial Design Based on Steady-State Assumptions

The safety concept for a distillation process involving a thermally sensitive substance utilizes multiple independent protection layers. One of the protection layers provided in this process is an automated dump system that transfers the contents of the distillation column into a quench vessel. Since the distillation process is normally operated under vacuum, it is desirable to increase the column pressure with nitrogen prior to opening the dump valves. This increased pressure assists in the column draining process and also prevents the introduction of atmospheric oxygen into the column during the dump process.

Nitrogen is supplied to the facility via a pipeline with a nominal diameter of two inches (50 mm). In order to monitor the normal consumption of nitrogen in the facility, the pipeline is equipped with an orifice flowmeter with an orifice diameter of one-half inch (13 mm). This flowmeter is not intended to measure the flow of nitrogen during the column pressurization process since the flow is much greater than the normal nitrogen consumption but infrequent and short in duration (less than two minutes). An evaluation of the maximum nitrogen flow rate attainable through the orifice flowmeter showed a steady-state flow rate of approximately 525 $kg\,h^{-1}$. As this flow rate was within the desired range, the remaining piping to each of the distillation columns was designed to provide minimal flow resistance. Based on this, a piping diameter of two inches (50 mm) was selected. An actuated ball valve was placed immediately upstream from the nitrogen inlet to each column. The distance from the flowmeter to each valve is approximately 40 m. In order to confirm the adequacy of the design, the steady-state flow rate through the piping was calculated based on the nitrogen supply pressure. The final design is summarized in Figure 17.1 below.

The distillation columns and nitrogen system were installed and the plant was brought on line. Over the first four years of operation, the columns were operated without any apparent problems.

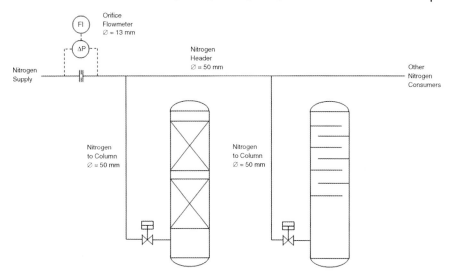

Figure 17.1 Nitrogen piping for distillation column pressurization.

17.2.2
Damage to Column Internals

Approximately four years into the life of the plant, the distillation columns were subjected to an internal inspection. The column inspections revealed that both columns had suffered damage to the internals. The column equipped with structured packing had a displaced packing support grid and damage to a liquid feed pipe and liquid distributor located above the bottom packed section. The column equipped with trays suffered from two displaced trays and deformed downcomers. Pictures of the damage are shown in Figures 17.2 and 17.3 below.

Once the column damage had been discovered, an analysis of the potential causes of the damage was conducted. Information from the plant data historian was reviewed to search for possible causes. No indication of liquid level above the reboiler return line or high column differential pressure during normal operation could be identified. The data historian did reveal that the column was experiencing high differential pressure across the packed sections during periods of nitrogen flow. Elevated differential pressure did not seem to be unusual for these periods of time, but the maximum pressure drop could not be ascertained due to the limited measurement range of the meters.

17.2.3
Dynamic Model of Nitrogen Flow Rates and Column Pressurization

Upon further investigation of the system design, it was realized that a large section of piping was located between the orifice flowmeter and the actuated valve at each column. Due to the decision to minimize flow resistance in the piping system downstream of the flowmeter, any nitrogen stored in this section of the piping would be pressurized but not subject to significant flow resistance upon opening of the valve. This compressed nitrogen represents a significant source of potential energy stored in the piping downstream of the flowmeter. It was suspected that the nitrogen in this section of piping could result in initial flow rates into the column that were significantly higher than the steady-state value.

In order to evaluate this theory, a dynamic model of the system was created in a custom modeling environment. The dynamic model includes the orifice flowmeter,

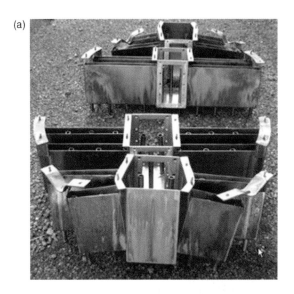

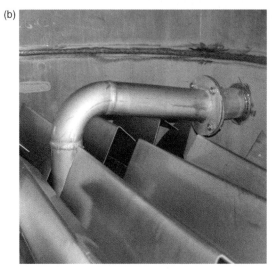

Figure 17.2 Damage to packed column internals. (a) liquid distributor damage, (b) feed pipe damage, (c) displaced support grid.

(c)

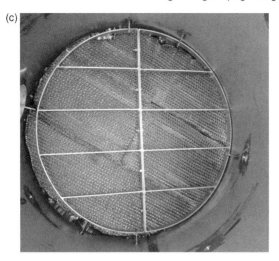

Figure 17.2 (*Continued*)

(a)

(b)

Figure 17.3 Damage to column trays. (a) displaced tray sections, (b) deformed distillation tray section.

the downstream piping, the actuated valve, and a simplified model of each distillation column vessel. In order to accurately represent the transient behavior of the system, the actuated valve was modeled utilizing the valve coefficient (K_v or C_v) provided by the valve manufacturer and the valve opening time measured prior to startup.

Figure 17.4 shows the nitrogen flow rate to the packed column as a function of time. When a dynamic model of the system is utilized, the flow rate begins at a very low value due to the opening characteristics of the actuated valve. As the valve moves toward the fully open position, the nitrogen downstream of the orifice flowmeter accelerates toward the column with minimal flow resistance. This results in a peak nitrogen flow rate of approximately 4750 kg h^{-1} (greater than nine times the steady-state value). It is very likely that these large nitrogen flow rates damaged the column internals after repeated column pressurization cycles.

In order to address the problem, an additional restriction orifice with a diameter of one-half inch (13 mm) was placed at the nitrogen inlet to both distillation columns. This resulted in a more constant nitrogen flow rate as shown in Figure 17.5. In addition, the coupled distillation column model shows that the extension of the column pressurization duration is minimal. The column pressure as a function of time is depicted in Figure 17.6. The addition of a restriction orifice increases the column pressurization time from 98 seconds to 109 seconds.

In this example, a simplified steady-state model was utilized in the original process design. Since a significant amount of pressurized nitrogen was located downstream of the components that provided a majority of the flow resistance, the nitrogen flow rate entering the column was not accurately estimated. As mentioned in the opening sections of this chapter, one reason for utilization of more rigorous methods is the violation of the assumptions of simplified approaches. In this example, the lack of recognition that the simplified model assumptions were being violated resulted in undesirable process behavior.

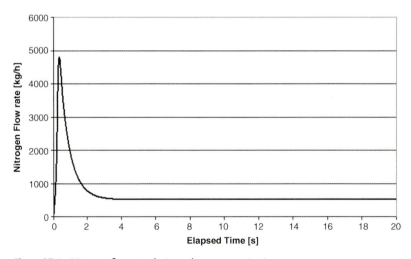

Figure 17.4 Nitrogen flow rate during column pressurization.

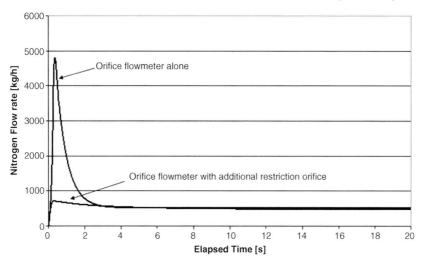

Figure 17.5 Effect of additional restriction orifice on nitrogen flow rate.

17.3
Tube Failure in a Wiped-Film Evaporator

17.3.1
Tube Failure – A Potentially Dangerous Overpressurization Scenario

When determining the required relieving rate for a relief device, Section VIII of the ASME Boiler and Pressure Vessel Code requires the consideration of internal failures in heat exchangers. The tube rupture scenario may be of particular interest in

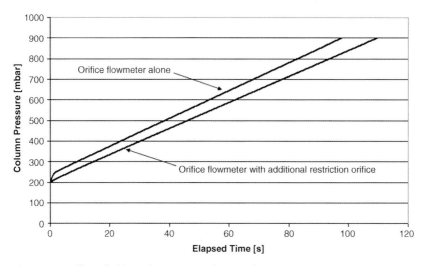

Figure 17.6 Effect of additional restriction orifice on column pressurization.

wiped-film evaporators equipped with internal condensers (also known as short-path evaporators).

In a short-path evaporator, the feed enters at the top of the evaporator and is distributed onto the heated walls where a thin film is created by rotating blades. Due to heat transfer from the jacket, vapors are generated that flow inward toward the condenser surface. The condensed vapors flow down the tube surface and exit the evaporator as a distilled product. The liquid feed continues to flow down the heated walls and ultimately exits the evaporator as the residue stream. Non-condensed vapors exit the short-path evaporator via the outlet to the vacuum system. A typical short-path evaporator design is shown in Figure 17.7.

As wiped-film evaporators are almost always operated under vacuum conditions, the design pressure of the evaporator shell is typically quite low (for example, 1 barg).

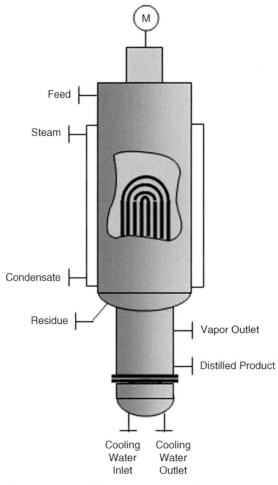

Figure 17.7 Typical short-path evaporator design.

If the wiped-film evaporator is normally operated at a temperature greater than the saturation temperature of the cooling medium at the design pressure, overpressurization of the evaporator can result from a tube failure.

In the design of a plant producing specialty esters, a short-path evaporator was incorporated into the product purification area of the process. The evaporator shell possesses a design pressure of 3.1 barg. The maximum normal operating temperature in the evaporator is 186 °C. The internal condenser is supplied with cooling water. The saturation pressure of water at 186 °C is 10.5 barg, which exceeds the vessel design pressure of 3.1 barg. Therefore, vessel overpressurization due to a tube failure is a credible scenario.

17.3.2
Required Relieving Rate Based on Steam Flow – An Unsafe Assumption

In many emergency relief scenarios, the design engineer is concerned with the maximum heat input resulting from the failure to the fully open position of a control valve that provides steam for heating. For many scenarios, the assumption of a fully open control valve is conservative. For the case of this evaporator, however, this assumption is not conservative and could easily lead to an undersized relief device.

Due to the construction of the evaporator and its relatively high operating temperature, a significant quantity of energy in the form of sensible heat is stored in the evaporator shell. For this particular evaporator, it was estimated that nearly 100 kg of the steel could quickly transfer energy in the form of sensible heat to the incoming water in the event of a tube failure. The transferred heat would result in vaporization of the incoming water and heating of the resulting steam.

As the thermal energy stored in the evaporator shell is located downstream of the steam control valve, the rate of release of this energy is not limited by the control valve. This is another example in which a simplifying assumption is not sufficiently conservative and must be replaced by satisfactory assumptions or more rigorous techniques.

17.3.3
Required Relieving Rate Based on Water Flow – An Expensive Assumption

The required relieving rate can be conservatively estimated by assuming that all water entering the vaporizer is instantaneously vaporized and must be vented. By assuming that the tube failure occurs as a fully split tube, an initial water flow rate of 29 100 kg h^{-1} enters the vaporizer. Under the assumption that sufficient heat is available from either the stored thermal energy in the evaporator shell or from additional steam condensing in the evaporator jacket, the required relieving rate can be set equal to the water flow rate. As shown in the next section, this is an example of an assumption that is excessively conservative and can be replaced with a more rigorous technique capable of calculating a less conservative relieving rate.

17.3.4
Dynamic Simulation of Wiped-Film Evaporator – An Optimal Solution

In order to yield an improved calculation of the required relieving rate, a dynamic simulation of the short path evaporator was developed. For purposes of the simulation, several important assumptions were required. It is assumed that the tube failure is the result of a clean break of a single tube. This results in an inlet area for water flow equal to twice the cross-sectional flow area of the tube. It is assumed that approximately 100 kg of steel in the evaporator shell is capable of transferring heat to the incoming water very quickly. For purposes of the simulation, it is assumed that the heat transfer coefficient is infinite. This is equivalent to an assumption in which the relevant portion of the evaporator shell and the evaporator contents remain at thermal equilibrium with one another throughout the tube failure event. In the dynamic model, the evaporator shell is protected by a rupture disk that opens at a specified pressure and allows steam to vent from the evaporator.

By solving differential equations for the mass and energy balances, the conditions in the evaporator can be calculated as a function of time. The figures below show the results for an appropriately sized rupture disk.

Figure 17.8 depicts the overall mass balance for the evaporator. Both the flow rate of water entering through the failed tube and the flow rate of generated steam exiting via the relief device are shown in this figure. It should be noted that the maximum relieving rate of steam in this figure (10 400 kg h^{-1}) is significantly lower than the maximum flow rate of water into the evaporator (29 100 kg h^{-1}). Therefore,

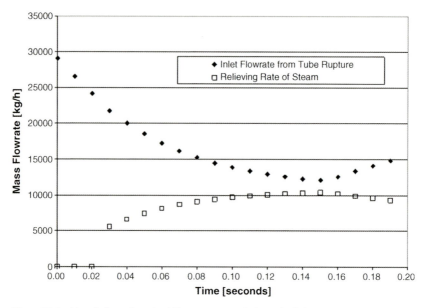

Figure 17.8 Mass balance for wiped-film evaporator during tube failure.

utilization of a more rigorous dynamic simulation allows the required relieving rate to be reduced by a factor of 2.8.

The temperature and pressure profiles during the tube failure event are depicted in Figure 17.9. As water begins to enter the evaporator, the temperature of the system immediately begins to decrease. The system temperature continues to decrease throughout the event. Due to vaporization of the incoming water and subsequent heating of the vapor, the pressure within the evaporator vessel increases rapidly. As the pressure increases, the relieving rate through the rupture disk begins to approach the flow rate of water entering through the failed tube. This results in a lower rate of pressure rise in the evaporator.

Although the steam generated in the evaporator is initially superheated, the incoming water quickly cools the steam to saturated conditions. Once saturated conditions have been reached, liquid water begins to accumulate in the evaporator. In addition to the liquid water entering through the failed tube, steam that was previously generated in the evaporator begins to condense. This condensation results in a decrease in the system pressure as seen in Figure 17.9. The formation of a liquid water phase in the evaporator can be observed in the graph of quality as a function of time in Figure 17.10.

In this case, utilization of a dynamic simulation is justified by the impact of the rigorous model on the required relieving rate (approximately a threefold reduction). It should be noted that the rate of pressure rise experienced when water is introduced into the wiped-film evaporator is extremely fast, with the peak pressure reached within 0.15 s of the tube failure. If economically feasible, the most desirable solution would be the utilization of a cooling medium with a lower vapor pressure that would not result in evaporator overpressurization in the event of a tube failure.

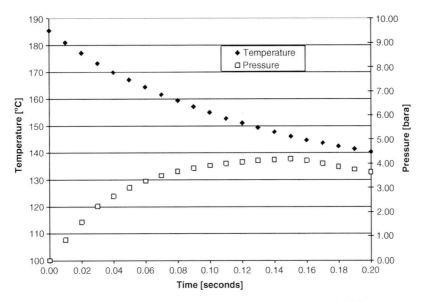

Figure 17.9 Temperature and pressure in wiped-film evaporator during tube failure.

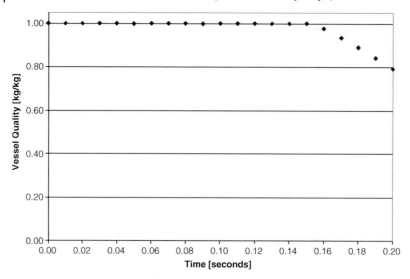

Figure 17.10 Quality of mixture in wiped-film evaporator during tube rupture.

17.4
Phenol-Formaldehyde Uncontrolled Exothermic Reaction

Leung [1] and D'Alessandro [2] both present example calculations pertaining to the uncontrolled exothermic reaction between phenol and formaldehyde.

17.4.1
Assumptions Regarding Single-Phase Venting

When evaluating emergency relief systems, the possibility of two-phase venting should be carefully considered. If two-phase flow does occur, the assumption of single-phase vapor venting can easily lead to undersized relief devices that will not be capable of maintaining the vessel pressure within acceptable limits. Based on a 10% overpressure in the example of the phenol-formaldehyde reaction, Leung shows that the required vent areas for vapor-only venting and homogeneous venting are 0.003 m^2 and 0.05 m^2, respectively. It is obvious that the failure to adequately consider the possibility of two-phase venting phenomena may lead to drastically undersized relief devices. The possibility of two-phase flow would dictate the use of methods that do not rely on the assumption of single-phase venting.

17.4.2
Will Two-Phase Venting Occur?

Based on fluid characteristics such as foaming behavior and viscosity, the design engineer can make basic assumptions regarding the nature of the two-phase system. For example, if the system has a strong foaming tendency, homogeneous venting behavior should be considered. If the system exhibits high viscosities, a viscous bubbly behavior is likely to be considered. For fluids that are well-behaved with low viscosities and no foaming tendencies, the assumption of churn-turbulent behavior is probably appropriate.

During homogeneous venting, the vapor void fraction is equal throughout the vessel. However, for fluids exhibiting churn-turbulent or viscous bubbly behavior, partial vapor-liquid disengagement will occur. Partial disengagement of the vapor and liquid results in a vapor void fraction in the relief device inlet that is greater than the average vapor void fraction within the vessel. Due to this partial disengagement, vessels exhibiting churn-turbulent or viscous bubbly behavior must contain a sufficient quantity of liquid in order for two-phase venting to occur.

In order to assess whether two-phase flow will occur for a given churn-turbulent or bubbly system, a two-phase disengagement diagram should be utilized. The axes of such diagrams are dimensionless superficial vapor velocity and average vapor void fraction. A curve on the diagram serves as the boundary between two-phase and single-phase venting regions. This curve is based on the average vessel void fraction at full disengagement as a function of the dimensionless superficial vapor velocity. D'Alessandro [3] presents these disengagement diagrams for both churn-turbulent and bubbly systems for various values of the so-called disengagement parameter (C_o).

17.4.3
Effect of Disengagement Behavior on Required Relieving Rate and Area

Once the design engineer has determined that some degree of vapor-liquid disengagement is likely to occur, the calculation of required relieving rates and relieving areas can proceed based on the appropriate partial disengagement behavior.

For the example of the phenol-formaldehyde reaction, the homogeneous calculation of Leung yields a required relieving rate of $150 \, \text{kg s}^{-1}$. When utilizing the most conservative values of the disengagement parameters, the required relieving rates presented by D'Alessandro for churn-turbulent and bubbly behaviors are $25 \, \text{kg s}^{-1}$ and $91 \, \text{kg s}^{-1}$, respectively. When utilizing the values recommended by the Design Institute for Emergency Relief Systems (DIERS) for the disengagement parameters, the required relieving rates presented by D'Alessandro for churn-turbulent and bubbly behaviors are $19 \, \text{kg s}^{-1}$ and $82 \, \text{kg s}^{-1}$, respectively. It is apparent that the selection of vessel disengagement behavior has a significant role in the required relieving rate.

For a system involving the polymerization of styrene, Leung [4] compares the required relieving areas for vapor-only venting with those for churn-turbulent, bubbly, and homogeneous models. Based on an overpressure of 20%, Leung reports that the churn-turbulent area is nearly twice the area for vapor-only venting and that the bubbly area is approximately three times the area for vapor-only venting. The required area for homogeneous venting is slightly higher than the area required for bubbly behavior.

In summary, the consideration of two-phase venting has a significant effect on relief device sizing. Failure to evaluate the assumptions necessary to ensure vapor-only venting may result in use of vapor-only methods for systems exhibiting two-phase flow. This will lead to undersized relief devices. Failure to consider the effects of partial disengagement (churn-turbulent or bubbly) when evaluating two-phase relief may lead to excessively high required relieving rates and excessively large relief devices.

17.5
Computational Fluid Dynamics – Is It Ever Necessary?

The problems in the previous examples in this paper could be adequately solved with either simplified methods or more rigorous methods such as dynamic simulation. However, none of the previous examples required the use of computational fluid dynamics (CFD). Does CFD have a role in process safety? The answer is clear – 'Yes.' CFD simulation is yet another tool in the process safety toolbox. Are there more process safety problems that can benefit from CFD than those that cannot? This answer is also clear – 'No.' Just as dynamic simulation finds limited use within the area of process safety, the situation is similar for CFD. These are not tools that should be used unnecessarily but should be used when they are the best or only tool for the job.

The following sections illustrate some process safety challenges in which computational fluid dynamics have been appropriately utilized.

17.5.1
Design of Storage Tanks for Thermally Sensitive Liquids

Temperature monitoring during storage of many thermally sensitive fluids is important because of the potential increase in temperature due to self-reaction (for example, polymerization) or decomposition. If a temperature increase due to reaction is detected promptly, actions can be taken to prevent the scenario from elevating to a more hazardous condition.

In this particular example, a thermally sensitive product is stored in a low-pressure tank. The tank is equipped with an external circulation pump that returns the circulated liquid back into the tank via mixing eductors. In order to detect a temperature rise in the vessel as early as possible, it is desirable to position the temperature measurement locations at points in the vessel that will experience the highest temperature. The optimal location is affected by the fluid circulation patterns, the heat generation rate based on reaction kinetics and reaction energy, and the loss of heat via external cooling or heat exchange with the environment. Figure 17.11 shows the temperature distribution in the tank based on CFD results and the recommended temperature measurement locations.

17.5.2
Dispersion of Sprayed Droplets during Application of a Surface Coating

Not only is computational fluid dynamics (CFD) a valuable tool for the chemical processing industry, but it may also assist in better understanding the applications of the

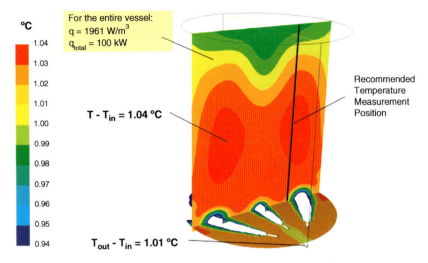

Figure 17.11 Temperature distribution in storage tank of thermally sensitive liquid.

products that the chemical industry produces. In this particular example, a product utilized as a surface coating is applied to a building exterior via a spray device. The spray device is designed to produce a spray of droplets with a defined droplet size distribution.

In order to better understand the efficiency of the product application and the amount of the product entering the environment, a CFD simulation was developed. This simulation utilizes the droplet size distribution, the building geometry, the presence of the person operating the spray device, and the wind characteristics to model the trajectories of sprayed droplets and to estimate the fraction of the sprayed droplets that reach the intended surface. Figure 17.12 shows typical results from this study.

17.5.3
Dispersion of Heat and Chemical Substances

Computational fluid dynamics is an ideal tool for studying many dispersion phenomena. These can include dispersion of chemical substances or heat. These tools are particularly important when the assumptions of simplified dispersion methods cannot be satisfied. For example, dispersion analysis via CFD is of particular interest when complex terrain or building structures are involved or when heavier-than-air gases are being dispersed. Figure 17.13 depicts the CFD results of a simulation involving thermal dispersion.

17.6
Computational Fluid Dynamics – The Future of Safety Technology?

Computational fluid dynamics (CFD) belongs not only to the future of process safety but also to the present. As one of many tools in the toolbox of the process

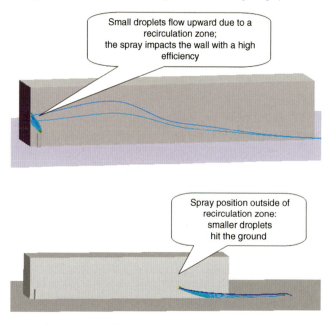

Figure 17.12 Simulation of sprayed droplet trajectories for surface coating application.

safety practitioner, it is already playing a valuable role in the field and should continue to grow. Increased utilization of CFD in the area of process safety will be driven by a number of factors. An increased interest in process safety problems particularly suited for analysis by CFD could drive an increase. As an example, the increasing focus on consequence analysis could lead to increased application of CFD for

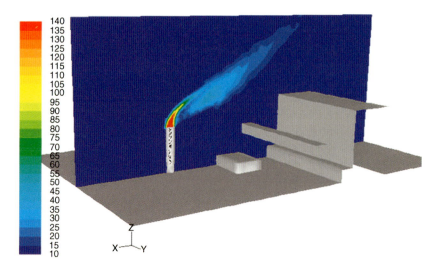

Figure 17.13 Temperature distribution resulting from thermal dispersion of stack effluent.

dispersion modeling. In addition, increasing computing power and improved CFD algorithms or codes should continue to increase the attractiveness for CFD as the tool of choice for process safety problems suitable to multiple solution techniques.

Acknowledgments

The author gratefully acknowledges the contributions of computational fluid dynamics examples that were originally prepared by Catrin Becker and Karsten Riest of Evonik Degussa.

References

1 Leung, J.C. (1986) Simplified vent sizing equations for emergency relief requirements in reactors and storage vessels. *AIChE Journal*, **32** (10), 1622–1634.

2 D'Alessandro, R. (2004) Emergency venting requirements for tempered systems considering partial vapor-liquid separation with disengagement parameters greater than unity – Part II: Application. *Process Safety Progress.*, **23** (2), 86–98.

3 D'Alessandro, R. (2004) Emergency venting requirements for tempered systems considering partial vapor-liquid separation with disengagement parameters greater than unity – Part I: Model development. *Process Safety Progress*, **23** (1), 1–15.

4 Leung, J.C., (1987) Overpressure during emergency relief venting in bubbly and churn-turbulent flows. *AIChE Journal*, **33**, **(6)**, 952–958.

18
CFD for Reconstruction of the Buncefield Incident*

Simon E. Gant and G.T. Atkinson

18.1
Introduction

On the morning of 11 December 2005 there was a large explosion and fire at the Buncefield Oil Storage Depot near Hemel Hempstead, UK[1]) (see Figure 18.1). The explosion was the largest in peacetime Europe, registering 2.4 on the Richter scale and breaking windows up to 1 km away. The subsequent fire burned for several days, destroying most of the site. Over 40 people were injured and there was significant damage to commercial and residential property, but fortunately no fatalities. The cost of the incident in terms of the damage to property and loss of business is estimated to have been in the region of £1 bn.

Examination of the pumping records for one of the pipelines importing fuel into the storage depot at the time of the incident indicated there was an overfilling of one of the large tanks (Tank 912) used to store unleaded petrol shortly before the explosion [1]. From the time that the tank started overflowing until the explosion at 06:01, approximately 260 m^3 or 180 tonnes of petrol was released from the tank. Images from CCTV cameras in the minutes leading up to the explosion showed a visible, low-lying mist flowing in all directions away from the bund in which the overflowing tank was situated. This mist flowed along the ground across the site and into neighboring areas, filling the car parks of two office and manufacturing units. It is thought that this mist coincided with a flammable mixture of hydrocarbon vapor, which had been generated by the overflowing tank. The cloud ignited, probably from an electrical source in an emergency pump room, leading to a severe explosion, and subsequently a large fire.

Since the Buncefield Incident took place in 2005, there have been two large-scale incidents with striking similarities: the Caribbean Petroleum Corporation fuel depot incident in Puerto Rico on 23 October 2009 and the Indian Oil Corporation fuel depot incident at Jaipur, India, on 29 October 2009. In the latter case, 11 people were killed and over 100 injured.

The present paper describes one element of HSL's investigation into the Buncefield Incident, namely the use of CFD to study the dispersion of flammable vapor

1) http://www.buncefieldinvestigation.gov.uk, accessed November 2010.
*Crown Copyright (2012).

Process and Plant Safety: Applying Computational Fluid Dynamics, First Edition. Edited by J. Schmidt.
© 2012 Wiley-VCH Verlag GmbH & Co. KGaA. Published 2012 by Wiley-VCH Verlag GmbH & Co. KGaA.

Figure 18.1 Aerial photo of the Buncefield site on fire after the explosion.

from the overfilling storage tank. This work was undertaken to understand how the vapor cloud spread over such a large area and to provide data that could be used for explosion modeling studies. Although the simulations were performed in 2006, they have not been presented publicly previously due to restrictions imposed by the legal proceedings. Other research undertaken by HSL into the source of flammable vapor from overfilling tank releases has been published by Atkinson et al. [2] and Coldrick et al. [3], and the investigation into the explosion mechanism is ongoing [4, 5].

The first section of this paper considers the incident CCTV records and presents maps showing the extent of the visible mist. The second section concentrates on the CFD modeling, describing the underlying physics of the model, sensitivity tests, source conditions, and the results from dispersion simulations. Finally, conclusions and recommendations for further work are presented.

18.2
Observations from the CCTV Records

18.2.1
Progress of the Mist

An aerial photograph of the Buncefield site prior to the incident is shown in Figure 18.2. The oil storage depot and neighboring office buildings were equipped with more than a dozen CCTV cameras that recorded images of a visible mist shortly before the explosion. The mist first appeared flowing over the north-western edge of Hertfordshire Oil Storage Limited (HOSL) Bund A towards the water tank at approximately 05:38 (Figure 18.3). Five minutes after its first appearance from Bund A, it arrived in the Fuji car park and the north-eastern corner of the Northgate car park (Figure 18.4). It then proceeded to spread across and fill the car parks. The mist front traveling down the road to the north of HOSL Bund B was estimated at 05:43 to have an average speed of approximately $0.6\,\mathrm{ms}^{-1}$. From the numerous cameras around the site, a map showing the arrival times of the mist has been produced (Figure 18.5).

18.2 *Observations from the CCTV Records* | 315

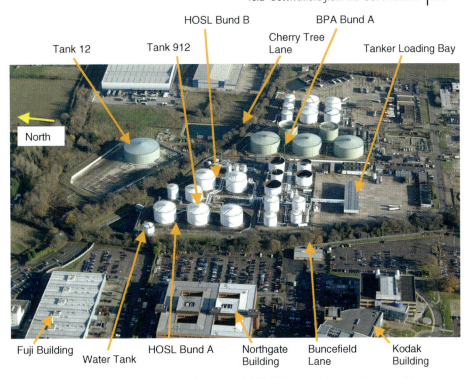

Figure 18.2 Aerial view looking east toward the Buncefield fuel depot and surrounding areas. The photo was taken in 2004, before the incident.

Figure 18.3 Images from CCTV Camera 5 looking north toward the water tank. The solid lines are added to help indicate the location of the mist front. Times shown at the top of each frame are approximately 17 min fast.

316 | *18 CFD for Reconstruction of the Buncefield Incident*

Figure 18.4 Images from Camera T8 looking due west showing the mist in the Fuji car park (to the right of the picture). The times shown at the top of each frame are correct.

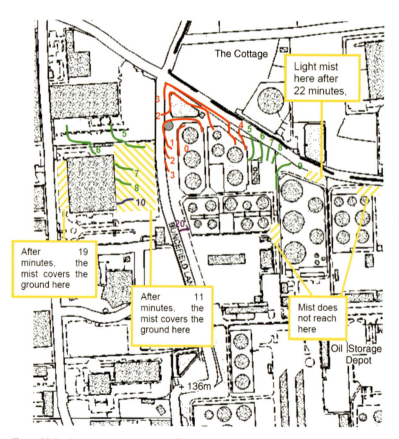

Figure 18.5 Approximate positions of the visible mist over time as indicated by the CCTV records. Numbers indicate the time in minutes from the instant when the mist started to overtop the bund wall.

18.2.2
Wind Speed

There were no measurements of wind speed taken on site at the time of the incident. The nearest meteorological measurements at Luton and Northolt airports recorded the weather conditions to be calm, cold, stable, and humid. The movement of the mist as indicated by the CCTV cameras across the Buncefield site appears to be consistent with completely zero-wind speed conditions.

18.2.3
Final Extent of the Mist

The final mist depths in the moments immediately before the explosion are shown on the map in Figure 18.6. These depths have been estimated from the relative size of neighboring objects in the CCTV footage and, due to the variation in illumination levels across the site, the values are subject to a degree of uncertainty. Mist depths are only given to the nearest meter and the results should be interpreted accordingly. The final images taken of the Northgate and Fuji car parks were recorded around 4 min before the explosion. Mist depths could therefore have increased here beyond the levels shown in Figure 18.6.

The area immediately adjacent to HOSL Bund A and to the north, around the water tank and the lagoon, was covered with a mist around 4 m deep moments before the explosion. Across the Northgate and Fuji car parks, the mist depth reached between

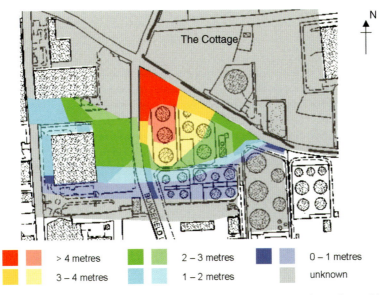

Figure 18.6 Approximate final depths of the visible mist. Solid colours indicate observed depths of the mist taken directly from the CCTV cameras. Transparent colors indicate cloud depths inferred from the surrounding levels.

2 and 3 m. The mist appears not to have extended to the control room at the south edge of the HOSL site, although witnesses at the tanker loading gantry reported wisps of mist being visible at their feet moments before the explosion.

The extent of the mist correlated reasonably well with the areas burned in the subsequent explosion and fire [1]. The exception is the car park area to the west of the Northgate building where the mist was approximately 2 m deep but there was no burn damage. This would suggest that the concentration of vapor here was below the Lower Explosive Limit (LEL).

18.2.4
What Was the Visible Mist?

When the liquid petrol cascaded from the roof of the tank, evaporation of the more volatile fractions will have lowered its temperature. Analysis of droplet evaporation indicates that this would probably have led to temperatures dropping below 0 °C and perhaps as low as −10 °C in the surrounding vapor phase. A sudden drop in temperature will have caused water vapor to condense out of the air and form very small droplets or perhaps ice crystals (essentially, fog). As the petrol vapor/mist spilled over the bund wall, it will have dispersed across the site with the visible mist appearing at the same time as the flammable vapor, the two being combined together in the gravity current.

18.3
CFD Modeling of the Vapor Cloud Dispersion

18.3.1
Initial Model Tests

To perform CFD simulations of the vapor cloud dispersion at Buncefield across the whole area covered by the visible mist is challenging because of the relatively large flow domain and the long duration of the release. In order to have good spatial and temporal resolution, the computing effort required is significant. Moreover, there are a number of uncertainties, including specifying the correct source conditions for the vapor and approximating the effect of obstructions to the vapor flow.

To address these issues, tests were first performed with a simplified CFD model of the whole site to assess the feasibility of performing detailed dispersion simulations for the full duration of the release. Results from this first assessment are not reported here, but they showed that it was possible to run the calculations, although they would be at the limit of what could be achieved using the available computing resources at HSL, with a calculation time in the order of weeks. In the second step, sensitivity studies were undertaken where each of the model uncertainties were assessed in isolation, to understand which factors were important and would require further attention and which factors could effectively be disregarded. Rather than run these tests using the full-scale model for the whole site, which would have taken many months, only a small region of the site was simulated. Having obtained the results from these tests, a final detailed CFD model was built for the whole site and simulations run for the full duration of the release.

18.3.2
Vapor Source Term

To calculate the amount of vapor produced by the overfilling tank, the method described by Atkinson et al. [2] was used. This approach assumes that the spray droplets released from the tank and the surrounding vapor reach equilibrium conditions. The rate at which petrol was released from the tank was known from the pumping records. The rate at which fresh air was entrained into the overflowing tank cascade was calculated using simple CFD simulations of a downward-directed spray with the same cross-sectional area as that observed in tank overfilling experiments. Having established the quantity of air entrained into the spray, equilibrium calculations were performed to arrive at the source conditions, based on the simplified petrol composition which was derived from analysis of the winter petrol that was being pumped into Tank 912 at the time of the incident.

The source conditions for the dispersion model were complicated by the fact that the filling rate to Tank 912 increased during the period of the release. Approximately 8 min before the explosion, the flow rate increased from an average of 550 $m^3 h^{-1}$ to around 920 $m^3 h^{-1}$. The vapor source for the CFD model was adjusted accordingly (see Table 18.1).

Table 18.1 Summary of petrol vapor source conditions.

Petrol pumping rate	550 $m^3 h^{-1}$	920 $m^3 h^{-1}$
Hydrocarbon vapor mass fraction		
Butane	5.9% (w/w)	7.5% (w/w)
Pentane	6.3% (w/w)	7.2% (w/w)
Hexane	2.2% (w/w)	2.3% (w/w)
TOTAL	14.4% (w/w)	17.0% (w/w)
Hydrocarbon vapor molar fraction		
Butane	3.2% (mol/mol)	4.2% (mol/mol)
Pentane	2.8% (mol/mol)	3.2% (mol/mol)
Hexane	0.8% (mol/mol)	0.9% (mol/mol)
TOTAL	6.8% (mol/mol)	8.3% (mol/mol)
Lower explosive limit of mixture	1.6% (mol/mol)	1.6% (mol/mol)
Upper explosive limit of mixture	8.1% (mol/mol)	8.1% (mol/mol)
Stoichiometric condition	2.7% (mol/mol)	2.7% (mol/mol)
Temperature	−9.1 °C	−7.6 °C
Density	1.46 kg m^{-3}	1.47 kg m^{-3}
Mass flow rate of air/vapor mixture	139 kg s^{-1}	174 kg s^{-1}
Volume flow rate of air/vapor mixture	95.7 $m^3 s^{-1}$	118 $m^3 s^{-1}$
Area covered to 2 m depth after 23 min (equivalent area)	66 000 m^2 (257 m × 257 m)	81 400 m^2 (285 m × 285 m)

As a check on the source term calculations, the area of ground that would be covered by a layer of vapor 2 m deep after a 23 min release was calculated, assuming that there was no further air entrained into the vapor cloud (see Table 18.1). For the 550 m^{3}h^{-1} and 920 m^{3}h^{-1} release rates, these areas were 257 × 257 m^2 and 285 × 285 m^2, respectively. The cloud sizes are smaller than the extent of the mist visible in the CCTV records (approx. 500 × 400 m^2). However, some further air entrainment into the cloud would have occurred, and the depth of the visible mist was non-uniform. The vapor production rates therefore seem reasonable from this simple comparison.

18.3.3
CFD Model Description

The source conditions described above were used to set the velocity, composition, and temperature of the vapor that was released in the CFD model. The gas was released uniformly from the floor of HOSL Bund A in the model to produce a steady upwelling flow. This approach did not capture any large-scale unsteady motion within the bund which may have resulted from the cascades of liquid petrol produced by the overflowing tank. However, it was assumed that this would not affect the vapor cloud dispersion in the far field.

The CFD model was developed using the commercial CFD code, ANSYS-CFX version 10. To model the effect of turbulence on the mean flow, the SST model [6] was used with a wall treatment that switched automatically from a low-Reynolds-number approach to a turbulent-wall function based on the local flow conditions and near-wall cell size. Since the vapor was initially 11% denser than the surrounding air, a non-Boussinesq treatment was adopted to account for density variations in the governing equations. The temperature of the ground was fixed at the ambient temperature of 0 °C, and zero-pressure entrainment boundaries were used for the top and sides of the flow domain to allow flow to enter or escape the computational domain unimpeded. Second-order accurate numerical schemes were used for both the spatial and temporal discretization, and the computational time-step chosen was 0.5 s.

To model the various hedges, such as those running alongside the Buncefield and Cherry Tree Lanes (Figure 18.2), it was not feasible to use a fine grid to resolve individual trees or shrubs. Instead, they were modeled as continuous porous regions with dimensions roughly equivalent to the hedges (details are provided below).

18.3.4
Sensitivity Tests

Sensitivity tests were conducted to assess the influence of the following parameters: grid resolution, turbulence, ground topology, the presence of hedges and obstacles, and ground surface roughness. These test simulations were performed using only a small region of the site, covering HOSL Bund A and the area to the west across the Northgate and Fuji car parks, as shown in Figure 18.7. Since the dense vapor cloud flowed along the ground and there was no wind, the domain only extended 5 m in the vertical direction.

inferred from the surrounding levels.

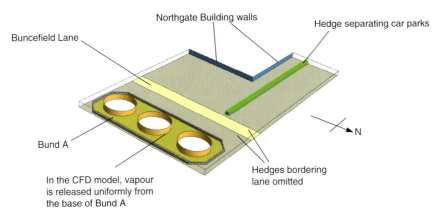

Figure 18.7 CFD model geometry used in the sensitivity tests.

18.3.4.1 Grid Resolution

Because of the very large domain required for the Buncefield simulations it was recognized that it would be very difficult or impossible to achieve a fully grid-independent solution. A pragmatic approach was therefore taken where the finest possible grid resolution was used given the available computing resources and time available, and tests were undertaken to assess the likely magnitude of errors due to possible under-resolution. A similar approach was adopted by Scargiali *et al.* [7] in their simulations of dense vapor dispersion over a $30 \times 30\,km^2$ area.

Two grids were used to assess the sensitivity to the grid. To achieve the best resolution possible of the vertical span, prism-shaped cells were used for the majority of the domain. The coarse mesh used 130,000 nodes with a first cell height of 15 cm. The finer mesh used 720,000 nodes with a first cell height of 7 cm. Results from the two simulations are shown in Figure 18.8. The dense cloud traveled across the Northgate car park faster with the coarse grid than with the fine, a difference in the distance traveled of around 15% after 4 minutes.

18.3.4.2 Turbulence

The release of liquid petrol from Tank 912 will have induced air currents and turbulence within Bund A. Characterizing this turbulence level for the vapor source is extremely difficult. To examine this uncertainty, two simulations were performed with different turbulence levels in the source flow of vapor. The first simulation used no turbulent fluctuations in the vapor source, that is, the flow was assumed to be laminar, and the second used an inlet turbulence intensity of 10% (considered as a relatively high level for many industrial turbulent flows). The results shown in Figure 18.9 indicated that the predicted maximum extent of the cloud was not affected strongly by the initial turbulence level, although the concentrations within

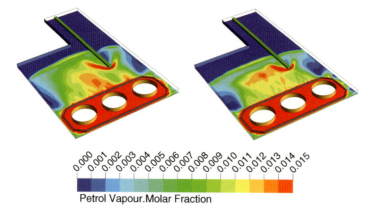

Figure 18.8 Petrol vapor concentrations on a plane 0.5 m above ground level for (a) the coarse and (b) fine grids, 4 min after the start of the release.

the cloud were modified slightly. In the final detailed CFD model, presented in Section 3.5, a turbulence intensity of 5% was used.

18.3.4.3 Ground Topology

The effect of a slope in the ground was assessed by tilting the whole of the CFD geometry to an angle of 2.86° down towards the north, equivalent to a slope of 1 : 20. This does not reflect the actual slope of the ground around HOSL Bund A, which was relatively flat. It does, however, represent a typical slope that could be encountered elsewhere on the site. Figure 18.10 compares results from using the 1 : 20 slope to those obtained using the previous flat-ground model. The difference is very significant, with the overwhelming majority of the cloud flowing directly northwards rather than west across the car parks as before. The importance of the slope of the ground was also reflected in the CCTV records of the incident. These results indicated that in

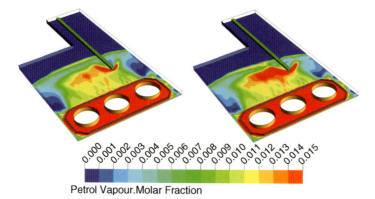

Figure 18.9 Petrol vapor concentrations on a plane 0.5 m above ground level for (a) the laminar treatment and (b) with an inlet turbulence intensity of 10%.

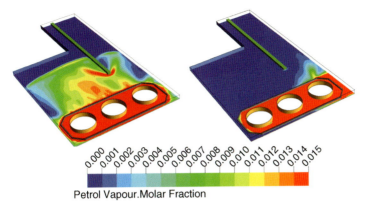

Figure 18.10 Petrol vapor concentrations on a plane 0.5 m above ground level for (a) completely flat ground and (b) for a 1:20 or 2.86° slope.

the final dispersion simulations it was important to represent the shape of the terrain as accurately as possible.

18.3.4.4 Hedges and Obstacles

The effect of hedges and obstructions on the vapor cloud flow was assessed by running simulations with and without hedges on either side of Buncefield Lane. The drag force produced by the trees and hedges was the subject of some uncertainty. CFD simulations were therefore performed using two different flow resistances that were calculated using the empirical Ergun correlation [8] with volumetric porosities, ε, and characteristic length scales, L, of $\varepsilon = 50\%$ and $L = 10$ cm (for the less porous hedge) and $\varepsilon = 70\%$ and $L = 30$ cm (for the more porous hedge). The results from these simulations (Figure 18.11) showed that the hedges clearly had a significant effect on the spreading of the dense cloud. This was also confirmed by the CCTV

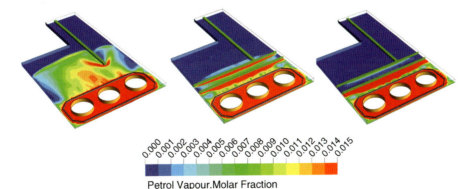

Figure 18.11 Petrol vapor concentrations on a plane 0.5 m above ground level for (a) an unobstructed release, (b) with porous hedges alongside Buncefield lane and (c) for less porous hedges.

records of the incident, where the cloud depth in the Fuji car park appeared at times to be significantly deeper than that in the adjoining Northgate car park, the two areas being separated by an evergreen hedge.

Having established that hedges might have a significant effect on the spreading of the vapor cloud, a more in-depth study of the literature was undertaken, which established that an alternative empirical correlation from Hoerner [9] might be more appropriate for hedges:

$$C_2 = \frac{1}{2}\left[\frac{3}{2\varepsilon} - 1\right]^2 \qquad (18.1)$$

where ε is the porosity, and the C_2 coefficient is used in a quadratic resistance term in the momentum equations. This model was previously used by Li et al. [10] to simulate the flow through windbreaks using CFD. In the dispersion simulations for the whole Buncefield site, presented in Section 3.5, Eq. (18.1) was used with porosities of 40 and 50%. These values were taken from Bean et al. [11] for 'medium porous' shelterbelts. The lower porosity of 40% was used for the hedge between the Northgate and Fuji car parks and the hedge in the northern half of the Fuji car park, which were composed of a dense evergreen laurel.

18.3.4.5 Ground Surface Roughness

The effect of surface roughness on the vapor cloud dispersion was assessed by performing simulations with different roughness heights of $h = 0.1$ mm and $h = 1.0$ mm. The roughness height does not refer here to the actual height of the roughness elements on the ground, but is a factor of around 10 times smaller than the real-life obstacles (see, for example, Blackadar [12]). Values of $h = 0.1$ mm and $h = 1.0$ mm therefore roughly correspond to the surface of tarmac and a lawn with coarse grass, respectively. The results from dispersion simulations using these surface roughnesses are compared to those obtained using perfectly smooth terrain in Figure 18.12. Unlike all of the previous tests, these simulations were performed using the k-ε turbulence model, rather than the SST model. There was hardly any perceptible

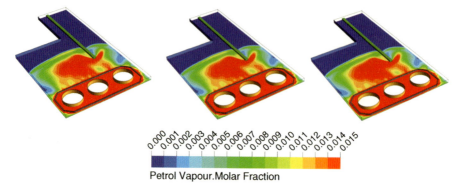

Figure 18.12 Petrol vapor concentrations on a plane 0.5 m above ground level for (a) smooth ground, (b) ground with 0.1 mm surface roughness and (c) 1 mm surface roughness.

difference between the results with different ground roughnesses, and for the final dispersion model the terrain was simply treated as smooth.

18.3.4.6 Summary of Sensitivity Tests

In summary, the sensitivity tests showed that there were two major factors influencing the flow of vapor over the ground: the slope of the terrain and the presence of obstacles such as hedges. Turbulence levels and the grid resolution had some limited effects, but predictions appeared to show that the flow was relatively insensitive to the surface roughness of the ground.

18.3.5
Final Dispersion Simulations

The sensitivity tests indicated that it was important to capture with as much detail as possible the slope of the terrain in order to obtain accurate predictions of the vapor cloud dispersion. To achieve this, topographical data from a photogrammetric survey of the site was used to build the CFD model geometry. The laborious process of converting the site surface levels from this survey data to the CFD geometry introduced some localized anomalies, such as some small pits and bumps. The general site surface levels were, however, faithfully reproduced. The contour data omitted details of the hedges, bund walls, and buildings. These were added subsequently by hand, using plans and photos of the site taken prior to the incident as a guide. Two porous regions representing the resistance to the flow due to pipework and obstacles in the loading bay area and the HOSL Bund C area were also included in the model.

The computational grid used for the final dispersion simulations is shown in Figure 18.13. The top surface of the domain was flat, and therefore the domain depth varied from approximately 5 m in the south to a maximum of just over 10 m in the east, where the ground level was lowest. The height of the first cell was 12 cm, and in total the grid comprised 1.5 million nodes. This was the maximum resolution that

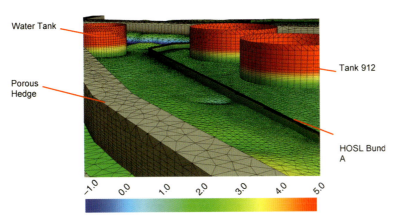

Figure 18.13 The computational grid near Bund A with the terrain and walls colored according to the height in meters.

Figure 18.14 Dispersion of the vapor cloud over time. The ground is colored according to elevation, where dark red to dark blue represents a difference in height of 6 m. Numbers shown indicate the time elapsed in minutes from the moment when the cloud first started flowing over the bund wall.

could be run given the available computing resources. A single simulation took 3 weeks, running in parallel on six 3.6 GHz and 3.8 GHz Xeon processors.

Figure 18.14 shows the CFD model results for the first 6 minutes of the release. The cloud shown in these plots is defined by an iso-surface at the LEL for the petrol mixture of 1.6% mol/mol. In order to compare to the observations from the CCTV records, the moment when the cloud started to flow over the bund wall has been called 'time zero' ($t = 0$). The predicted progress of the cloud for the full duration of the release is summarized in Figure 18.15.

The spreading rate of the vapor cloud was predicted by the CFD model to be generally slightly faster than observed in the CCTV records. In particular, across the Northgate car park the CFD model predicted the cloud to reach the Northgate building at $t = 4$ min, whereas the CCTV records showed it reaching there at $t = 6$ min. The shape of the cloud was also slightly different across the Northgate and Fuji car parks: the CFD model did not capture the observed faster progress of the cloud across the Fuji car park than across the Northgate car park, but instead predicted the speed to be approximately the same in the two areas.

Despite these differences, the CFD model predictions produced generally good qualitative agreement with the CCTV observations. The modeled cloud dispersed slowly in the southerly direction, similar to that observed in the CCTV records. In the north-east of the site, although the cloud arrived slightly too early in the CFD simulations, it was predicted to flow through the hedge into Cherry Tree Lane and towards the lagoon rather than propagate further east along the north of BPA Bund A, similar to the behavior observed in the CCTV records.

In the analysis of the CCTV records, it was estimated that the mist traveled at around 0.6 ms^{-1} as it flowed along the road to the north of HOSL Bund B. The CFD model predicted the front speed in this area to be between 0.4 ms^{-1} and 0.6 ms^{-1}.

18.3 CFD Modeling of the Vapor Cloud Dispersion | 327

Figure 18.15 CFD predictions of the progress of the vapor cloud across the Buncefield site. Times shown are in minutes from the moment the mist appeared over the wall of Bund A.

The predicted final depths of the vapor cloud across the site are shown in Figure 18.16. The depth of the cloud immediately adjacent to Bund A was underpredicted somewhat, mainly towards the north near the water tank, where the mist level was more than 4 m deep in the CCTV records but only around 3 m deep in the CFD model. However, there was generally good agreement between the CFD simulations and the observed behavior elsewhere, particularly in the south and east. In the west, the CFD model predicted that the vapor cloud would fill the small car park area on the western side of the Northgate building to a depth of between 1 and 2

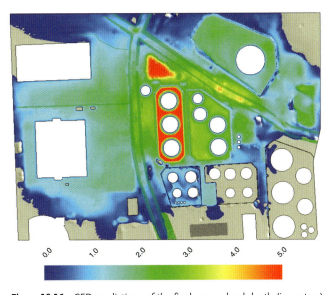

Figure 18.16 CFD predictions of the final vapor cloud depth (in meters).

meters. Although the observed mist level in this area reached around 2 meters, the burn damage suggested that the vapor concentration was below the LEL.

Care should be exercised when comparing the extent of the vapor cloud in the CCTV records with the CFD results. Firstly, estimating cloud depths from the CCTV records was subject to some uncertainty due to the variation of the lighting level across the site. Secondly, the CFD results shown in Figures 18.14–18.16 are based on a definition of the cloud as an iso-surface at 1.6% mol/mol petrol vapor. It is not possible to establish at what vapor concentration the mist became visible in the CCTV footage. However, results that were produced with different iso-surfaces of 1% and 2.7% mol/mol showed similar trends in overall behavior.

18.4
Conclusions

This paper has presented an analysis of the dispersion of mist in CCTV footage of the Buncefield Incident and CFD simulations of the vapor dispersion. The CCTV records showed that the mist produced by the petrol cascade from the overflowing tank spread over a period of approximately 23 minutes to cover an area measuring approximately 500 m by 400 m. In the final moments before the explosion, it was over 4 m deep immediately adjacent to HOSL Bund A and between 2 m and 3 m deep in the main Northgate and Fuji car park areas.

The CFD simulations showed that the main factors affecting the dispersion of vapor were topography and the presence of obstacles, such as hedges. The dispersion behavior was found to be relatively insensitive to the surface roughness of the ground, while turbulence levels and the computational grid resolution used in the simulations had limited effects. The model results showed similar qualitative behavior to that observed in the incident in terms of the arrival time of the cloud and its final depth.

Whilst CFD simulations such as those presented here provide a useful tool for incident investigation, it would be impractical to apply the same methodology for the purposes of risk assessment. A range of possible release scenarios and weather conditions would need to be considered in any risk assessment, and such analysis would not be feasible with CFD, due to the lengthy computing time involved. The Buncefield Incident took place in zero wind speed conditions, and dispersion of the dense vapor was affected strongly by topography and the presence of obstacles. Simple models that could be used routinely in risk assessment are currently unable to account fully for these effects. It would be beneficial for this issue to be addressed in future work. A potential solution would be to generalize a shallow-layer model, such as SPLOT [13, 14], to simulate dispersion of dense vapor over complex terrain in nil wind.

18.5
CFD: The Future of Safety Technology?

Over the last 30 years, CFD has been demonstrated to be an invaluable tool for the investigation of fluid flows. For some analyses, it remains the only way in which a problem can be solved. For example, in the present work, no other dispersion models currently exist that could simulate the flow. However, there remain some types of fluid flow where CFD models are not yet sufficiently mature to be used with confidence in an industrial context. This includes flows featuring primary liquid breakup and laminar-turbulent transition. Here it may be necessary to use experiments or empirical correlations in combination with CFD models. A good example of such a collaborative link between experiments and CFD is in the recent work on overfilling tank releases by Coldrick *et al.* [3]. Experiments also still have a very important role to play in validating CFD models. Finally, it should be conceded that there are some areas of study where, although CFD could be used in theory, it does not provide the most cost-effective solution. An example of this is in risk assessment, where a very wide range of scenarios needs to be investigated quickly, and where simpler models may offer more appropriate solutions.

The next decade will probably see the continued increase in overall computer speed, driven by machines becoming highly parallel. If this is combined with a reduction in the cost-per-processor of commercial CFD software, it should enable CFD practitioners to perform more extensive sensitivity tests faster and more efficiently, helping to improve understanding and confidence in predictions. This is particularly important in process safety applications, which, as demonstrated by the present study, are often characterized by significant uncertainty in boundary conditions. In the past, many simulations have used coarse meshes or long time steps as a matter of necessity. Gradually, over time, this should become less of an issue. There is also likely to be a move toward the use of time-resolving turbulence models, although this may be driven more by the desire to make simulations appear realistic rather than necessarily to improve the accuracy of the mean flow predictions.

Acknowledgments

This publication and the work it describes were funded by the Health and Safety Executive (HSE). Its contents, including any opinions and/or conclusions expressed, are those of the authors alone and do not necessarily reflect HSE policy.

References

1 Buncefield Investigation Progress Report. Major Incident Investigation Board (MIIB) 21 February, 2006 (available from http://www.buncefieldinvestigation.gov.uk).

2 Atkinson, G., Gant, S.E., Painter, D., Shirvill, L., and Ungut, A. (2008) Liquid dispersal and vapour production during overfilling incidents *Proc. IChemE Hazards XX Symposium & Workshop*, Manchester, UK, 15–17 April. Also published in *IChemE Loss Prevention Bulletin* 207.

3 Coldrick, S., Atkinson, G.T., and Gant, S.E. (2011) Large scale evaporating liquid cascades – an experimental and computational study. Proc. IChemE Hazards XXII Symposium & Workshop, Liverpool, UK, 11–14 April.

4 Health and Safety Executive Buncefield Explosion Mechanism Phase 1 Research Report RR718, 2009 (available from http://www.hse.gov.uk/research/rrpdf/rr718.pdf).

5 Atkinson, G.T. and Cusco, L. (2010) Unsteady deflagration of large low lying vapour clouds. 6th International Seminar on Fires and Explosions, Leeds, UK.

6 Menter, F.R. (1994) Two-equation eddy-viscosity turbulence models for engineering applications. *AIAA Journal*, **32**, 1598–1605.

7 Scargiali, F., di Rienzo, E., Ciofalo, M., Grisafi, F., and Brucato, A. (2005) Heavy gas dispersion modelling over a topographically complex mesoscale: a CFD based approach. *Process Safety and Environmental Protection*, **83**, 242–256.

8 Ergun, S. (1952) Fluid flow through a packed column. *Chemical Engineering Progress*, **48**, 89–94.

9 Hoerner, S.F. (1993) Fluid-Dynamic Drag. Hoerner Fluid Dynamics (publisher), ISBN 9993623938.

10 Li, W., Wang, F., and Bell, S. (2003) Windbreak sheltering effects on an outdoor open space. Proc. 8th International IBPSA Conference, Eindhoven, Netherlands.

11 Bean, A., Alperi, R.W., and Federer, C.A. (1975) A method for characterizing shelterbelt porosity. *Agricultural Meteorology*, **14**, 417–429.

12 Blackadar, A.K. (1998) *Turbulence and Diffusion in the Atmosphere*, Springer-Verlag.

13 Webber, D.M. and Ivings, M.J. (2010) Modelling bund overtopping using shallow water theory. *Journal of Loss Prevention in the Process Industries*, **23**, 662–667.

14 Webber, D.M. and Ivings, M.J. (2004) *Feasibility Study of Modelling gas Dispersion using a Shallow Layer Approach with a Robust and Rigorous Numerical Solution Technique*, Report CM/04/14 Health and Safety Laboratory, Buxton, UK.

Part Six
Contributions for Discussion

19
Do We Really Want to Calculate the Wrong Problem as Exactly as Possible? The Relevance of Initial and Boundary Conditions in Treating the Consequences of Accidents

Ulrich Hauptmanns

19.1
Introduction

Uncertainties are a characteristic of engineering calculations. Yet they are frequently neglected (cf. [1]). They are either due to lacking or insufficient knowledge of the phenomena to be described (epistemic uncertainties) or they derive from their stochastic nature (aleatory uncertainties). Both types of uncertainties play an important role in the assessment of accident consequences. Whilst the epistemic uncertainties may, at least in principle, be reduced by further research, the aleatory uncertainties are intrinsic.

The present focus is on stochastic (aleatory) uncertainties, which chiefly affect the boundary conditions of the models used. These derive largely from the fact that the moment of the accident cannot be predicted. Thus, for example, the weather conditions then prevailing cannot be foretold. Additionally, the calculations are affected by modeling uncertainties, that is, models do not perfectly represent the phenomena addressed.

Accident consequence calculations usually require the deployment of a chain of models. Therefore inaccuracies caused by the interfacing between models are to be borne in mind. For example, the time-dependent discharge rate as an output of the discharge calculation cannot be input in complete detail into the model for atmospheric dispersion, as in the widely used model of Ref. [2].

In view of the above, the question arises whether complex and more exact models like, for example, those of computational fluid dynamics (CFD) (cf. [3]) are needed for accident consequence assessment. Or is the potential improvement they provide clouded by other uncertainties, thus making it more useful to adequately represent the stochastic boundary conditions (and possibly other uncertainties)?

This question is addressed here by considering by way of example an accident consisting of the release of a toxic substance in the gaseous state in unobstructed surroundings. In order to assess its consequences, the following sequence of model calculations has to be performed:

- assessment of the leak (i.e., size and shape, frequency of occurrence, elevation),
- determination of the discharge rate as a function of time,
- calculation of the atmospheric dispersion,

- assessment of the toxic load and of the conditional probability of damage (e.g., death) at one or several points of reference, and
- by combination with the frequency of occurrence of the leak, assessment of the location risk, which becomes the better known individual risk if a person remains for 24 h at the location in question.

The findings obtained apply *mutatis mutandis* to accident sequence chains which involve more calculation steps than are required here, for example, to the release of a liquid with the additional modeling of the evaporation process or flash releases with consequential explosions or fires.

19.2 Models

19.2.1 Leaks

In assessing a leak, the following stochastic parameters have to be accounted for:

- size of the opening,
- geometry of the opening,
- location of the opening (e.g., elevation above the ground, orientation),
- time necessary for successful isolation.

In what follows, some of these topics are commented upon.

19.2.1.1 Leak Size
The following relationships to assess leak sizes in process plants are in use:

- determination according to Brötz [4]

$$D_L = 0.11284 \cdot DN \quad DN < 100 \tag{19.1}$$

- determination according to Strohmeier [5]

$$D_L = 0.02111 \cdot DN^{1.1} \tag{19.2}$$

- determination according to Moosemiller [6]

$$D_L = \frac{0.00635}{f \cdot DN} \tag{19.3}$$

In Eqs. (19.1)–(19.3) D_L denotes the leak diameter in mm (a circular opening is assumed) and DN is the nominal pipe width (approximately the inner diameter in mm). Eq. (19.3) is the only one to establish a relationship between leak size and the corresponding annual expected frequency of occurrence, f; the latter refers to a pipe length of 1 m. The values are typical for steel pipes in oil and chemical process operations.

Table 19.1 Leak cross sections obtained with different rules of calculation.

Equation	Leak diameter D_L in mm	Leak cross section F_L in mm²	Expected annual frequency for 1 m of pipe length
(19.1)	9.03	64	not considered
(19.2)	2.62	5.38	not considered
(19.3)	9.03	64	$8.79 \cdot 10^{-6}$
(19.3)	2.62	5.38	$3.05 \cdot 10^{-5}$

A comparative calculation of leaks in a pipe of size DN 80 is shown in Table 19.1. Additionally, in Ref. [7], a full bore opening of a pipe DN 25 is considered, which leads to a leak cross section of $F_L = 581.1$ mm².

The relationship between the leak cross section and its expected annual frequency of occurrence is illustrated in Figure 19.1.

19.2.1.2 Geometry of the Aperture

In all cases a circular geometry is assumed, which does not necessarily correspond to reality. Contraction of flow and different degrees of smoothness of the leak edges are accounted for by the discharge factor C_d. Numerical values used are 0.595–0.62; they increase with the roundness of the edges [8].

19.2.2
Discharge of a Gas

Assuming an isentropic expansion of an ideal gas the mass flow rate is calculated in the case of sub-sonic flow by (cf. [9])

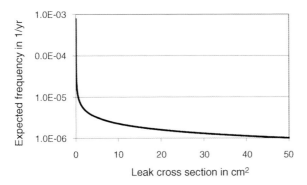

Figure 19.1 Leak cross sections and expected annual frequencies of occurrence according to Eq. (19.3) for a pipe length of 1 m.

$$\dot{m}_L = C_d \cdot F_L \cdot \left\{ \frac{2 \cdot \kappa}{\kappa - 1} \cdot \varrho_0 p_0 \left(\frac{p_1}{p_0}\right)^{\frac{2}{\kappa}} \left[1 - \left(\frac{p_1}{p_0}\right)^{\frac{\kappa-1}{\kappa}}\right] \right\}^{1/2} \qquad (19.4)$$

In Eq. (19.4) C_d is the discharge coefficient, F_L the leak cross section, κ the isentropic coefficient, ϱ_0 the gas density, p_0 the upstream pressure and p_1 the downstream pressure (in the present case inside, respectively outside the tank). Eq. (19.4) applies if

$$\frac{p_1}{p_0} \leq \left(\frac{2}{\kappa+1}\right)^{\frac{\kappa}{\kappa-1}},$$

If this is not the case the flow becomes choked and the mass flow is calculated by

$$\dot{m}_L = C_d \cdot F_L \cdot \left\{ \varrho_0 p_0 \kappa \left(\frac{2}{\kappa+1}\right)^{\frac{\kappa+1}{\kappa-1}} \right\}^{1/2} \qquad (19.5)$$

The above equations are solved using small time steps and establishing new values for pressure and density based on the ideal gas law after each of them. The calculation ends when the upstream pressure equals the downstream pressure or after the leak is isolated.

The resulting time-dependent discharge flow rates may well be represented by the following relationship:

$$\dot{m}_L(t) = A \cdot \exp(-\lambda \cdot t) \qquad (19.6)$$

In Eq. (19.6) the constants A and λ are determined by regression analysis of the results obtained according to the procedure described above; λ is a time constant, which represents the speed of discharge.

Some of the relevant stochastic parameters involved in the discharge calculation are:

19.2.2.1 Filling Ratio
As a consequence of the unforeseeable moment of the accidental opening, the filling ratio of the tank is a stochastic variable. It affects the upstream pressure and therefore the discharge rate. For the present purpose it is assumed to lie in the range from 60% to 100%.

19.2.2.2 Duration of Release
The duration of release strongly affects the quantity of gas discharged. Without intervention the release lasts until the outside and the inside pressures are equalized. In many cases isolation is possible. Time periods t^* of 5 min, 10 min [7] and 30 min [10] are claimed to be necessary for isolation but, depending on the circumstances, much more time may be needed, for example, a day.

19.2.2.3 Ambient Temperature and Pressure
Both ambient temperature and pressure are stochastic variables. They influence the discharge process, as can be inferred from Eqs. (19.4) and (19.5).

19.2.3
Atmospheric Dispersion

Starting point is the atmospheric diffusion equation (see Appendix) with a constant eddy coefficient K and a constant wind speed u. Since both quantities vary with height and time the results of this approach may not be meaningful. However, the theory is used most profitably to determine which variables are most important in the diffusion process and to obtain order-of-magnitude results [11].

The solution for the concentration c in the case of an exponentially decreasing discharge mass flow rate (see Eq. (19.6)), which drops to 0 at point of time t^* (either because of equalization of pressures or leak isolation) is

$$c(x_1, y_1, z_1, t) = \frac{2 \cdot A}{8 \cdot \sqrt{\pi^3 \cdot K^3}} \cdot \int_0^t z^{-3/2} \cdot \exp\left(-\lambda(t-z) - \frac{(x_1 - u \cdot z)^2 + y_1^2 + z_1^2}{4 \cdot K \cdot z}\right) dz$$

$$- U(t-t^*) \cdot \frac{2 \cdot A}{8 \cdot \sqrt{\pi^3 \cdot K^3}} \cdot \int_0^{t-t^*} z^{-3/2} \cdot \exp\left(-\lambda(t-z) - \frac{(x_1 - u \cdot z)^2 + y_1^2 + z_1^2}{4 \cdot K \cdot z}\right) dz$$

(19.7)

In Eq. (19.7) x_1, y_1 and z_1 are the coordinates of the reference point; u is the wind speed, K is the eddy diffusion coefficient, and U the unit step function. The latter is equal to 1 for $t-t^* > 0$ and 0 otherwise.

Relevant stochastic parameters besides the duration of release, t^*, are the following:

19.2.3.1 Wind Speed
Wind speed may be represented by a Weibull distribution (cf. [12]), whose probability density function (pdf) is given by

$$f_U(u) = a \cdot b(a \cdot u)^{b-1} \cdot \exp(-a \cdot u)^b \quad u, a, b > 0 \tag{19.8}$$

In Eq. (19.8) u is the wind speed in m s^{-1} and a and b are distribution parameters; $a = 0.27778$ s m^{-1} and $b = 1.77$ are chosen here. They would represent the situation in the area of Cologne [12] and refer to the anemometer height of $z_A = 10$ m. The conversion of the resulting wind speed to the reference height z_1 is made by using the relation

$$u(z_1) = u(z_A) \cdot \left(\frac{z_1}{z_A}\right)^m \tag{19.9}$$

with $m = 0.2$ for unstable layers and $z_1 = 1.8$ being chosen here.

19.2.3.2 Eddy Coefficient
According to Ref. [11] we have

$$K = \frac{k^2 \cdot u \cdot (z_1 + z_0)}{\ln \frac{z_1}{z_0}} \tag{19.10}$$

In Eq. (19.10) k is von Karman's constant (0.4), u the wind speed at height z_1 in m s^{-1}, and z_0 the roughness length. The latter describes the characteristics of the terrain surrounding the point of release (0.02...1.2 m) [2]. It is evident that K is linearly coupled with the stochastic variable 'wind speed'.

19.2.4
Health Effects

Health effects of toxic substances are often represented by means of a Probit (probability unit) equation [9]. This relates a causative factor, for example, a toxic load, with a probability of suffering harm. Its general form is

$$Y = a + b \cdot \ln(c^n \cdot t) \tag{19.11}$$

In Eq. (19.11) a, b, and n are empirical coefficients representing the underlying evidence of harm caused by the substance; $c \cdot t$ (t being the time of exposure) is the dose, that is, the integral of the concentrations over the time of exposure. If n is not equal to 1, the term 'toxic load' is used for $c^n \cdot t$ instead of toxic dose. The numerical values of the coefficients depend on the substance involved (see Table 19.2).

In order to obtain the probability of damage, Y from Eq. (19.11) is introduced into

$$P_{\text{damage}} = \frac{1}{\sqrt{2\pi}} \int_{-\infty}^{Y-5} \exp\left(-\frac{x^2}{2}\right) dx = \phi(Y-5) \tag{19.12}$$

where ϕ is the standard normal distribution.

Damage may mean death, as with the coefficients of Table 19.2, or reversible or irreversible health effects.

It should be borne in mind that the exposure depends on the stochastic parameter of being indoors or outdoors. In the case of being indoors the quality of the corresponding building plays a role as well (i.e., number of air exchanges per hour), just as the characteristics (e.g., duration) of possible evacuation measures.

Table 19.2 Coefficients of the probit equation for death from different substances with c in mg m^{-3} and t in min [13].

Substance	a	b	n
Acrolein	−4.1	1	1
Ammonia	−15.6	1	2
Carbon monoxide	−7.4	1	1
Chlorine	−6.35	0.5	2.75
Ethylene oxide	−6.8	1	1
Hydrogen sulfide	−11.5	1	1.9
Phosgene	−10.6	2	1

19.3
Case Study

In order to show the impact of the different stochastic parameters, the following case is examined in more detail. A tank of a volume of 35.6 m³ contains 100 kg of a toxic substance at $T = 290$ K and 199,5 kPa. Discharge and consequential atmospheric diffusion are calculated for reference points at distances of x_1. The health impact is assessed for a substance described by

$$Y = -6.8 + \ln\left(\int_0^t c^2(t')dt'\right) \tag{19.13}$$

Both deterministic and probabilistic calculations are performed. In the latter some of the deterministic boundary conditions are replaced by the corresponding stochastic characteristics.

19.3.1
Deterministic Calculations

The mass flow rates resulting from leaks of different sizes are presented in Figure 19.2. The calculations are performed for leaks of different sizes in a pipe of 1 m length connected to the tank at ground level using $C_d = 0.62$. Because of the shortness of the pipe, friction losses in the pipe are neglected.

The corresponding conditional probabilities of death (all at $z_1 = 1.8$ m and $y_1 = 0$ m with the wind speed u referring to a height of 1.8 m) as a function of the distance x_1 from the source are presented in Figure 19.3.

It is evident that the leak size largely affects the outcome and valuation of the calculation. The slower the discharge the milder are the health effects.

19.3.2
Sensitivity Studies

In order to assess the sensitivity of the final result to possible changes of input parameters, a number of stochastic input parameters are examined deterministically.

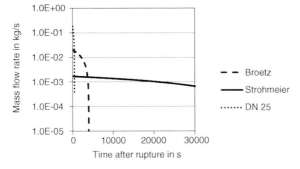

Figure 19.2 Mass flow rates for different leak sizes.

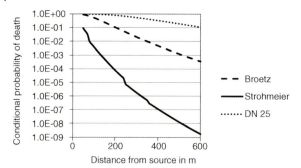

Figure 19.3 Distance-dependent conditional probability of death for different leak sizes ($u = 3.0\,\text{m s}^{-1}$, $z_0 = 0.2\,\text{m}$, $K = 0.4369\,\text{m}^2\,\text{s}^{-1}$).

In the first place changes of the probability of death as a function of the toxic load are shown in Figure 19.4.

It is evident that in the range of low toxic loads (small leaks and/or large distances from the source) the response to small changes is large, while for high toxic loads (large leaks and/or distances close to the source) large increases of the toxic load make only a small difference to the probability of death.

The sensitivity of the probability of death with respect to variations of the time constant of discharge, λ, is shown in Figure 19.5. Obviously the impact is stronger the smaller is the mass flow rate from the leak.

The impact of variations of the wind speed is shown in Figures 19.6 and 19.7. It is evident that the variation with wind speed is much larger than that caused by possible variations of the discharge rate despite the relatively small wind speed range (1.5–4.5 m s^{-1}) considered. Hence, possible errors in the discharge calculation, which could be reduced by using the more exact CFD procedure, have a comparatively smaller impact on the probability of death.

Again there is a dependence on leak size, and larger variations occur with lower probabilities of death, that is, for smaller leak sizes (Strohmeier).

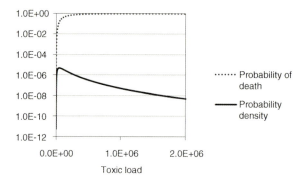

Figure 19.4 Probability of death and its probability density as a function of the toxic load.

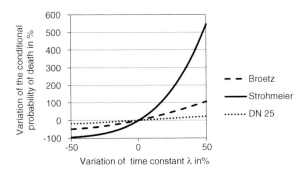

Figure 19.5 Variation of the conditional probability of death with variation of the time constant of discharge λ for a distance of 300 m from the source ($u = 3.0\,\text{m s}^{-1}$, $z_0 = 0.2\,\text{m}$, $K = 0.4369\,\text{m}^2\,\text{s}^{-1}$).

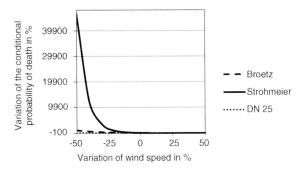

Figure 19.6 Variation of the conditional probability of death with variation of the wind speed at a distance of 300 m from the source; $u = 3.0\,\text{m s}^{-1}$, $z_0 = 0.2\,\text{m}$, K according to Eq. (19.10).

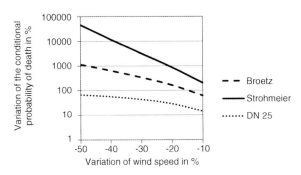

Figure 19.7 Variation of the conditional probability of death with variation of the wind speed at a distance of 300 m from the source; $u = 3.0\,\text{m s}^{-1}$, $z_0 = 0.2\,\text{m}$, K according to Eq. (19.10) (excerpt).

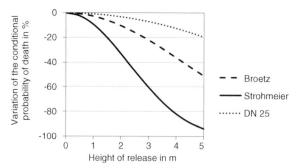

Figure 19.8 Variation of the conditional probability of death with variation of the height of release at a distance of 300 m from the source; $u = 3.0\,\mathrm{m\,s^{-1}}$, $z_0 = 0.2\,\mathrm{m}$, K according to Eq. (19.10).

The influence of the elevation of the point of release is demonstrated in Figure 19.8. The smaller the leak the larger is the impact to be expected.

Another parameter which changes stochastically is the degree of filling, that is, stored mass, at the point in time of pipe rupture. Its impact is shown in Figure 19.9.

Finally, the influence of different durations of release is shown in Figure 19.10. The differences are considerable, thus highlighting the importance of quick isolation.

In summary it may be stated that there are a number of parameters which change stochastically depending on the point in time of release and have a large impact on the result. In several cases the percentage changes are much higher than those that would result from possible lacks of precision of the discharge calculations.

19.3.3
Probabilistic Calculations

The stochastic nature of some of the input variables is accounted for by embedding the sequence of models in a Monte-Carlo environment (cf. [14]). The sequence is

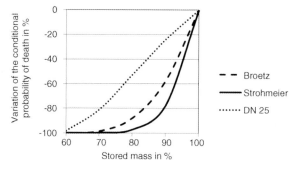

Figure 19.9 Variation of the conditional probability of death with variation of the reference mass of 100 kg at a distance of 300 m from the source; $u = 3.0\,\mathrm{m\,s^{-1}}$, $z_0 = 0.2\,\mathrm{m}$, K according to Eq. (19.10).

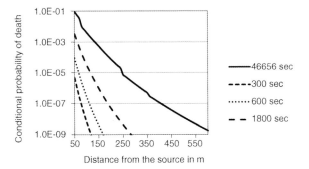

Figure 19.10 Conditional probability of death for the Strohmeier leak and different durations of release; $u = 3.0\,\text{m s}^{-1}$, $z_0 = 0.2\,\text{m}$, K according to Eq. (19.10).

calculated 10000 times (trials), each time using a set of input variables randomly chosen from the corresponding probability distributions, e.g., Eq. (19.8) in combination with Eq. (19.9) (referring the result from Eq. (19.8) to $z_1 = 1.8\,\text{m}$) for the wind speed u, and for the other input quantities from

- the rectangular distribution with pdf (cf. [15])

$$f_X(x) = \begin{cases} \dfrac{1}{b-a} & \text{if } a \le x \le b \\ 0 & \text{otherwise} \end{cases} \tag{19.14}$$

which is used if no specific details of the underlying distribution are known, the reason being either that there is no empirical evidence or that such evidence has not been collated in the present context, or

- the triangular distribution with pdf (cf. [15])

$$f_X(x) = \begin{cases} \dfrac{2}{b-a} \cdot \dfrac{x-a}{c-a} & \text{if } c \ge x \ge a \\ 0 & \text{otherwise} \\ \dfrac{2}{b-a} \cdot \dfrac{b-x}{b-c} & \text{if } b \ge x \ge c \end{cases} \tag{19.15}$$

which is used if little data is known, as is the case here with the times until the isolation of the leak; a is the lowest value, b the modal value, and c the highest.

The input quantities along with the probability distributions and parameters used are given in Table 19.3.

Figure 19.11 shows a comparison between the results of the deterministic calculations using the input data given above and the mean as well as the 95th centile (the value below which 95% of the results are expected to lie) of the probabilistic approach. In the latter the stochastic nature of the input variables is accounted for according to the procedure described above.

Table 19.3 Input quantities and corresponding probability density distributions for the Monte-Carlo calculations.

Input quantity	Equation used	Rationale	a	b	c
C_d	(19.14)	[7]	0.595	0.62	
t^*	(19.15)	[6, 10] + estimate	300 s	3600 s	28 800 s
Filling ratio	(19.14)	Estimate	0.6	1.0	
Storage temperature	(19.14)	Estimate	$-10\,°C$	$+40\,°C$	
u	(19.8)	[12]	$0.27778\,\mathrm{m\,s^{-1}}$	1.77	

Figure 19.12 finally presents the location risk in the surroundings of the point of release including the 95th centile. It is obtained by integrating over all possible leak cross sections, that is

$$F = \int_0^{f(D_{L,max})} f' \cdot P_{damage}\, df' \qquad (19.16)$$

In Eq. (19.16) F is the expected frequency of death and f is calculated according to Eq. (19.3); P_{damage} stems from Eq. (19.12), which is combined with Eq. (19.13) to give the conditional probability of death. The concentration needed for Eq. (19.12) is calculated with Eq. (19.7) where A and λ are obtained from Eqs. (19.4) through (19.6). The integration of Eq. (19.16) is done using the trapezoidal rule; the lower limit is taken to be 0, although Eq. (19.3) would give infinity for $D_L = 0$. On the other hand $\lim_{D_L \to 0} f \cdot P_{damage} = 0$ because $P_{damage} = 0$ for a pipe without leak.

Both figures show that the deterministic approach gives only a partial view of a much more complex reality, which is characterized by an even larger number of

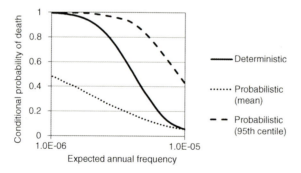

Figure 19.11 Conditional probability of death versus annual frequency of occurrence of different leak sizes according to Eq. (19.3) (input data for the deterministic calculations: $u = 3.0\,\mathrm{m\,s^{-1}}$, $z_0 = 0.2\,\mathrm{m}$, $K = 0.4369\,\mathrm{m^2\,s^{-1}}$) compared with the mean and 95th centile of the probabilistic calculation at a distance of 300 m from the point of release ($z_0 = 0.2\,\mathrm{m}$).

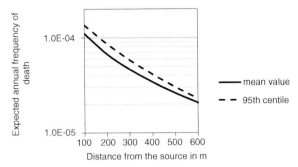

Figure 19.12 Location risk of death for the case study accounting for the stochastic character of the input variables ($z_0 = 0.2$ m).

stochastic input variables than the ones accounted for or mentioned here. For example, the roughness length z_0 has a large impact. Usually it is assessed on the basis of a natural language characterization of the terrain surrounding the point of release, which lends itself to misjudgments.

19.4
Conclusions

It was shown that the calculation of accident consequences has to take into account a number of stochastic input variables. Possible variations of these have a much larger impact on the final result than potential inaccuracies in discharge calculations accompanying the use of a simple model. Hence, CFD is not essential for this type of calculation, and the analyst is well advised to devote more attention to the stochastic nature of the input variables. A combination of CFD with stochastic input variables would require prohibitive times for calculation without providing a substantial improvement in the results. In fact, employing high-quality models like CFD and correspondingly a more sophisticated approach to atmospheric diffusion without paying due attention to the stochastic character of the boundary conditions involves the danger of calculating the wrong problem as exactly as possible, making an enormous effort in computing time and laboriously preparing the input for the CFD calculations, without any benefit to the safety assessment.

Appendix

The atmospheric diffusion equation is (cf. [11])

$$\frac{\partial c}{\partial t} = K \left(\frac{\partial^2 c}{\partial r^2} + \frac{2}{r} \frac{\partial c}{\partial r} \right)$$

Its solution in the Laplace domain, which keeps the concentration finite for $r \to \infty$, is

$$c(r,s) = \frac{B}{r} \cdot e^{-r \cdot \sqrt{\frac{s}{k}}}$$

The constant of solution B is determined from the following initial condition

$$\lim_{r \to 0} -K \cdot 2\pi r^2 \cdot \frac{dc(r,s)}{dr} = \frac{A}{\lambda + s} \cdot [1 - \exp(-(\lambda + s) \cdot t^*)]$$

It accounts for the finite duration of release $[0, t^*]$ from a source described by Eq. (19.6) and for the impervious ground by allowing a release only into the upper hemisphere (multiplication by a factor of 2). The complete solution in the Laplace domain then is

$$c(r,s) = \frac{2 \cdot A}{4 \cdot \pi \cdot K} \cdot \frac{1 - \exp(-(\lambda + s) \cdot t^*)}{\lambda + s} \cdot \frac{1}{r} \cdot e^{-r \cdot \sqrt{\frac{s}{k}}}$$

The inversion of the Laplace Transform (cf. [16]), replacing r^2 by $x^2 + y^2 + z^2$ and introducing a co-ordinate system moving with velocity u in the x-direction gives Eq.(19.7).

References

1 Knetsch, T. and Hauptmanns, U. (2005) Integration of stochastic effects and data uncertainties into the design of process equipment. *Risk Analysis*, **25** (1), 189–198.

2 Institut für Meteorologie der TH Darmstadt, Störfall-Programm zur VDI-Richtlinie 3783 Blatt 1 (1987) 'Ausbreitung von störfallbedingten Freisetzungen; Sicherheitsanalyse' und Blatt 2 (1990) 'Dispersion of heavy gas emissions by accidental releases - Safety study', Beuth Verlag, Berlin.

3 (2005) FLUENT 6.2 User's Guide. Fluent Inc., Lebanon, NH 03766.

4 Brötz, W. (1979) *Gutachten Sicherheitstechnik NRW im Auftrag des MAGS*, Stuttgart.

5 (1994) Abschlußbericht des Arbeitskreises 'Novellierung der 2. StörfallVwV', TAA GS-03.

6 Moosemiller, M.D. (2009) Development of Algorithms for Predicting Ignition Probabilities and Explosion Frequencies. paper 00813, 8th World Congress of Chemical Engineering, Montreal, August 23rd -27th.

7 (November 2010) Kommission für Anlagensicherheit beim Bundeministerium Umwelt, Naturschutz und Reaktorsicherheit, Leitfaden 'Empfehlungen für Abstände zwischen Betriebsbereichen nach der Störfall-Verordnung und schutzbedürftigen Gebieten im Rahmen der Bauleitplanung- Umsetzung §50 BImSchG', 2. Überarbeitete Fassung, KAS-18.

8 Perry, R.H. and Green, D.W. (eds) (1998) *Perry's Chemical Engineering Handbook*, McGraw Hill, New York.

9 Lees, F.P. (1996) *Loss Prevention in the Process Industries*, vol. 1–3, Butterworth Heinemann, Oxford.

10 (2009) Rijksinstitut voor Volksgezondheid en Milieu (RIVM), Centrum Externe Veiligheid (Hrsg.), Handleiding Risicoberekeningen Bevi, Juli.

11 Sellers, W.D. (December 1965) *Physical Climatology*, University of Chicago..

12 Quaschning, V. (2010) *Understanding Renewable Energy Systems*, John Wiley & Sons Ltd., Chichester.
13 Uijt de Haag, P.A.M. and Ale, B.J.M. (2005) *Guideline for Quantitative Risk Assessment 'Purple Book'*, CPR 18E, VROM.
14 Ripley, B.D. (1987) *Stochastic Simulation*, John Wiley & Sons, New York.
15 Hartung, J. (1991) *Statistik: Lehr- und Handbuch der Angewandten Statistik*, R. Oldenbourg Verlag, München.
16 Abramowitz, M. and Stegun, I. (1972) *Handbook of Mathematical Functions with Formulas, Graphs, and Mathematical Tables*, U.S. Department of Commerce, Washington.

20
Can Software Ever be Safe?
Frank Schiller and Tina Mattes

20.1
Introduction

In recent decades, software became the main carrier of functionality in almost all common industries. There is neither a process engineering plant nor a manufacturing cell nor a car nor a train that is capable of operating without software. Lots of medical equipment – as well as household appliances – is controlled by software. The major advantages of software are

- its flexibility, as software can be changed easily and adopted for changing applications, and
- its copyability and spreadability, as software is not strongly connected to its physical memory.

Obviously, the principles of operation that were formerly realized by more or less fixed and inflexible hardware are now realized by software executed on microcontrollers or processors utilizing its advantages.

As the intelligence of technical systems is more and more represented by software, failures caused by software have increasingly to be faced. All important characteristics of technical systems like quality, reliability, maintainability, usability, and safety depend on the condition of the software.

The process of design and development of software can only roughly be compared to corresponding development processes of other typical products. The relationship between the properties of the final product and the steps involved in the design and development phases are not always evident in software engineering.

The soaring performance of processor hardware (memory, processing time) is heavily utilized, and new applications are continuously being identified. Therefore, the complexity of software is increasing simply because of the increasing number of lines of code and data to be processed and the increasing extent of interaction between different software components. Pragmatic unstructured approaches are no longer acceptable.

To deal with this, several software development methods like the well-known V-Model have been developed and successfully applied. These regulate the complete software development process starting with the analysis of the requirement and continuing with the design phases up to the implementation and test procedures [1].

Applying these software development methods usually leads to reasonably functioning software. But as soon as it comes to safety-critical applications, where safety has not only to be assured but to be proved, it is not at all enough to ensure a reasonable standard of functioning operation. Safety functions have to be executed correctly or their incorrect execution has to be detected and appropriate action has to be taken in order to reliably avoid hazards.

Whereas random faults like hardware faults are relatively well dealt with by redundancy and coding techniques, software errors that are systematic are difficult to treat. In contrast to random faults, systematic faults and their effects cannot be acceptably described by probabilities.

The chapter continues with relevant definitions and categorizations of general faults and general corresponding countermeasures in Section 20.2. Section 20.3 gives a description of different types of errors in software as well as methods to correct them. The main goal of these methods is the so-called perfect software that is completely free of errors. Potential future approaches are discussed in Section 20.4, and Section 20.5 summarizes the chapter.

20.2
Basics

This section contains some relevant definitions as well as general strategies to handle erroneous systems in a safe way. More information about the basics can be found in Ref. [1].

20.2.1
Definitions

To avoid any misconceptions, some definitions according to IEC 61 508 [2] are given below:

- Fault: 'abnormal condition that may cause a reduction in or loss of the capability of a functional unit to perform a required function' (e.g., the calculation `int result = 100/x;` within the software code, where x is a user input)
- Error: 'discrepancy between a computed, observed or measured value or condition and the true, specified or theoretically correct value or condition' (e.g., the calculation `int result = 100/x;` gets executed with the user input 0)
- Failure: 'termination of the ability of a functional unit to perform a required function' (e.g., the function containing the calculation above returns no valid result, as the execution stops with an exception)

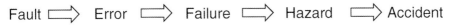

Figure 20.1 Relationship between the different relevant terms.

- Hazard: 'potential source of harm' (e.g., the software supervising the pressure of a tank stops executing)
- Accident (see Ref. [2]): 'unintended event or sequence of events that causes death, injury, environmental or material damage' (e.g., the high pressure of a tank greater than the specified one causes a leakage and further damage).

Figure 20.1 visualizes the relationship between these terms. If a fault is brought into action, for example, in a computation process, then an error occurs. If the error is not detected early enough it can affect the required function and hence cause a failure. A function that operates incorrectly is a potential risk and with that a hazard. Under certain circumstances, the hazardous situation can result in an accident.

According to the definitions, incorrect software is firstly considered to be a fault (incorrect software stored in the memory but without any effect initially), then to be an error (there might be effects during compilation or at the beginning of the execution, but the specification from an external view point is still fulfilled since the environment has not yet been affected). After activating the software of, for example, the controller of a plant, failures may occur.

20.2.2
General Strategies

The overall goal of safe software systems is to avoid hazards and accidents. According to the relationships in Figure 20.1, there are several possibilities whereby measures of safe software systems can break the cause–effect chain. Depending on the location within this chain, these measures constitute specific properties of the resulting system. These different approaches are categorized in Figure 20.2 and will be further described below.

20.2.2.1 Perfect Systems
Following the perfect systems approach, faults should be entirely avoided. The goal is to achieve safety by shielding the system from any kind of faults, so that, in consequence, no external cause can affect the system (see Figure 20.3). The output (y) should be correct.

To avoid faults, three general measures are applied:

- Fault prevention: Preventing the introduction or occurrence of faults (e.g., by formal methods in software engineering)
- Fault removal: Detecting and correcting existing faults by extensive analysis and diagnosis
- Fault forecasting: Identification of faults by imagining future incidents (e.g., on the basis of statistics or by fault tree analysis). Afterwards, fault prevention and removal strategies have to be applied.

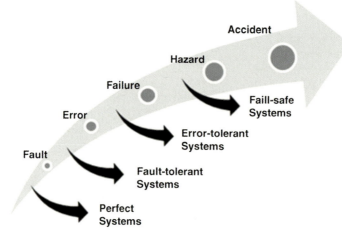

Figure 20.2 Approaches to prevent hazards.

20.2.2.2 Fault-Tolerant Systems

In contrast to perfect systems, the intention of fault-tolerant systems is not to avoid faults but to accept inevitable ones and to define and implement measures of fault detection and corresponding reactions to prevent errors (Figure 20.4). The output (y) should again be correct.

This can be achieved by controller-like approaches or (more related to software) by redundancy in various ways:

- Redundant components: for example, if the required function is to measure a certain value, two sensor units instead of one can be installed, and if one fails the other one is still able to send the required data.

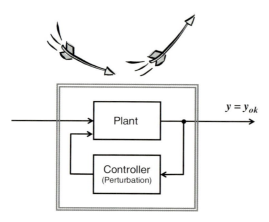

Figure 20.3 Schematic illustration of the perfect systems approach.

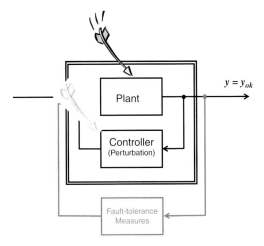

Figure 20.4 Schematic illustration of fault-tolerant systems.

- Redundant data: for example, error-correcting codes in communication; that is, out of the data to be sent a certain checksum is calculated that enables bits that have been corrupted during transmission to be determined and corrected.
- Redundant computations: The calculation of critical values can be implemented two or more times with various algorithms and on different processors.

20.2.2.3 Error-Tolerant Systems

The aim of fail-operational systems is to enable the system to keep certain availability in spite of errors with respect to the avoidance of hazardous situations (Figure 20.5).

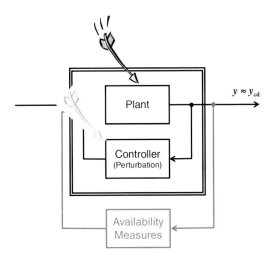

Figure 20.5 Schematic illustration of fail-operational systems.

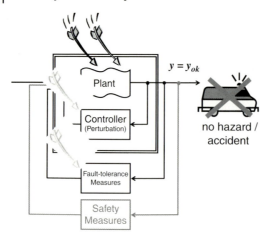

Figure 20.6 Schematic illustration of fail-safe systems.

In contrast to fault-tolerant systems, where the environment does not notice faults of the system since they do not lead to errors (Figure 20.4), error-tolerant systems accept certain errors that affect the environment (Figure 20.5, the output y is approximately y_{ok}) but strategies to cope with errors have to be implemented. For example, if the considered function is to measure a value and the sensor unit fails then the system can use a default value under certain circumstances.

20.2.2.4 Fail-Safe Systems

The only goal of fail-safe systems is the avoidance of hazards and accidents (Figure 20.6).

To achieve this kind of safety it is necessary to detect failures (usually by redundancy and/or supervision) and implement a corresponding reaction (transition into a safe state of the system). The safe state depends on the application and the kind of failure. Usually, switching off the power supply would be a safe reaction but in manufacturing this may not be so in many cases. In general, the reaction has to be adjusted according to the application (e.g., if a magnet carries a heavy weight, it could be better just to stop the movement and let the magnet hold the weight).

In the next section, errors concerning software as well as general methods to avoid or treat them are presented.

20.3
Software Errors and Error Handling

In software development, two kinds of errors have to be considered: errors during the software development (see Section 20.3.1) and implementation errors (see Section 20.3.2). These two kinds of errors and methods to prevent them are described below.

20.3.1
Software Development Errors

Software development is a very complex issue, as there are many steps from the idea of an application to the final product. During this process, a lot of errors can occur, and these are described in the next section. Finally, process models providing possible control mechanisms for software development and therefore helping to avoid errors are discussed.

20.3.1.1 Errors in Software Development
This section concentrates on the various kinds of errors occurring in software development.

20.3.1.1.1 Problem Analysis Error
A problem analysis error occurs when parts of the problem are wrongly interpreted.

20.3.1.1.2 Requirement Definition Error
A requirement definition error manifests itself in the wrong formulation of a requirement.

20.3.1.1.3 Design Error
Design errors are the wrong interpretation of requirements.

20.3.1.1.4 Implementation Error
When parts of the specification are not implemented as intended, implementation errors occur.

20.3.1.1.5 Test Definition Error
A very important phase in software development is the definition of test cases. In this phase, the most common error is a test definition error, meaning that parts of the specification are not correctly mapped to the corresponding test cases.

20.3.1.2 Process Models for Software Development
Errors mentioned in the preceding section are very hard to detect, as the corresponding erroneous behavior of the software product can only be detected in the final phases of the software development. Therefore, those errors are very expensive, because correction often means going back at least one phase in software development and because one error can propagate into several software modules. Generally speaking, the later an error is revealed, the more expensive is the correction. For this reason, several process models for software development have been established, allowing a controlled development to avoid errors mentioned in the preceding section. Further information about the different models can be found in Ref. [5].

20.3.1.2.1 Waterfall Model
The waterfall model (Figure 20.7) is a sequential design process that was adapted from hardware manufacturing industries where subsequent changes are extremely expensive.

There are various modifications of the waterfall model. The original model consists of seven phases that are passed through sequentially:

1) Requirements specification
2) Design

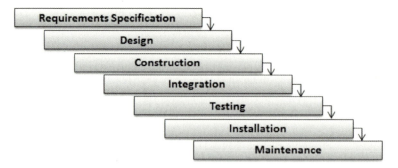

Figure 20.7 Waterfall model.

3) Construction (i.e. coding or implementation)
4) Integration
5) Testing (i.e. validation)
6) Installation
7) Maintenance

The process is constantly monitored by reviews, making sure that only when a phase is fully and correctly completed is the next phase started. Every phase has to be documented before proceeding to the next phase. In the original model, it was forbidden to revisit a previous phase. Newer versions of the waterfall model schedule contain intended skips to previous phases in order to improve the overall quality of the software product. However, this model is still too inflexible for many applications.

20.3.1.2.2 **V-Model** The V-Model is an approach intended to facilitate understanding of the complexity of the system to be designed. The model (see Figure 20.8) can be

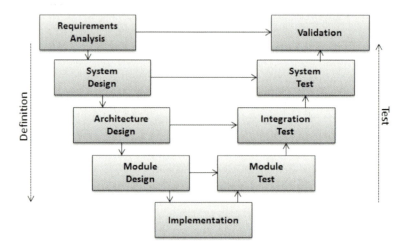

Figure 20.8 V-Model.

divided into two main phases, the definition phase and the test phase. Each step in the definition phase has a corresponding step in the test phase.

The definition phase consists of the following steps:

1) Requirement Analysis
2) System Design
3) Architecture Design
4) Module Design.

Each step produces the basis of the corresponding steps within the test phase. The definition phase peaks in the center step of the development process, the implementation. Moving forward from implementation, the steps of the definition phase are executed in the reverse order from a test point of view.

20.3.1.2.3 Sawtooth Model The sawtooth model can be considered as a modification of the V-Model, where instead of or in addition to the definition of the test cases, prototypes are developed (see Figure 20.9). These prototypes are very helpful to include the customer in early steps and to enable flexibility since the requirements may change during the development process.

The first prototypes are mainly demonstrators where many functions are not executed yet but prepared proxy data is used.

Especially for simulation software like CFD this approach, starting with abstract models and stepwise detailed definition is recommended.

20.3.1.2.4 Spiral Model The spiral model is an iterative process (see Figure 20.10). Each iteration contains the following activities that are assigned to specific quadrants:

1) Definition of goals, alternative solutions, and framework requirements
2) Evaluation of alternative solutions and analysis and reduction of risk
3) Realization and test of the intermediate product
4) Scheduling the next cycle

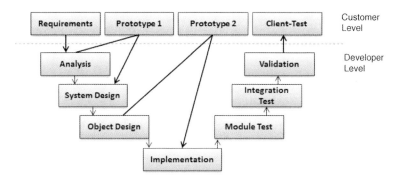

Figure 20.9 V-Model.

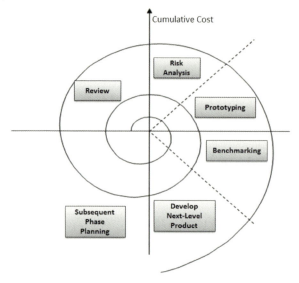

Figure 20.10 Spiral Model.

The spiral model is a further development of the waterfall model, in which the phases are run through repeatedly. The new idea and main characteristic of the spiral model is the risk analysis.

20.3.2
Errors and Methods concerning Errors in Source Code

This section is dedicated to errors occurring within the source code of a software product and methods specifically designed to avoid or find those errors. First, an overview of the most common errors in source codes is given. Then, different strategies to deal with those errors are classified and explained. Further information about possible errors in source code and methods to prevent these can be found in Refs [2–4].

20.3.2.1 Errors in Source Code
Modern programming languages provide a large functionality and flexibility for users. Accompanying those advantages are potential sources of errors, which are described in this subsection.

20.3.2.1.1 Syntax Errors In general, the expression 'syntax error' refers to an error in the syntax of a sequence of characters or tokens that is intended to be written in a particular programming language. An example of a syntax error is the wrong use of brackets within C-source, as shown in the following expression.

```
void test [string input];
```

This line of code would produce a syntax error within the syntax check of the compiler, as parameters for functions are listed within round brackets in standard C syntax. Thanks to the rapid development in the sector of IDEs (Interactive Development Environments), simple cases of syntax errors are already dealt with by the user interface. The IDE constantly checks the source code for syntax errors during user input, so that those errors can be corrected even before the compiling process is started. Errors which cannot be recognized by the IDE can still be found at compilation time, as the compiler performs more intensive checks of the syntax.

Apart from these examples, there are also syntax errors which cannot be recognized at implementation or compilation time and lead to an unexpected behavior of the resulting software. These syntax errors are misconceptions concerning syntactical expressions. These errors cannot be recognized at compilation time, as the syntax of the source code is correct, but the user's understanding of the written source code is different to the interpretation of the compiler. For example, the following line of code is written in C:

```
int c = a- - -b;
```

Without specific understanding of the used compiler, this simple line allows the following interpretations:

```
int c = (a - -) - b;
int c = a- (- -b);
```

Depending on the mechanisms of the compiler, one of the possibilities is chosen. Most modern compilers work in a greedy fashion, meaning that the biggest, syntactically correct token is taken and interpreted. In this example, the compiler first evaluates the partial expression 'int c='. After that, the compiler takes the token 'a--', as the next expression, as it is the biggest token providing a syntactically correct token.

Another example of a syntactical error leading to a misconception is the use of semicolons. An extra semicolon in C may be harmless, as it usually leads to a null statement or a warning message by the compiler, which makes it easy to remove. But in combination with causal statements, it could provide a very surprising functionality. Considering the following example:

```
if (a &gt; 0);
c = b;
```

The compiler will process the semicolon, as it is a valid syntactical expression. But the result of those two lines will be quite different from the user's intended meaning shown in the following expression:

```
if (a > 0);
c = b;
```

The first example with the semicolon ending the 'if line' is interpreted like the following snippet:

```
if (a > 0) {}
c = b;
```

This code snippet is of course equivalent to the single line 'c=b;'. Apart from additional semicolons, missing semicolons can cause trouble, too. The following code snippet is considered:

```
if (a > 0)
return
c = b;
e = f;
g = h;
```

Even though the return statement is missing the semicolon at the end, this code snippet will successfully compile, but will definitely lead to wrong behavior. As mentioned before, compilers work in a greedy fashion. After evaluating the conditional expression, the compiler takes the longest valid token. The result of this interpretation is the following source code:

```
if (a > 0)
return c = b;
e = f;
g = h;
```

This example will always return the result of the assignment 'c=b;'. Depending on the result type of the surrounding function, this error might not even lead to an error message within the source code.

20.3.2.1.2 **Semantic Errors** This subsection deals with so-called semantic errors. A semantic error occurs if a source code actually shows functionality different to the one expected by its textual representation. An example of such unexpected behavior is array indexing, which is very specific to the used programming language. While FORTRAN, PL/I, and Snobol4 start subscripting arrays from 1, Algol and Pascal force the user to explicitly give lower and upper bounds for every array. In standard Basic, the declaration of an array with 10 elements really allocates an array with 11 elements using index 0 to 10 inclusive. In C, the array index runs from 0 through 9. For further explanation of semantic errors based on array indexing, the following example in C is considered:

```
int index, testArray[10];
for (index = 1; index <= 10; index++) {
testArray[index] = 0
}
```

In this example, all values of 'testArray' are expected to be set to 0. In C, the index of the array runs from 0 through 9. As the condition of the for-loop proofs if the index is <= 10, the loop also sets the value on position 'testArray[10]' to 0. The result of this line is that the nonexistent element number 10 is set to zero, meaning that the word that follows 'testArray' in the memory is deleted, which is obviously not the desired functionality of that code section. Furthermore, if this program is run on a compiler that allocates memory for variables at decreasing addresses, the word after 'testArray' turns out to be the memory space of index, leading to an infinite loop, as the call 'testArray[10]' sets the loop back to its start.

Another example of a semantic error is shown in the following code segment:

```
int main()
{
printf('Hello World\n');
}
```

This code segment is designed to display the chosen text as message and shows the desired functionality at every run. The semantic error only manifests itself when this code segment is used within a bigger software architecture starting this program as subprogram and testing if the program has been executed correctly. In C, whenever no return type is declared for a function, the return type int is presumed. As this example fails to return a value, it usually returns some garbage integer, meaning that the test to find if the program has executed correctly (by indicating 0 as return value) always fails (except for the case that the garbage integer is 0).

As one can see in these examples, semantic errors can be very hard to find, and therefore the goal should be the avoidance of such errors.

20.3.2.1.3 **Numerical Errors** Programming languages offer various data types for the processing of numerical values. Apart from problems with overflow, integer operations are processed with full precision. But if the source code has to handle float and double values, possible numerical errors have to be considered, as all values including continuous values are stored discretely. As float and double data types can only be stored with limited precision, operations could lead to imprecise results, as they are only an approximation of the real results. If these results are used in an iterative fashion, the extent of the imprecision increases and could eventually lead to a wrong functionality of the source code.

A very infamous example of numerical errors is the incident with the American Patriot surface-to-air missile (SAM) system in 1991. This system is designed to provide defense against aircraft and cruise missiles. When the system recognizes an object in the monitored airspace, the software calculates the next assumed position of the object, according to which the rocket is fired. This algorithm showed a numerical imprecision leading to a 50% deviation of the optimal result after a runtime of 8 h. This problem was discovered within the test phase of the system. The reaction to this error was the simple statement that the usual usage time of the system must not exceed 8 h. A software update resolving the problem was released, but it did not reach the concerned system in time. On February the 25th in 1991 the system failed and missed an Iraqi scud rocket, which hit American barracks leading to the deaths of 28 soldiers and hundreds of casualties.

Like the errors in the earlier sections, numerical errors can be very hard to find and deal with.

20.3.2.1.4 **Portability Errors** In modern software engineering, the principle of modularity is of great importance. Modularity allows the simplified porting of existing code snippets to a new software product. However, despite this big advantage, there is always the risk of a portability error, as there are many aspects which have to be considered when a software module is ported. The most important is the interface between new and existing source code. The data types and range of passed variables have to be checked before usage of the software. An infamous example of a portability error involving the interface between two software parts is the case of the Ariane 5 incident in 1996. After the successful start of the rocket, it drastically changed its course and finally exploded. For the Ariane 5 rocket, software modules of the Ariane 4 rocket were used. As the earlier rocket did not reach the speed of the Ariane 5 rocket, smaller data types could be used. But if the rocket reached a higher speed, the calculations of a subsystem led to an overflow and to the self-destruction of the rocket, as the engines were about to break off. In addition, the error-causing software module was not required for the flight phase, as it was only part of the flight preparation software.

Apart from interface errors, further portability errors are possible. A simple example is the use of an existing source code with a different compiler. As mentioned in Section 20.3.2.1.1, compilers usually work in a greedy fashion. When the code fragment 'int c = a---b;' is ported to another compiler, it is possible, that this compiler will not work in the same manner, so that the line is interpreted in another way, leading to unexpected behavior.

20.3.2.2 Methods for Preventing Software Errors

As the preceding section shows, many possible errors have to be dealt with in software engineering. The importance of methods to prevent such errors constantly increases, especially for safety-critical applications. This section classifies current methods for providing software quality and gives a short overview of the most common approaches.

The methods described in this section can be divided into the two groups, static and dynamic testing. They are described in the following subsections. Figure 20.11 gives an overview of various categories of methods of avoiding and revealing errors in code. Further information about various test methods can be found in, for example, Ref. [4].

20.3.2.2.1 **Static Testing** The methods associated with static testing are applied to the existing source code and the implementation of the source code, meaning that the code is not executed in order to avoid and find errors.

20.3.2.2.1.1 **Coding Guidelines** A simple method to prevent most common errors in software implementation is the use of coding guidelines. These guidelines provide

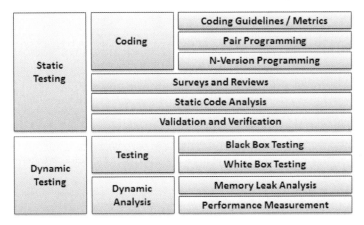

Figure 20.11 Overview of the classification of methods to assure a certain level of software quality.

rules for the syntax and semantics of the source code. Syntax rules include, for example, name conventions, meaning that used variable names have to follow certain patterns so that the readability of the code is maximized, thereby preventing errors caused by misconception of variables by the originator of the software and by users maintaining the software. Apart from these fundamental syntax rules, semantic rules are also included. The syntactical error in Section 20.3.2.1.1 in the code segment 'int c = a---b;' can be resolved by the coding guideline, that increments and decrements are only allowed as standalone statements, resulting in the following code segment:

```
int c = a-b;
a- -;
```

In this way, the resulting code is unambiguous and can be compiled with the same result in machine code by all compilers.

Of course, coding guidelines can only help to prevent errors if these rules are checked. Apart from the manual examination of the code, modern IDEs allow the definition of custom rules to be checked at implementation time, so that the coding guidelines can be automatically checked.

Very widespread versions of guidelines are the MISRA (Motor Industry Software Reliability Association) guidelines, which are mostly used in automation industry.

One of the methods to improve the quality of the created source code is by so-called software metrics, which are intended to measure certain aspects of the source code as characteristic numbers. These numbers are defined in the guidelines (meaning and lower/upper bounds). Apart from simple metrics like lines of code, there are also more complex metrics measuring the length of single functions, the ratio between lines of code and comments, and so on. More information about code metrics can be found in Ref. [5].

20.3.2.2.1.2 Pair Programming Pair programming affects the process of creating the source code. Two programmers work on the same work station. One person (called the driver) types the code, while the other person (the observer or navigator) reviews each line of code and scans it for errors. The roles are switched frequently.

Of course, this implementation method is more cost-intensive, as two programmers are needed while only one person is typing code. But studies have shown that the resulting code is more effective and – as many errors are instantly found – has fewer errors.

20.3.2.2.1.3 **N-Version Programming** N-Version programming is an approach introduced in 1977. Several functional equivalent versions of the same specification are implemented in order to use the resulting versions of the software in a system based on redundancy.

The problem of this approach is the fact that all versions are based on the same specification and are often implemented by the same programming language or by use of identical components of libraries, so that the probability that identical errors occur increases.

20.3.2.2.1.4 **Surveys and Reviews** The method of surveys and reviews uses the already created software product as the basis. The source code is reviewed by a single person or a whole committee in order to find potential errors within the code.

Apart from the fact that this approach is very time consuming, the question arises whether each single line of code is really reviewed. Furthermore, the probability that more persons will miss the same error is high, as the human brain tends to map incomplete fragments to a known complete picture, leading, for example, to the impression that a return statement is complete when in fact a semicolon is missing.

20.3.2.2.1.5 **Static Code Analysis** In addition to the manual methods described in the preceding sections, static code analysis is an approach to guaranteeing software quality by automatically reviewing the source code. The goals of the analysis are

- to test whether the source code complies with the coding guidelines
- to identify (potential) bugs
- to identify design and implementation problems.

The biggest disadvantage of this approach is obviously the number of rules to be implemented within the analysis, as every rule that has to be checked has to be mapped to the analysis logic. However, on the one hand there are already numerous frameworks for static code analysis which already support the most common rules, and on the other hand, the results of the static code analysis are undeniable, as the user can be sure that each single line of code is checked and every aspect which might embody a potential risk is identified.

20.3.2.2.1.6 **Validation and Verification** The methods of validation and verification are intended to assure that the software product meets the user's needs. While validation tries to find the answer to the question 'Are we building the right product?' (the software should do what the user really requires), verification asks the question 'Are we building the product right?' (the software should conform to its specification).

Validation and verification have to be applied to every step of the development process. The goal of the approach is to discover defects in the system and assess whether the system is usable in an operational situation.

20.3.2.2.2 Dynamic Testing Apart from the methods which are based on the static source code, there are also approaches using the executing source code in order to find defects and reveal other forms of problems. These can be classified into the two groups testing and dynamic analysis, which are further described in the following sections.

20.3.2.2.2.1 **Testing** There are two main types of testing, which differ in the goals they try to achieve: Black Box and White Box testing.

20.3.2.2.2.1.1 **Black Box Testing** Black box testing is a method to test the functionality of an application by feeding different inputs (valid and invalid inputs) to the test object and observing if the output meets the expectations of the specification. The internal structure or workings of the application are totally ignored.

20.3.2.2.2.1.2 **White Box Testing** As the name suggests, white box testing is the exact complement of black box, meaning that white box testing has the goal of testing the internal structure and workings of the application. There are several techniques that are used in white box testing, for example, data flow testing (to test whether data is correctly handled and passed within the application) and branch testing (to test whether every branch of the software is reachable and executed correctly).

20.3.2.2.2.2 **Dynamic Analysis** Dynamic analysis approaches are intended to ensure that the executed application meets all requirements concerning simply the execution.

20.3.2.2.2.2.1 **Memory Leak Analysis** A very widespread problem within applications written in C, for example, is so-called memory leaks. A memory leak occurs when an application allocates memory but does not release it after usage. Such an unintended increase in memory usage of applications is often called a memory leak. A memory leak can lead to unacceptable memory usage of applications, disturbing the time functionality of the source code.

A method to find such potential risks is the reference counting method, which keeps track of how many references to certain objects are still present and releases the memory if the reference counter is 0 (a method which is implemented in most common garbage collectors).

20.3.2.2.2.2.2 **Performance Measurement** Performance measurements are methods to test whether the application meets the time requirements given in the specification. This is particularly interesting for embedded systems, as these systems work in a cyclic pattern and therefore need to produce the output for certain inputs within tight time limits.

All these methods just try to avoid errors in the software. But one can never be sure that they succeed a hundred percent. The main problem is not that these methods have to be improved or that further error detection techniques are needed, but strategies to handle 'non-perfect' software have to be applied.

20.4
Potential Future Approaches

In the future, much more emphasis will be put on models, since they can be validated and verified by means of Formal Methods and are easier to understand by interdisciplinary developing groups. Models of technical systems will be extensively used as the basis for industrial software specifications.

Nowadays, all specifications developed for specific tasks are considered as models (of the software) in typical Computer Science. If the task changes then the model has to be adjusted. It is much more effective to distinguish between the model of the technical system and a formal description of the tasks.

More emphasis will be put on

- *model-based development* with models of the technical process to be controlled (plant): since the requirements of the controller software may contain faults, connections to a simulation model in the first stages of the development process is proposed. Additionally, it is easier and therefore less fault-prone to define requirements with respect to the final combined system of plant and controller instead to only the controller software. In safety-related systems, the combination of plant and controller has to fulfill requirements which have to be proved.
- *model-based formal methods*: instead of starting the development process with the explicit requirements with respect to the software, a model of the plant is to be developed. Properties of this formal model can be proved, and necessary additional model parts (the later software of the controller) are deduced automatically.
- *hierarchical modeling*: in order to reduce complexity, different models with different levels of abstraction are necessary. If the safety proof succeeds on an upper level, a detailed modeling of lower levels of the plant is not necessary.
- *diverse software*: a combination of the above formal approaches on the one hand with regular software development on the other hand reduces the possibility of fault tremendously. Two or more very different solutions will be developed and not simply different interpretations of the same solution.
- *reduction of safety-relevant parts* (encapsulation): instead trying to ensure correct behavior it is much more effective only to supervise functionality and to react in a safe way. Many software components are correct but their correctness cannot be proved. A small piece of software with proved correctness supervises these components and activates safe reactions.

All these approaches still suffer from complexity. The challenge is to develop less complex models that are suitable for the description of safety characteristics.

In the future, these approaches will help to prove and to certify software solutions to safety-relevant applications – and not only solutions to the development process (as is the case today).

20.5
CFD - The Future of Safety Technology?

Up to now, the main goal of software development is to avoid faults and errors. But their absence cannot neither be 100% achieved nor be proved. Therefore, the main focus concentrates on the development process at present.

Model-based approaches, including models of the technical plant such as CFD, are necessary. They are easier to understand, and they support fault and error avoidance from the first stages of the development process in this way.

The inclusion of CFD in software for safety-relevant applications will improve safety as well as its proof. Additionally, the CFD approaches will have an impact on other model-based approaches and encourage scientists and developers working on these.

References

1 Motet, J.-C.G.A.G. (2002) *Design of Dependable Computing Systems*, Kluwer Academic Publishers, Dordrecht.
2 Lyu, M.R. (1995) *Software Reliability Engineering*, IEEE Computer Society Press, Los Alamitos.
3 Hoffmann, D.W. (2008) *Software-Qualität*, Springer-Verlag, Heidelberg.
4 Liggesmeyer, P. (2002) *Software-Qualität*, Spektrum, Akad. Verl., Heidelberg.
5 Kan, S.H. (2007) *Metrics and Models in Software Quality Engineering*, 2nd. edn, [S.l.]: Addison Wesley.

21
CFD Modeling: Are Experiments Superfluous?

B. Jörgensen and D. Moncalvo

In this brief text the major strengths and weaknesses of a computational fluid dynamics (CFD)-oriented development of safety valves against the traditional intensive testing are exposed. Before this, a short introduction on the subject of safety valves is given.

Safety valves are mounted on pressurized items to protect them from excessive overpressure. A safety valve commences to open when the pressure in the protected item equals the predetermined set pressure and discharges the foreseen mass flow rate when the relieving pressure is reached. The choice of safety valve depends on, among other things, its opening behavior and its discharge coefficient. The opening of a safety valve can be either stepwise, in the so-called proportional safety valves, or suddenly, in popping full-lift valves. The discharge coefficient is the ratio between the mass flow rate in the safety valve and a reference mass flow rate at the same relieving condition and back pressure. The current international standards ISO 4126-1 [1] and ASME Boiler and Pressure Vessel Code Section VIII Division 1 [2] describe the procedures for certifying the discharge coefficient for gas and liquid flows. For safety valves with adjustable lift or for those facing a high back pressure at the outlet, ISO 4126-1 requires the manufacturer to provide an additional curve illustrating the dependence of the discharge coefficient on either the lift or the back pressure. Measurements for certification of the discharge coefficient must be made in stationary or quasi-stationary conditions.

Certification testing is surely the most crucial task in the development of a new valve family. The usual development practice of many manufacturers still relies massively on in-house experience to achieve a combination of desired technical features with the most economical method of manufacture. Prototypes of safety valves are optimized on the basis of experimental results gained on the test rig until they discharge so much flow rate to achieve or even exceed the design discharge coefficient in the foreseen range of test pressures. The development of a new product family is usually an expensive and time-consuming task for many manufacturing firms [3].

Process and Plant Safety: Applying Computational Fluid Dynamics, First Edition. Edited by J. Schmidt.
© 2012 Wiley-VCH Verlag GmbH & Co. KGaA. Published 2012 by Wiley-VCH Verlag GmbH & Co. KGaA.

The optimization of development procedures is therefore the target of many valve manufacturers. The intrinsic difficulty of the traditional optimization routine is due to the fact that experimental measurements deliver only 'integral' values – like mass flow rates and forces – although 'local' distributions of pressure or velocity would definitely be more useful to the R&D engineers, enabling them to identify areas for improvement in the contour of the valve. For example, the impact of small cavities or protrusions between the disc and the valve nozzle are responsible for the generation of small scale transient swirls during the opening of the disc, which cannot be foreseen with precision during the design phase.

In this concern it came into question if pressure and velocity contours calculated using CFD software packages may help engineers to design safety valves more efficiently. In recent years, CFD has spread from large multinationals R&D units to midsize manufacturing companies. One factor favoring its success is surely the coupling with 3D computer aided design (CAD) tools, facilitating the import and the discretization of complex geometries like the contours of safety valves. The first published simulations [3–5] of quasi-steady air and liquid flows using CFD tools are encouraging, since they prove that it is possible to approximate the flow rate of both compressible and incompressible media in safety valves within an error of a few percent in a broad range of operating conditions. However, larger deviations have been observed in the case of gaseous flows at relieving pressures of several hundred bars [4] or in the case of very viscous non-Newtonian media [6]. For these flows the worsening in the accuracy may also be related to the difficulty of accurately describing the properties of the medium in a broad range of conditions with simplified physical models. Another factor affecting the result is surely the difficulty in quantifying the amount of impurity in the medium and the intensity of thermal and mechanical exchange between these impurities and their surroundings during the flow.

Some practical aspects must be addressed to illustrate the strengths and the weaknesses of a CFD-based design of safety valves. In our experience at LESER, CFD has helped us to reduce the time required for the development of new valve types by selecting the most promising prototypes on the basis of numerical calculations and testing them on the rig. As an example of an optimization of a valve body, out of fourteen calculated prototypes only two were tested in our test rig and they both were successful in reaching the maximum lift while discharging the required mass flow rates. The error between numerical and experimental values of the discharge coefficient was approximately 0.03 using both the above-mentioned sizing standards. At the moment, LESER has begun to investigate the opening of safety valves using CFD, and physically plausible transient distributions of pressure and velocity have been calculated. For full-lift safety valves the calculation starts from an almost closed valve, ideally with an initial lift slightly above the value at which it pops. If air flows are considered, at these small lifts the turbulence modeling must include corrections for low-Reynolds-number turbulent flows, which become inconsistent and must be dropped at a certain lift, which is usually a few tenths of millimeter above the initial value. The identification of the most suitable initial lift and of the lift where corrections for low-Reynolds-number turbulence models can be dropped is very complex and is often done using rule-of-thumb criteria. Surface roughness inhomogeneity and the deviation between the cast product and the 3D design are additional factors which can not be accounted for in CFD, although they both affect the accuracy of the calculation of the disc acceleration at small lifts. Furthermore, the initial velocity of the disc must be known in order to calculate the opening characteristic. An uncertain or a rough estimation of this quantity may generate qualitatively plausible but quantitatively incorrect opening dynamics. A stable and automatic grid update routine throughout the CFD calculation requires that the disc rise is confined to few micrometers for each time step of few microseconds, which would require sophisticated measurement techniques and good expertise in the test rig in order to have reliable measurements to validate the calculation.

Back to the main question. According to our experience we can conclude that numeric development is a tool to save time and money in the development phase, but a good design practice cannot avoid physical testing of few selected prototypes. The numerical calculation of the opening of safety valves is a very recent topic of investigation. In the opinion of both authors of this paper, further research must go in this direction.

References

1 (2004) ISO 4126-1, Safety devices for protection against excessive pressure – Part 1: safety valves.
2 (2010) ASME Boiler and Pressure Vessel Code. Section VIII Division 1. Rules for construction of pressure vessels.
3 Moncalvo, D., Friedel, L., Jörgensen, B., and Höhne, T. (2009) Sizing of safety valves using ANSYS CFX Pro. *Chemical Engineering and Technology*, **32** (2), 247–251.
4 Schmidt, J., Peschel, W., and Beune, A. (2009) Experimental and theoretical studies on high pressure safety valves: sizing and design supported by numerical calculations. *Chemical Engineering and Technology*, **32** (2), 252–262.
5 Stoffel, B., Beinert, C., and Tembrink, H. (2007) Strömungssimulation als entwicklungswerkzeug für Armaturen. *Industriearmaturen*, **1**, 50–54.
6 Moncalvo, D., Friedel, L., and Jörgensen, B. (2010) A computational investigation on the flow of non-Newtonian polymers in safety valves. *Chemical Engineering and Technology*, **33** (11), 1825–1832.

Index

a

aerosol behavior 59, 60
– application of DNS and LES methods 59
– CFD calculations 59
– modeling of aerosol flows 59
– PHEBUS test facility 60
– RANS simulations 59
– validation 60
aleatory uncertainties 333
ALE coupling, with multi-block grid 18
ALE grid 20, 21
Algol/Pascal force 360
ANSYS-CFX version 10 320
APOLLO code 13, 14
APOLLO simulation 15, 16, 23
– surfaces of fireball, and leading shock wave 16
ASME Boiler 301, 369
atmospheric diffusion equation 245, 246
atmospheric dispersion 337
– eddy coefficient 337, 338
– wind speed 337
automotive crash simulation 9

b

balance equations
– dynamic modeling of disturbances in distillation columns 164
– for filtered quantities 248
– integral form of 163
– for mass, momentum, and heat transfer 162, 163
– Navier–Stokes equations, in combination with 234
– used in form of ordinary differential equations (ODEs) 162
beyond-design-based accidents (BDBA) analyses 53
– catalytic recombiners 55
– CFX and REACFLOW, for simplified EPR geometry 58
– characterized by damage to fuel rods 53
– coarse-mesh code GOTHIC 58
– EUproject ECORA 56
– formation of stable density stratification 56
– passive recombiners (PAR) 54, 55
– post-test calculations of HYJET experiments 57
– production of hydrogen by zirconium-water reaction 54
BFRL's experiment, to test response of smoke detectors on 166. *See also* fire modeling
– benefits from CFD calculations 175
– buoyant non-reacting flow over heated surface 168–170
– isothermal turbulent flow through 166–168
– – Smagorinsky factor 168
– simulation of incipient fire 170–173
– simulation of smoke migration 174
– uncertainties, to be assessed for fire simulations 175
blast parameters, computed with APOLLO and AUTODYN 15
blow-down system 269
boiler feed water (BFW) 292
boundary conditions 2, 38, 39
– of CFD pool fire simulation 145
– for CFD simulation 51
– geometrical 40
– influence of 207, 208
– periodic 55, 245
– refining of mesh/changes of 115
– significant uncertainty in 329
– in treating consequences of accidents, relevance of 333
Buncefield fuel depot, aerial view 315

Buncefield incident, CFD for reconstruction 313–328
– extent of mist 317, 318
– final dispersion simulations 325–328
– fire after explosion, aerial photo of 314
– ground topology 322, 323
– hedges and obstacles 323, 324
– sensitivity tests 320, 325
– vapor cloud dispersion, CFD modeling of 318–328

c

case studies
– future application of CFX 103–105
– to show, impact of different stochastic parameters 339
cavitation hammer effects 95. *See also* water hammer, effects
C3/C4 rectification system 278
C3/C-separation system
– dynamic UniSim model of 280, 281
– mass flows, on cooling water failure 286
– process flow diagram of 278, 284
certification testing, of new valve family 369
CFD applications, on explosions and blast waves 12
– APOLLO code 13
– conservation equations of continuum mechanics 12
– verification and validation 12
CFD codes. *See also* commercial codes
– categories, applications in explosive safety and 10
– model used as options for radiation models in 203
– quality assessment procedures 200
– risk analysis, categories 10
– validity of 20, 210
CFD-FLO Module 113
CFD in nuclear reactor safety
– application, state-of-art of CFD modeling 42–53 (*See also* CFD modeling)
– deboration transients in pressurized water reactors 42–47
CFD modeling 369–371
– for prediction
– – of final vapor cloud depth 327
– – of gas dispersion, within urban canopy layer 213–215
– – of progress of vapor cloud across 327
– safety valves 369, 370
– software packages 370
– weaknesses of 369
CFD program results, reliability of 200
– boundary conditions, influence of 207, 208
– grid dependency, influence of 206, 207
– SMARTFIRE program, benchmark test 201
– user's influence 200, 201
– validation 200
– verification 200
CFD simulations. *See also* direct numerical simulation (DNS); large eddy simulation (LES); RANS simulations
– adaptive code 195
– applied for safety engineering tasks 3
– commercial software 146
– 3D CFD simulations 11, 36
– to determine critical thermal distances 151
– in fire safety applications 198
– flow around catalyst plates 54
– initial and boundary conditions 51
– of large pool fires 139
– quantification of area of uncertainty for 191
– two-phase CFD simulation 49
– of vapor cloud dispersion at 322
CFD tools 1–3, 7
– license/quality labels 4
Chapman-Jouget model 14
chemical reactors, for high-temperature reactions 251–256
coarse-mesh code GOTHIC 58, 59
column trays, damage 299
combustion modeling 141, 142
combustion systems, characteristic properties of 234
– ignition delay times 237–239
– ignition of flammable mixtures 234–237
– laminar flame velocities 239–242
– turbulent flame velocities 242–247
combustion zone, in symmetry plane 15
commercial codes
– ANSYS AUTODYN 14
– AUTOREAGAS 19
– equipped with MpCCI interfaces 18
– FLACS 19
composite model, for disturbances in distillation columns 263
computational safety engineering 4
computational structural dynamics (CSD) code 17, 18
computer aided design (CAD) tools 370
computer-aided engineering (CAE) environment 9
condensation-induced water hammer 97
conservation laws 139
containment systems, for carriage of LNG 123

cooling water failure 285
core melting behavior 60–64
– CFD to calibrate models 63
– coarse-mesh approach 63
– concept of effective convectivity 62, 63
– Generation III LightWater Reactor 63
– KERENA, solution applied 63
– modeling of external RPV cooling 63
– SIMECO tests 61
– temperature field in core melt 64
coupled fluid–structure simulations 18

d

deflagration-to-detonation transition (DDT) 25
design-based accidents 37
design driven, by probabilistic approaches 11
Design Institute for Emergency Relief Systems (DIERS) 307
design of safe complex system, challenges 12
diffusion coefficient 139, 240, 337
direct numerical simulation (DNS) 3, 41, 140, 181, 192, 214, 244, 246, 248, 251, 256
discretized reactor, for synthesis of sulfur dioxide 251
dispersion
– CFD, application in 309, 310
– of heat and chemical substances 309
– of sprayed droplets, during application of a surface coating 308, 309
– of vapor cloud over time 326
displaced support grid, in packed column 298
distillation column pressurization, nitrogen piping 296–301
– damage to column internals 297
– nitrogen flow rates, dynamic model of 297–301
– steady-state assumptions 296, 297
distillation columns 262, 268, 300
– column head pressure during failure scenario 271
– disturbances, dynamic modeling (See dynamic simulation model)
– pilot plant flowsheet 270
DNS. See direct numerical simulation (DNS)
droplet evaporation, analysis of 318
DTBP (di-tert-butyl peroxide) pool fires 139
– block structured grid, for simulation 144
– CFD predicted instantaneous temperature fields 147
– ignition behavior of 235
– measured and CFD-predicted SEP of 151
– measured and CFD-predicted time-averaged irradiances E_{CFD} from 152

– time-averaged axial temperature profiles 147
Dufour effect 140
dynamic process simulation, applications 277
– hydrogen plant 288–293
– model building, and verification of 289–293
– process safety-related application of 284–287
– rectification systems 277
– verification of 278–284
dynamic simulation model 262, 278
– additional equations 266
– balance equations 264
– CFD calculations 262
– CFD for safety technology 269–272
– column stages 263
– discharged mass during pressure relief 271
– as hybrid model 271
– incoming vapor flow 265
– outgoing liquid flow 265, 266
– phase equilibrium 265
– pilot plant flowsheet 270
– relief device 266–268

e

eddy diffusion coefficient 337
eddy viscosity turbulence models 3
energy conservation equation 140
engineering methods
– evaluation 22
– TNT equivalence (See TNT equivalence)
equilibrium stage model. See dynamic simulation model
ERCOFTAC knowledge base Wiki 181, 182
– application challenges (AC) 182, 183
– – index 186
– – test cases for different application areas 182
– content of knowledge base 184
– interaction with users 185
– navigation tree of Wiki 183
– underlying flow regimes (UFR) 183, 184
– – flow types 183
– – index 187
Ergun correlation 323
erroneous systems, general strategies to handle in safe way 350–354
error-causing software module 362
errors avoidance, software development 355
explosive safety, and risk analysis 13

f

fail-operational systems   353
fail-safe systems   354
fast-acting valves   95. *See also* water hammer
– benchmark test, for numerical diffusion   107–109
– boundaries, characteristics of material and medium   108
– case studies, future application of CFX   103–105
– development of pressure, over time with linear closing of control surface
– – breakup of 1st cavitational hammer   107
– – within 10 s   105, 106
– experimental results   100–103
– experimental set-up   99, 100
– Kaplan turbine failure, 1D modeling of   105
– multi-phase flow test facility   97, 98
– PPP extension, for software validation   99
– use of CFX for calculations, difficulties in   106, 107
faults avoidance, general measures   351
fault-tolerant systems   352, 353
feed pipe damage   298
Fick's law   139
fireball surface
– cauliflower shape caused by instabilities   15
– comparison, of simulation results   17
– effects of combustion   16
– and shock front in simulated detonation   16
fire dynamics simulator (FDS)   167, 199, 201, 202, 206
fire modeling
– analytical models   161
– empirical model   161
– zone models   161, 162
flow force   113
flow simulations on uncertain parameters, dependency of   196–198
flow testing   114
fluid dynamics modeling
– coupled to chemical reaction kinetics   35
fluid–structure coupling   9, 18, 20
fluid–structure interaction (FSI)   97
Fourier's law   140
Francis weir formula   265
Friedlander function

g

gas discharge   335
gas dispersion, within urban canopy layer
– analysis of data from urban monitoring station   218–227
– CFDs model for   214, 215
– validation data requirements, of model   215–218
– validation of mathematical flow, and dispersion models   215
Gaussian dispersion models   295

h

hazards prevention, approaches for software systems   351, 352
HAZOP (hazard and operability study)   10
heat flux   124, 127, 134–136
heat transfer, evaluation of   125
– burning insulation   128
– deterioration, different phases of   126, 127
– film boiling, possibility of   127, 128
– simplified steady-state model   125, 126
Hertfordshire Oil Storage Limited (HOSL)   314, 320, 322, 325, 326
Heskestad model. *See* plume models
hydrocarbonl fires, large
– CFD simulations   144, 145
– – boundary conditions   145
– radiation modeling   142–144
– turbulence modeling   140, 141
hydrocarbon rectification, dynamic simulation of   277
hydrogen plants   288
– dynamic UniSim model of   290
– process flow diagram of   288
HYJET experiments   57

i

ideal gas, isentropic expansion   335
industrial turbulent flows   321
Interactive Development Environments (IDEs)   359

j

jet-in-cross-flow set-ups, mixing of fuels with air in   247–251
– filtered momentum equations   248
– turbulent flame velocity   247

k

Kaplan turbine failure. *See* fast-acting valves
Karman's constant   338
Kolmogoroff dissipation scale   189

l

Laplace transform, inversion of   346
large eddy simulation (LES)   3, 39, 47, 48, 64, 140, 166, 168, 181, 191, 194, 195, 214, 248, 251, 256

leak cross sections 335
leak diameter 334
leaks
– atmospheric diffusion equation 345, 346
– atmospheric dispersion 337, 338
– deterministic calculations 339
– distance-dependent conditional probability of death for 340
– gas discharge 335, 336
– geometry of aperture 335
– location risk of death 345
– mass flow rates for 339
– probabilistic calculations 342, 345
– sensitivity studies 339, 342
– size 334, 335
LES. *See* large eddy simulation (LES)
light water reactors, safety and safety analysis 33–36
– barriers against release of radioactive substances 34
– beyond-design based accidents 35
– design-based accidents 34
– emergency core cooling (ECC) system 34
– loss-of-coolant accidents 34
– probabilistic safety analysis (PSA) 35
– reactivity-induced accidents (RIA) 36
– severe accident management guidelines 35
liquefied natural gas (LNG) 123, 124
– transportation 124
liquid distributor damage 298
liquid layer, hydrostatic pressure of 265
liquid petrol cascaded 318
liquid–vapor distribution 262
lower explosive limit (LEL) 318
low-pressure valves. *See* low-pressure venting
low-pressure venting 113
– CFD modeling 113
– design of vent 114
– different orientation of geometries 117
– effect of rotated T-pipe part 116
– end-of-line vent 115
– flow force calculation 113, 114
– flow testing 114
– lift as function of force and alpha value 116
– measured mass flow 113
– mesh and prism layer 114
– optimized geometry 115
– piping structure, influencing vent's flow and 115
– volume flow tank pressure diagram 118
low-Reynolds-number turbulence models 370

m

mass balance 83, 85, 162, 176, 237, 244, 264, 266, 304
mass flux density 139
mathematical modeling, of physical phenomena 159
– balance equations for 159, 160, 176
– principles 159
– verification and validation 159
membrane tanks 123
– advantages 124
memory leaks 365
message passing interface (MPI) 18
model's limits of applicability, influence of 202–204. *See also* grid dependency
– fire in closed room 204
– fire in leaky room 205, 206
– fire in open room 204, 205
momentum conservation equation 140
Monte-Carlo calculations 344
Moss type LNG carrier 124
– boundary conditions 129
– CFD-calculations, of response to fire exposure 128
– checking CFD model 129–131
– melting of insulation caused by heat flux 134
– model of solution domain 129
– results of CFD calculation 133–136
– weather cover, evaluation 129
– – ABS FEA prediction 129
Motor Industry Software Reliability Association (MISRA) guidelines 363
multiphysics 162
Murphree efficiency 263, 265

n

natural gas 123
Navier–Stokes equations 9, 63, 194, 214, 234, 244, 246, 252
nitrogen flow rate
– additional restriction orifice, effect of 301
nitrogen piping, for distillation column pressurization 297. *See also* process safety toolbox
non-linear behavior, of devices 12
nuclear power plants (NPPs) 35, 48, 54, 99
nuclear safety 33, 35, 37, 38
numerics, and under-resolved simulations
– large eddy simulation (LES) models 194–196
– numerical discretizations, and under-resolution 190, 191

– Reynolds-averaged Navier–Stokes (RANS) models   192, 193
– turbulence modeling   191, 192

o

oxides   35, 54
– coefficients of probit equation   338
– mechanically stability   54

p

packed column internals, damage   298, 299
pair programming   363
PANDA drywell vessel   57
Peng-Robinson equations   279
perfect systems approach   352
peroxide pool fires. *See also* hydrocarbonl fires, large
– CFD predicted temperature fields, of DTBP pool fires   147
– CFD simulations   144, 145
– – block structured grid for   144
– – boundary conditions   145
– combustion modeling   141, 142
– JP-4 pool fire dynamics, quantitative description of   145, 146
– – critical thermal distances   150–154
– – flame temperature   146
– – irradiance E 149, 150
– – surface emissive power   147–149
– radiation modeling   142–144
– turbulence modeling   140, 141
petrochemical plants   277
petrol vapor concentrations, above ground level   322, 323, 324
petrol vapor source conditions   319
phenol-formaldehyde reaction   307
phenol-formaldehyde uncontrolled exothermic reaction   306–308
– disengagement behavior, effect of   307, 308
– single-phase venting, assumptions   306
– two-phase venting   306, 307
plume models   164. *See also* fire modeling
– equations   165
Pointing correction   265
polystyrene   123
pressure relief valves (PRVs)   261
pressure swing adsorption unit (PSA)   288
pressure vessel code   301
pressurized thermal shock (PTS)   49–53
– CATHARE code   51
– free surface flow, and stratification in cold legs   51
– high-end two-phase flow simulations   51

– initial and boundary conditions, taken from   51
– instantaneous surface deformations   52
– interfacial heat transfer and condensation   52
– NEPTUNE-CFD   50, 51
– steam condensation rate contours   52
– temperature distribution as result of mixing in liquid   52
– thermal hydraulic phenomena, to reactor pressure vessel wall   50
– TOPFLOW-PTS experimental setup   53
– turbulence modeled by   51
ProcessNet Safety Engineering Section   5–7
process safety toolbox   295–311

q

quantitative risk analysis (QRA), for explosive safety   10, 25–27
– CFD tools, role in QRA   26
– consecutive events following some original cause   26
– effects of single event   25
– event probability, expression   25
– modeling of event require   26
– models, currently being developed   27
– process of   25, 26
– quantitative measure of risk   25

r

radiation modeling   142–144
RANS simulations   39, 41, 48, 59, 252, 256
Rayleigh–Taylor instabilities   15
REACFLOW, combustion code   58
reactor, for producing sulfur dioxide   252–256
– boundary conditions   253
– droplet trajectories   255
– equation for movement of droplets in   254
– mass fraction field   254
– mass transfer coefficient   255
– probability of wall hitting in dependence of droplet sizes   256
– simulation of turbulent reacting flows   256
– temperature field   253
– velocity fields   253
reboiler
– heat transfer   287
– steam supply to   282
relief device model   266, 267
– closing characteristic   268
– opening characteristic   268
Richtmyer–Meshkov instabilities   15
RNG k-eps model   58

ROCOM reactor pressure vessel model, nodalization of 44

S
safe software systems. *See also* software, safety system
safety analysis 10
safety engineering 1
safety functions, software for 350
safety-related simulations, limitations of CFDs 208–210
safety valves 71
– computational domain, for flow simulation 87
– contour of flow 72
– – coefficients of 79
– 3D numerical simulation (CFD) 85, 91
– nozzle discharge coefficient models 80–82
– nozzle/discharge coefficient sizing procedure 72
– – throat *vs.* reduced inlet stagnation pressure 82
– nozzle flow models, for sizing coefficient C_g 80–82
– numerical model, and discretization 86, 87
– numerical results 87–90
– – acceleration of spindle with disk during 89
– – fluid–structure interaction (FSI) 89
– numerical sizing, for real gas flow 77, 78
– phenomenological description, of flow through 71, 72
– prototypes of 369
– valve sizing (*See also* sizing of safety valves, applying CFD)
– – according to ISO 4126-1 73, 74
– – limits of standard valve sizing procedure 74
– – method, for real gas applications 74–77
sawtooth model, for software errors handling 357
self-supporting tanks 123, 124
– *vs.* membrane systems 124
semantic errors 360
sensitivity tests 320, 321
– CFD model geometry 321
shear stress transport (SST) model 86
short-path evaporator
– design, typical 302
– dynamic simulation of 304
sizing of safety valves, applying CFD 82
– BASF high pressure safety valve 83
– disk lift measurement 84

– – linear displacement sensor 84
– high pressure test facility 82–86
Society of International Gas Tankers and Terminal Operators (SIGTTO) 123
software, developed for code coupling 18
software development methods 350
software errors, and error handling 354–366
software quality, classification of methods to assure 363
software, safety system
– advantages of 349
– black box testing 365
– CFD 367
– coding guidelines 362, 363
– definitions 350, 351
– development errors 355
– development methods 350
– dynamic analysis 365
– errors preventing methods 362
– errors within source code 358
– error-tolerant systems 353, 354
– fail-safe systems 354
– fault-tolerant systems 352, 353
– memory leak analysis 365
– numerical errors 361, 362
– N-Version programming 364
– pair programming 363, 364
– perfect systems 351, 352
– performance measurements 365, 366
– portability errors 362
– process models for 355, 356
– safety-critical applications 350
– sawtooth model 357
– semantic errors 360, 361
– spiral model 357, 358
– static code analysis 364
– static testing 362
– syntax errors 358–360
– user's needs 364
– validation and verification 364, 365
– V-Model 356, 357
– waterfall model 355, 356
– white box testing 365
Soret effect 139
spiral model, for software errors handling 357, 358
spontaneous condensation hammer 97
steady-state simulation, use of 287
steam reformer 288
– electric power failure
– – simulation results of 292
– emergency shutdown 291

– – pressure in cooling train and steam drum   292
– – steam and air flows   291
– – temperature profile of cooling train   291
stochastic input parameters   339
stochastic uncertainties   333
storage tank's design, for thermally sensitive liquids   308
Strohmeier leak   343
surface emissive power (SEP)   139
surface roughness   370
surface-to-air missile (SAM) system   361
syntax error   358

t

tank design   124
thermal energy   303
thermal fatigue, due to turbulent mixing   47, 48
thermal–hydraulic system codes   36
thermal radiation   34, 62, 139, 145
thermo-physical interaction   262
TNT equivalence   22
– blast parameters of spherical detonation   23
– deflagration-to-detonation transition (DDT), influencing factors   25
– for gas explosions, APOLLO simulation   23, 24
– Graham–Kinney correlations for   24
– hydrocarbon–air mixtures, rules for   25
– methane–air mixtures   24
– normalizing distance with   23
– peak overpressures and positive durations   24
– pressure–time curves   24
– scaled distance, defined   22
– scaling property   23
TOPFLOW-PTS test   54
Torricelli equation   265
total electric power failure   285, 292
toxic substances
– health effects of   338
– probability of death   340, 341, 342
transient cavitating flow, in piping systems   95
transient gaseous cavitation   95
transient vaporous cavitation   96, 97
tube failure, wiped-film evaporator during
– mass balance for   304
– quality of mixture in   306
– temperature and pressure   305
tunnel grid   20
turbulence modeling   86, 140, 141, 191, 194, 370

turbulent burning velocities   242, 245, 246
– of hydrogen mixtures with air   244
TUV-certified flow test facility   113

u

uncertainties data, for CFD simulations in fire safety applications   196
UniSim steady-state model   279

v

vapor cloud dispersion, CFD modeling of
– description of   320
– final dispersion simulations   325–328
– initial model tests   318, 319
– sensitivity tests   320
– – grid resolution   321
– – ground surface roughness   324, 325
– – ground topology   322, 323
– – hedges and obstacles   323, 324
– – turbulence   321, 322
– vapor source term   319, 320
VDI-GVC (the German society of engineers and society for chemical and process engineering)   5
vessel dimensions   11
viscous bubbly behavior   307
V-model, software errors handling   357

w

wash water flow   285, 286
– temperature profile in reboiler   287
waterfall model, modifications of   355, 356
water hammer
– difficulties in use of CFX for calculations   106–109
– effects   95
– impact of increased temperature on   98
– improvement, case studies   103–105
– induced by fast-acting valves   95
– and modeling of fluid dynamics processes   97
– modeling, use of CFX in   109, 110
– wave propagation   96
Weibull distribution   337
wind tunnel experiments   227–229
– generation of wind tunnel boundary layer   227
– time series of wind velocity, measurement   227
– validation data for RANS CFD-models   228
– vortex generators and roughness elements   227
– wind speeds, chosen to ensure   228

wiped-film evaporator, tube failure   301–306.
   See also process safety toolbox
– dynamic simulation of   304–306
– potentially dangerous overpressurization scenario   301–303
– required relieving rate
– – based on steam flow   303
– – based on water flow   303

z

zone models, as simplified form of fire models   164
– CFAST, computer code   164
– ordinary differential equations (ODEs)   164
– thermodynamic properties   164